Introduction to Geometry

Introduction to Geometry

Lydia McFarland

Willford Press
www.willfordpress.com

Published by Willford Press,
118-35 Queens Blvd., Suite 400,
Forest Hills, NY 11375, USA

ISBN: 978-1-68285-735-9

Cataloging-in-Publication Data

Introduction to geometry / Lydia McFarland.
p. cm.
Includes bibliographical references and index.
ISBN 978-1-68285-735-9
1. Geometry. 2. Mathematics. I. McFarland, Lydia.
QA444 .I58 2019
516--dc23

For information on all Willford Press publications visit our website at www.willfordpress.com

Willford Press

Printed in China.

Contents

Preface

Geometry falls under the domain of mathematics. It is concerned with the study of shape, size and the properties of space. Geometry is built on the fundamental concepts of points, lines, curves, planes, angles and symmetries and has applications across different branches of mathematics, in art, architecture and physics. Modern geometry can be classified into Euclidean geometry, differential geometry, topology and algebraic geometry, among others. This book is a compilation of chapters that discuss the most vital concepts in the field of geometry. It attempts to understand the multiple branches that fall under the discipline of geometry and how such concepts have practical applications. Coherent flow of topics, student-friendly language and extensive use of examples make this book an invaluable source of knowledge.

A foreword of all chapters of the book is provided below:

Chapter 1- Geometry is a branch of mathematics that is concerned with the study of shape and size of figures and the properties of space. Some of the prominent branches of geometry are Euclidean geometry, differential geometry, algebraic geometry and topology. This is an introductory chapter, which will introduce briefly the fundamentals of geometry and its important branches;

Chapter 2- An example of synthetic geometry is Euclidean geometry. The field develops logically from axioms that describe the basic properties of points, lines and other geometric objects, without the use of coordinates to describe these objects. This chapter has been carefully written to provide an easy understanding of the varied concepts of Euclidean geometry, such as Euclidean space, Euclidean distance as well as the Butterfly and Ceva's theorem, among others.;

Chapter 3 - Differential geometry is a branch of geometry that applies the techniques and principles of differential and integral calculus along with linear and multilinear algebra, to study geometry. The aim of this chapter is to explore the central aspects of differential geometry, such as Riemannian geometry, Finsler geometry, Symplectic geometry, Contact geometry, Euler's theorem, Cohn-Vossen's inequality, Darboux's theorem, etc.;

Chapter 4- The branch of mathematics that studies the zeros of multivariate polynomials is known as algebraic geometry. Abstract algebraic techniques, such as commutative algebra, are used for the solution of geometrical problems. This chapter closely examines the key concepts of algebraic geometry such as algebraic curve, algebraic variety, affine variety, Zariski's main theorem, Bézout's theorem, etc. to provide an extensive understanding of the field;

Chapter 5- Discrete geometry is a branch of geometry that studies discrete geometrical objects, such as points, lines, circles, spheres, planes, etc. In order to completely understand discrete geometry, it is necessary to understand the fundamentals of polyhedral combinatorics, structural rigidity, K-set, convex lattice polytope, Helly's theorem, McMullen problem, etc. which have been extensively discussed in this chapter;

Chapter 6- Convex geometry is a branch of geometry that studies convex sets in a Euclidean space. The concepts of convex functions, convex sets, convex bodies, etc as well as the principles of convex analysis, Busemann-Petty problem and Carathéodory's theorem have been carefully analyzed in this chapter;

Chapter 7 - Geometry is a vast subject that branches out into a number of significant sub-disciplines. Some of the prominent branches discussed in this chapter include spherical geometry, incidence geometry, non-Euclidean geometry, Projective geometry and Taxicab geometry.

At the end, I would like to thank all the people associated with this book devoting their precious time and providing their valuable contributions to this book. I would also like to express my gratitude to my fellow colleagues who encouraged me throughout the process.

Lydia McFarland

1

Geometry and its Branches

Geometry is a branch of mathematics that is concerned with the study of shape and size of figures and the properties of space. Some of the prominent branches of geometry are euclidean geometry, differential geometry, algebraic geometry and topology. This is an introductory chapter, which will introduce briefly the fundamentals of geometry and its important branches.

Geometry is an original field of mathematics, and is indeed the oldest of all sciences, going back at least to the times of Euclid, Pythagoras, and other "natural philosophers" of ancient Greece. Initially, geometry was studied to understand the physical world we live in, and the tradition continues to this day. Witness for example, the spectacular success of Einstein's theory of general relativity, a purely geometric theory that describes gravitation in terms of the curvature of a four-dimensional "space time". However, geometry transcends far beyond physical applications, and it is not unreasonable to say that geometric ideas and methods have always permeated every field of mathematics.

It is primarily developed to be a practical guide for measuring lengths, areas, and volumes, and is still in use up to now. Euclid turned the study of geometry into an axiomatic form at around 3rd century BC, and these axioms are still useful up to the present day. An important evolution for the science of geometry was created when Rene Descartes was able to create the concept of analytical geometry. Because of it, plane figures can now be represented analytically, and is one of the driving forces for the development of calculus. In addition, the rise of perspective gave rise to projective geometry. Nowadays, modern geometry has strong ties with physics, and is an integral part of new physical concepts such as relativity and string theories.

In real life, geometry has a lot of practical uses, from the most basic to the most advanced phenomena in life. Even the very basic concept of area can be a huge factor in how you do your daily business. For example, space is a huge issue when planning various construction projects. For instance, the size or area of a specific appliance or tool can greatly affect how it will fit in to your home or workplace, and can affect how the other parts of your home would fit around it. This is why it is essential to take account of areas, both of your space, and the item that you are about to integrate in there. In addition, geometry plays a role in basic engineering projects. For example, using the concept of perimeter, you can compute the amount of material (ex, paint, fencing material, etc.) that you need to use for your project. Also, designing professions such as interior design and architecture uses 3 dimensional figures. A thorough knowledge of geometry is going to help them a lot in determining the proper style (and more importantly, optimize its function) of a specific house, building, or vehicle.

Those are some of the more basic uses of geometry, but it doesn't end there. As some more professions use geometry in order to do their job properly. For example, computer imaging, something that is used nowadays for creating animations, video games, designing, and stuff like that, are created using geometric concepts. Also, geometry is used in mapping. Mapping is an essential element in professions such as surveying, navigation, and astronomy. From sketching to calculating distances, they use geometry to accomplish their job. In addition, professions such as medicine benefit from geometric imaging. Technologies such as CT scans and MRIs are used both for diagnosis and surgical aids. Such methods enable doctors to do their job better, safer, and simpler.

Fundamental Geometric Figures

When studying geometry, we will deal with several fundamental figures, including points, lines, and planes. A point can be thought of as an infinitesimally small sphere; that is, if you consider a "round" object such as a tennis ball, imagine that ball getting smaller and smaller until it is so small that it cannot be measured or even seen--this is akin to a point. A point has no width, length, or depth. A point can also be thought of as simply a location; a location does not have any physical dimension such as length or width, and yet we know intuitively how to speak of locations: "no trespassing beyond this point," for instance. Even though a point has no measurable size, we still need a way to represent it when studying geometry. Thus, we typically represent points as small dots, such as those shown below.

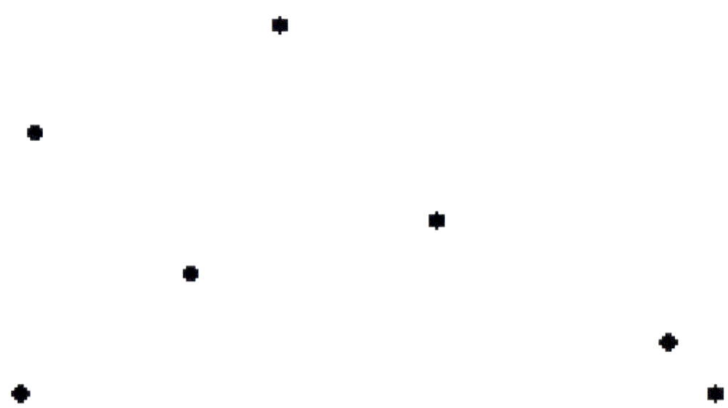

A line has slightly more character than a point: it is a figure that extends infinitely in one dimension. (A dimension can simply be thought of as a pair of opposite directions: forward and backward, left and right, or up and down) A line has infinite length, but it has no thickness. Imagine a piece of string that is pulled tight from two points that are immeasurably far away, and imagine that the thickness of the string decreases until the string is invisible- this string would then be a line. Because we can't show infinite distances on a piece of paper (or anywhere else, for that matter), we use arrows when drawing lines to indicate that the figure extends indefinitely. Below is an example of a line:

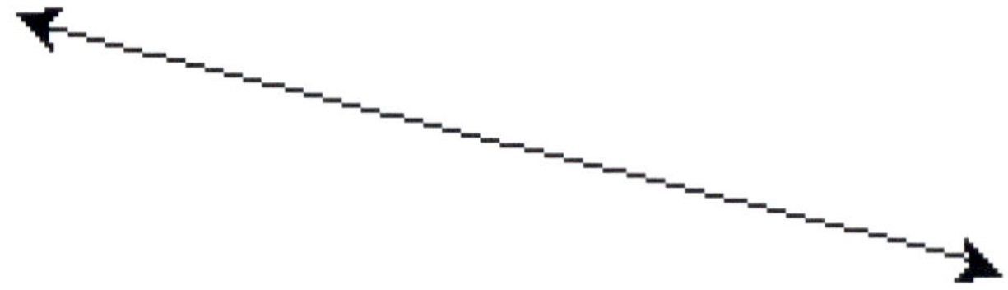

Naturally, we can conceive of a line that extends to infinity (that is, extends indefinitely) in only one direction but has an end point in the other direction; this figure is called a ray. Imagine the sun shining in the sky--a beam of light that originates from the sun can extend indefinitely away from it. The word "ray" in this case is closely linked with the geometric figure called a "ray."

If the line instead has two endpoints, we call it a line segment, which is illustrated below:

In some cases, it is helpful to show the endpoints of the ray or line segment more clearly using dots. The figures are the same:

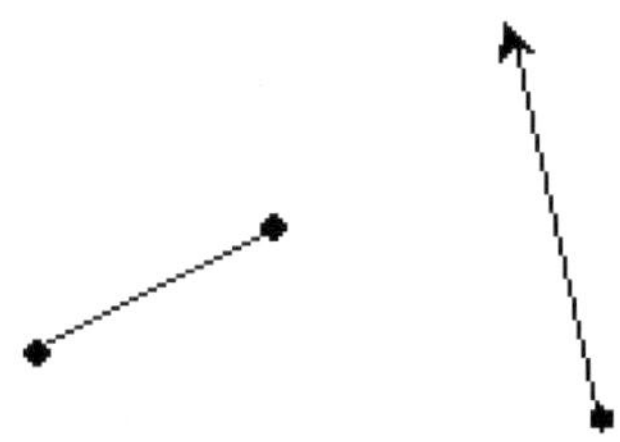

A plane is a geometric figure that extends to infinity in two dimensions. As such, we must show it in three dimensions, but you can imagine a plane as an even surface whose edges are so far away that they could never be reached. Planes are often drawn as parallelograms. Since, their infinite extent cannot be represented. We will draw the plane with a dashed border to differentiate it from a parallelogram.

Because we will mostly deal with (two-dimensional) planar geometry, we will not deal much with planes. Note, however, that we are in a sense dealing with a plane in planar geometry- we do all our drawing, moving, and analysing of figures located entirely on that one plane. Consider, for instance, the collection of basic figures shown on the plane below:

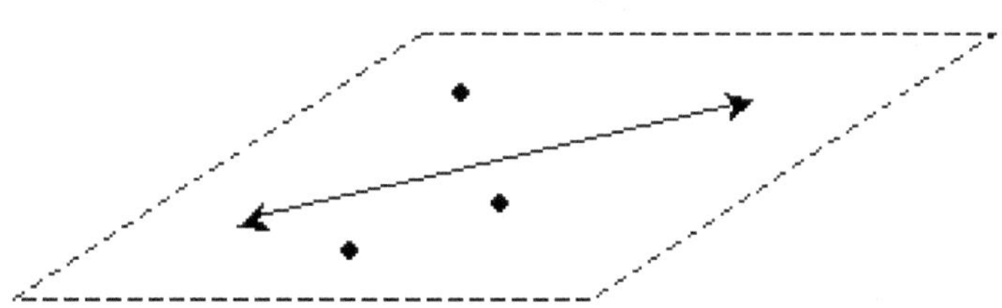

Branches of Geometry

Contemporary geometry has many subfields which are as follows:

1. Algebraic geometry – Algebraic geometry is a branch of geometry studying zeros of

multivariate polynomial. It includes linear and polynomial algebraic equation used for solving about the sets of zeros. The application of this type include Cryptography, string theory etc.

2. Discrete geometry – It is concerned with relative position of simple geometric object, such as points, lines, triangles, circles etc.
3. Differential geometry – Differential geometry uses techniques of algebra and calculus for problem solving. The various problems include general relativity in physics etc.
4. Euclidean geometry – The study of plane and solid figures on the basis of axioms and theorems including points, lines, planes, angles, congruence, similarity, solid figures. It has a wide range of application in Computer Science, Modern Mathematics problem solving, Crystallography etc.
5. Convex Geometry – It includes convex shapes in Euclidean space using techniques of real analysis. It has application in optimization and functional analysis in number theory.
6. Topology – This is concerned with properties of space under continuous mapping. Its application includes consideration of compactness, completeness, continuity, filters, function spaces, grills, clusters and bunches, hyperspace topologies, initial and final structures, metric spaces, metrication, nets, proximal continuity, proximity spaces, separation axioms, and uniform spaces.

Another way of classification of geometry includes plane and solid geometry:

- Plane geometry – This wing of geometry deals with flat shapes which can be drawn on piece of paper. These include lines, circles & triangles of two dimensions.
- Soid geometry – It deals with 3-dimensional objects like cubes, prisms, cylinders & spheres.

2

Euclidean Geometry

An example of synthetic geometry is Euclidean geometry. The field develops logically from axioms that describe the basic properties of points, lines and other geometric objects, without the use of coordinates to describe these objects. This chapter has been carefully written to provide an easy understanding of the varied concepts of Euclidean geometry, such as Euclidean space, Euclidean distance as well as the Butterfly and Ceva's theorem, among others.

Euclidean geometry is the study of plane and solid figures on the basis of axioms and theorems employed by the greek mathematician Euclid. In its rough outline, Euclidean geometry is the plane and solid geometry commonly taught in secondary schools. Indeed, until the second half of the 19th century, when non-Euclidean geometries attracted the attention of mathematicians, geometry meant Euclidean geometry. It is the most typical expression of general mathematical thinking. Rather than the memorization of simple algorithms to solve equations by rote, it demands true insight into the subject, clever ideas for applying theorems in special situations, an ability to generalize from known facts, and an insistence on the importance of proof. In Euclid's great work, the Elements, the only tools employed for geometrical constructions were the ruler and the compass—a restriction retained in elementary Euclidean geometry to this day.

In its rigorous deductive organization, the Elements remained the very model of scientific exposition until the end of the 19th century, when the German mathematician David Hilbert wrote his famous Foundations of Geometry (1899). The modern version of Euclidean geometry is the theory of Euclidean (coordinate) spaces of multiple dimensions, where distance is measured by a suitable generalization of the Pythagorean Theorem.

Axioms

The term 'postulate' was used by Euclid for an assumption in geometry, while the word "axiom" on the other hand, were the assumptions used throughout mathematics and not specifically linked to geometry. Some of Euclid's axioms, not in his order, are given below:

(1) Things which are equal to the same thing are equal to one another,

If $a = b$ and $a = c$ then $b = c$

(2) If equals are added to equals, the wholes are equal,

If $a= b$ and $c = d$, then $a + c = b + d$

(3) If equals are subtracted from equals, the remainders are equal,

If $a= b$ and $c = d$, then $a - c = b - d$

(4) Things which coincide with one another are equal to one another,

X coincide with X, hence X = X

(5) The whole is greater than the part,

$$a\rangle\frac{a}{2}$$

(6) Things which are double of the same things are equal to one another,

If a = b, then $2a = 2b$

(7) Things which are halves of the same things are equal to one another,

If a = b, then $\frac{a}{2} = \frac{b}{2}$

Theorems are statements which are proved.

Euclid's Postulates

The postulates stated by Euclid are the foundation of Geometry and are rather simple observations in nature. 'Euclid' was a greek mathematician regarded as the 'Father of Modern Geometry'.

Postulate – I

A straight line segment can be formed by joining any two points in space:

In Geometry, a line segment is a part of a line that is bounded by 2 distinct points on either end. It consists of a series of points bounded by the two endpoints. Thus a line segment is measurable as the distance between the two endpoints. A line segment is named after the two endpoints with an over bar on them:

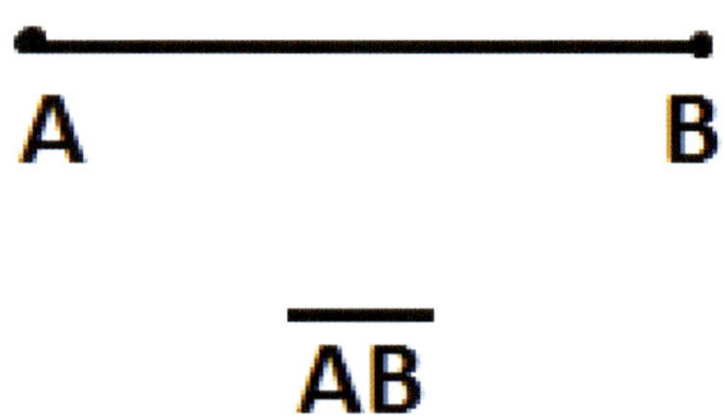

Postulate – II

Any straight line can be extended indefinitely on both sides. Unlike a line segment, a line is not bounded by any endpoint and so can be extended indefinitely in either direction. A line is uniquely defined as passing through two points which are used to name it:

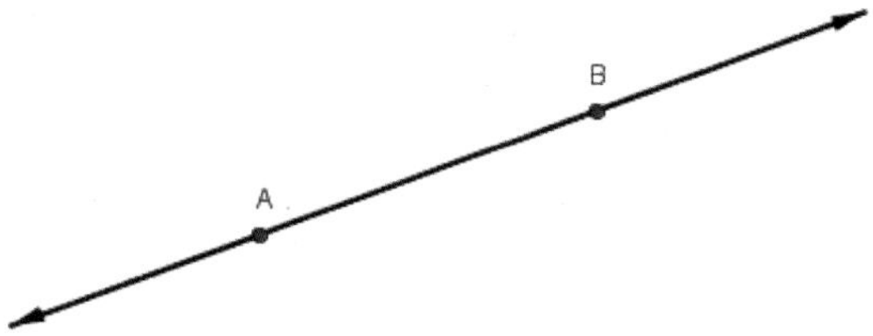

Postulate – III

A circle can be drawn with any centre and any radius. For any line segment, a circle can be drawn with its centre at one endpoint and the radius of the circle as the length of the line segment. Consider a line segment bounded by two points. If one of these points is taken as the centre of a circle and the radius of the circle is taken as equal to the length of the segment, a circle can be drawn with its diameter twice than the length of the line segment:

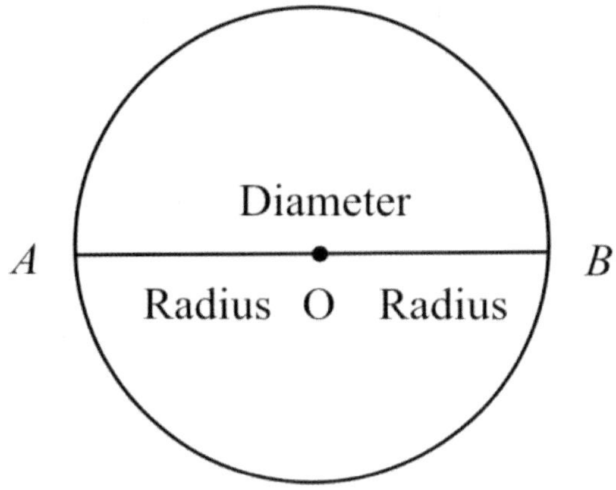

In the above example, the line segment AO serves as the radius of a circle with centre at point O and a diameter equal to AB.

Where, $l(AB) = 2l(AO)$.

Postulate – IV

All right angles are congruent or equal to one another. A right angle is an angle measuring 90 degrees. So, irrespective of the length of a right angle or its orientation all right angles are identical in form and coincide exactly when placed one on top of the other.

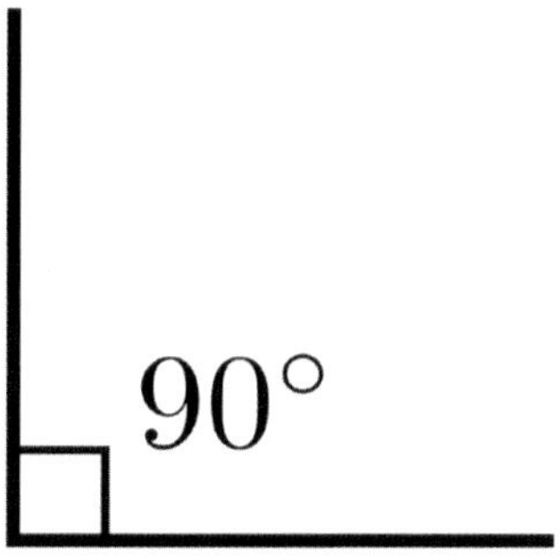

A right angle

Postulate – V

Two lines are parallel to each other if they intersect the third line and the interior angle between them is 180 degrees:

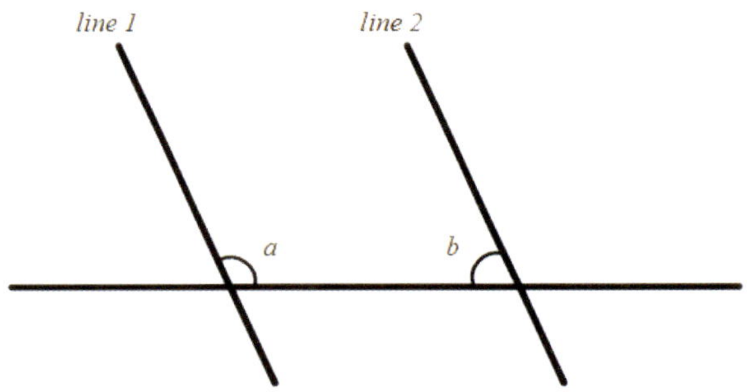

If: $a + b = 180°$

Then: *line 1 and line 2 are parallel*

'Parallel lines' are a set of 2 or more lines that never cross or intersect each other at any point in space if they are extended indefinitely. As you can see in the above image, line 1 and line 2 are parallel if and only if the sum of angles 'a' and 'b' they make with the transversal is 180 degrees.

Notation and Terminology

Naming of Points and Figures

Points are customarily named using capital letters of the alphabet. Other figures, such as lines, triangles, or circles, are named by listing a sufficient number of points to pick them out unambiguously from the relevant figure. For example, triangle ABC would typically be a triangle with vertices at points A, B, and C.

Complementary and Supplementary Angles

Angles whose sum is a right angle are called complementary. Complementary angles are formed when a ray shares the same vertex and is pointed in a direction that is in between the two original rays that form the right angle. The number of rays in between the two original rays is infinite.

Angles whose sum is a straight angle are supplementary. Supplementary angles are formed when a ray shares the same vertex and is pointed in a direction that is in between the two original rays that form the straight angle (180 degree angle). The number of rays in between the two original rays is infinite.

Modern Versions of Euclid's Notation

In modern terminology, angles would normally be measured in degrees or radians.

Modern school textbooks often define separate figures called lines (infinite), rays (semi-infinite), and line segments (of finite length). Euclid, rather than discussing a ray as an object that extends to infinity in one direction, would normally use locutions such as "if the line is extended to a sufficient length," although he occasionally referred to "infinite lines". A "line" in Euclid could be either straight or curved, and he used the more specific term "straight line" when necessary.

Some Important or Well-known Results

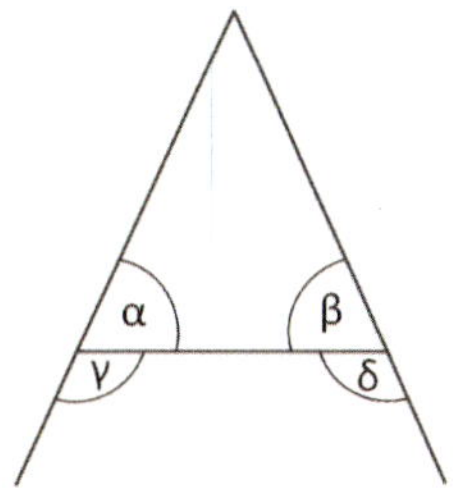

The Pons Asinorum or Bridge of Asses theorem states that in an isosceles triangle, α = β and γ = δ.

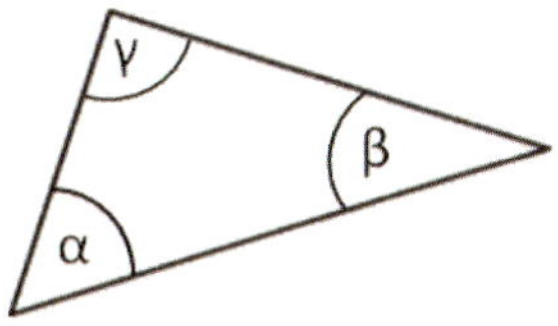

The Triangle Angle Sum theorem states that the sum of the three angles of any triangle, in this case angles α, β, and γ, will always equal 180 degrees.

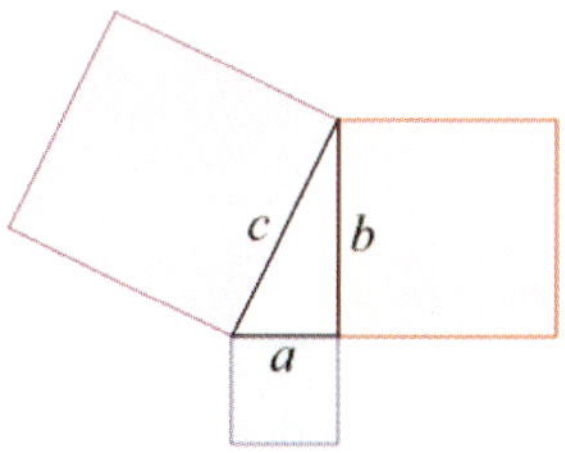

The Pythagorean Theorem states that the sum of the areas of the two squares on the legs (*a* and *b*) of a right triangle equals the area of the square on the hypotenuse (*c*).

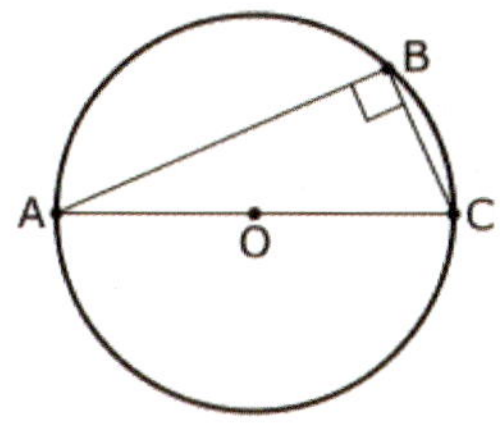

Thales' theorem states that if AC is a diameter, then the angle at B is a right angle.

Pons Asinorum

The Bridge of Asses states that *in isosceles triangles the angles at the base equal one another, and, if the equal straight lines are produced further, then the angles under the base equal one another*. Its name may be attributed to its frequent role as the first real test in the *Elements* of the intelligence of the reader and as a bridge to the harder propositions that followed. It might also be so named because of the geometrical figure's resemblance to a steep bridge that only a sure-footed donkey could cross.

Congruence of Triangles

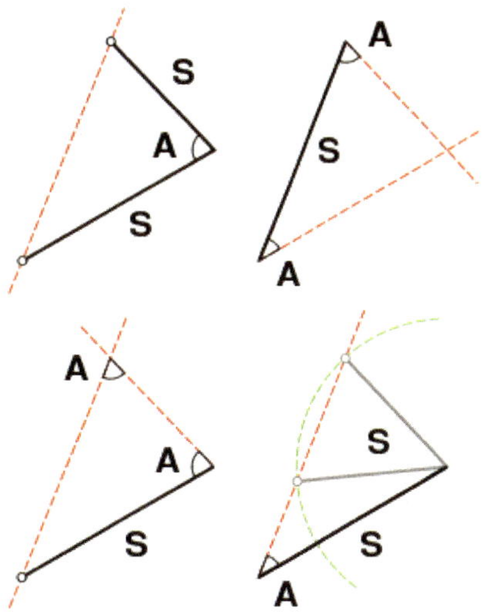

Congruence of triangles is determined by specifying two sides and the angle between them (SAS), two angles and the side between them (ASA) or two angles and a corresponding adjacent side (AAS). Specifying two sides and an adjacent angle (SSA), however, can yield two distinct possible triangles unless the angle specified is a right angle.

Triangles are congruent if they have all three sides equal (SSS), two sides and the angle between them equal (SAS), or two angles and a side equal (ASA). Triangles with three equal angles (AAA) are similar, but not necessarily congruent. Also, triangles with two equal sides and an adjacent angle are not necessarily equal or congruent.

Triangle Angle Sum

The sum of the angles of a triangle is equal to a straight angle (180 degrees). This causes an equilateral triangle to have three interior angles of 60 degrees. Also, it causes every triangle to have at least two acute angles and up to one obtuse or right angle.

Pythagorean Theorem

The celebrated Pythagorean theorem states that in any right triangle, the area of the square whose side is the hypotenuse (the side opposite the right angle) is equal to the sum of the areas of the squares whose sides are the two legs (the two sides that meet at a right angle).

Thales' Theorem

Thales' theorem, named after Thales of Miletus states that if A, B, and C are points on a circle. where the line AC is a diameter of the circle, then the angle ABC is a right angle.

Scaling of Area and Volume

In modern terminology, the area of a plane figure is proportional to the square of any of its linear dimensions, $A \propto L^2$, and the volume of a solid to the cube, $V \propto L^3$. Euclid proved these results in various special cases such as the area of a circle and the volume of a parallelepiped solid. Euclid determined some, but not all, of the relevant constants of proportionality.

For example, it was his successor Archimedes who proved that a sphere has 2/3 the volume of the circumscribing cylinder.

Applications

Because of Euclidean geometry's fundamental status in mathematics, it is impractical to give more than a representative sampling of applications here:

A survey or uses a level

Sphere packing applies to a stack of oranges.

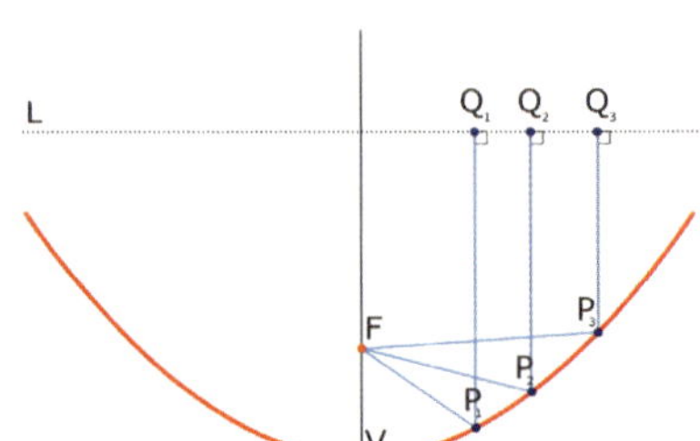

A parabolic mirror brings parallel rays of light to a focus.

As suggested by the etymology of the word, one of the earliest reasons for interest in geometry was surveying, and certain practical results from Euclidean geometry, such as the right-angle property of the 3-4-5 triangle, were used long before they were proved formally. The fundamental types of measurements in Euclidean geometry are distances and angles, both of which can be measured directly by a surveyor. Historically, distances were often measured by chains, such as Gunter's chain, and angles using graduated circles and, later, the theodolite.

An application of Euclidean solid geometry is the determination of packing arrangements, such as the problem of finding the most efficient packing of spheres in n dimensions. This problem has applications in error detection and correction.

Geometric optics uses Euclidean geometry to analyse the focusing of light by lenses and mirrors:

Geometry is used in art and architecture.

Geometry can be used to design origami.

The water tower consists of a cone, a cylinder, and a hemisphere. Its volume can be calculated using solid geometry.

Geometry is used extensively in architecture:

Geometry can be used to design origami. Some classical construction problems of geometry are impossible using compass and straightedge, but can be solved using origami.

Quite a lot of CAD (computer-aided design) and CAM (computer-aided manufacturing) is based on Euclidean geometry. Design geometry typically consists of shapes bounded by planes, cylinders, cones, tori, etc. CAD/CAM is essential in the design of almost everything, nowadays, including cars, airplanes, ships, and the iPhone. A few decades ago, sophisticated draftsmen learned some fairly advanced Euclidean geometry, including things like Pascal's theorem and Brianchon's theorem. But now they don't have to, because the geometric constructions are all done by CAD programs.

As a Description of the Structure of Space

Euclid believed that his axioms were self-evident statements about physical reality. Euclid's proofs depend upon assumptions perhaps not obvious in Euclid's fundamental axioms,in particular that certain movements of figures do not change their geometrical properties such as the lengths of sides and interior angles, the so-called *Euclidean motions*, which include translations, reflections and rotations of figures. Taken as a physical description of space, postulate 2 (extending a line) asserts that space does not have holes or boundaries (in other words, space is homogeneous and unbounded); postulate 4 (equality of right angles) says that space is isotropic and figures may be moved to any location while maintaining congruence; and postulate 5 (the parallel postulate) that space is flat (has no intrinsic curvature).

The ambiguous character of the axioms as originally formulated by Euclid makes it possible for different commentators to disagree about some of their other implications for the structure of space, such as whether or not it is infinite and what its topology is. Modern, more rigorous reformulations of the system typically aim for a cleaner separation of these issues. Interpreting Euclid's axioms in the spirit of this more modern approach, axioms 1-4 are consistent with either infinite or finite space (as in elliptic geometry), and all five axioms are consistent with a variety of topologies (For example, a plane, a cylinder, or a torus for two-dimensional Euclidean geometry).

Treatment of Infinity

Infinite Objects

Euclid sometimes distinguished explicitly between "finite lines" and "infinite lines". However, he typically did not make such distinctions unless they were necessary. The postulates do not explicitly refer to infinite lines, although for example some commentators interpret postulate 3, existence of a circle with any radius, as implying that space is infinite.

The notion of infinitesimal quantities had previously been discussed extensively by the Eleatic School, but nobody had been able to put them on a firm logical basis, with paradoxes such as

Zeno's paradox occurring that had not been resolved to universal satisfaction. Euclid used the method of exhaustion rather than infinitesimals.

Later ancient commentators, such as Proclus, treated many questions about infinity as issues demanding proof and, e.g., Proclus claimed to prove the infinite divisibility of a line, based on a proof by contradiction in which he considered the cases of even and odd numbers of points constituting it.

At the turn of the 20th century, Otto Stolz, Paul du Bois-Reymond, Giuseppe Veronese, and others produced controversial work on non-Archimedean models of Euclidean geometry, in which the distance between two points may be infinite or infinitesimal, in the Newton–Leibniz sense. Fifty years later, Abraham Robinson provided a rigorous logical foundation for Veronese's work.

Infinite Processes

One reason that the ancients treated the parallel postulate as less certain than the others is that verifying it physically would require us to inspect two lines to check that they never intersected, even at some very distant point, and this inspection could potentially take an infinite amount of time.

The modern formulation of proof by induction was not developed until the 17th century, but some later commentators consider it implicit in some of Euclid's proofs, e.g., the proof of the infinitude of primes.

Supposed paradoxes involving infinite series, such as Zeno's paradox, predated Euclid. Euclid avoided such discussions, giving, for example, the expression for the partial sums of the geometric series without commenting on the possibility of letting the number of terms become infinite.

Logical Basis

Classical Logic

Euclid frequently used the method of proof by contradiction, and therefore the traditional presentation of Euclidean geometry assumes classical logic, in which every proposition is either true or false, i.e., for any proposition P, the proposition "P or not P" is automatically true.

Modern Standards of Rigor

Placing Euclidean geometry on a solid axiomatic basis was a preoccupation of mathematicians for centuries. The role of primitive notions, or undefined concepts, was clearly put forward by Alessandro Padoa of the Peano delegation at the 1900 Paris conference:

> When we begin to formulate the theory, we can imagine that the undefined symbols are *completely devoid of meaning* and that the unproved propositions are simply *conditions* imposed upon the undefined symbols.

Then, the *system of ideas* that we have initially chosen is simply *one interpretation* of the undefined symbols, this interpretation can be ignored by the reader, who is free to replace it in his mind by *another interpretation* that satisfies the condition.

Logical questions thus become completely independent of *empirical* or *psychological* questions.

The system of undefined symbols can then be regarded as the *abstraction* obtained from the *specialized theories* that result when the system of undefined symbols is successively replaced by each of the interpretations.

That is, mathematics is context-independent knowledge within a hierarchical framework. As said by Bertrand Russell:

> If our hypothesis is about *anything*, and not about some one or more particular things, then our deductions constitute mathematics. Thus, mathematics may be defined as the subject in which we never know what we are talking about, nor whether what we are saying is true.

Axiomatic Formulations

- Euclid's axioms: In his dissertation to Trinity College, Cambridge, Bertrand Russell summarized the changing role of Euclid's geometry in the minds of philosophers up to that time. It was a conflict between certain knowledge, independent of experiment, and empiricism, requiring experimental input. This issue became clear as it was discovered that the parallel postulate was not necessarily valid and its applicability was an empirical matter, deciding whether the applicable geometry was Euclidean or non-Euclidean.
- Hilbert's axioms: Hilbert's axioms had the goal of identifying a *simple* and *complete* set of *independent* axioms from which the most important geometric theorems could be deduced. The outstanding objectives were to make Euclidean geometry rigorous (avoiding hidden assumptions) and to make clear the ramifications of the parallel postulate.
- Birkhoff's axioms: Birkhoff proposed four postulates for Euclidean geometry that can be confirmed experimentally with scale and protractor. This system relies heavily on the properties of the real numbers. The notions of *angle* and *distance* become primitive concepts.
- Tarski's axioms: Alfred Tarski and his students defined *elementary* Euclidean geometry as the geometry that can be expressed in first-order logic and does not depend on set theory for its logical basis, in contrast to Hilbert's axioms, which involve point sets. Tarski proved that his axiomatic formulation of elementary Euclidean geometry is consistent and complete in a certain sense: there is an algorithm that, for every proposition, can be shown either true or false. (This doesn't violate Gödel's theorem, because Euclidean geometry cannot describe a sufficient amount of arithmetic for the theorem to apply). This is equivalent to the decidability of real closed fields, of which elementary Euclidean geometry is a model.

Constructive Approaches and Pedagogy

The process of abstract axiomatisation as exemplified by Hilbert's axioms reduces geometry to theorem proving or predicate logic. In contrast, the greeks used construction postulates, and emphasized problem solving. For the greeks, constructions are more primitive than existence propositions, and can be used to prove existence propositions, but not *vice versa*. To describe problem solving adequately requires a richer system of logical concepts. The contrast in approach may be summarized:

- Axiomatic proof: Proofs are deductive derivations of propositions from primitive premises that are 'true' in some sense. The aim is to justify the proposition.

- Analytic proof: Proofs are non-deductive derivations of hypotheses from problems. The aim is to find hypotheses capable of giving a solution to the problem. One can argue that Euclid's axioms were arrived upon in this manner. In particular, it is thought that Euclid felt the parallel postulate was forced upon him, as indicated by his reluctance to make use of it, and his arrival upon it by the method of contradiction.

Andrei Nicholaevich Kolmogorov proposed a problem solving basis for geometry. This work was a precursor of a modern formulation in terms of constructive type theory.This development has implications for pedagogy as well.

If proof simply follows conviction of truth rather than contributing to its construction and is only experienced as a demonstration of something already known to be true, it is likely to remain meaningless and purposeless in the eyes of students.

Euclidean Space

Euclidean space is defined as two and even more dimensional space that follows the postulates and axioms explained by Euclid. In modern times, Euclidean space is more commonly referred as Cartesian space or Cartesian plane. Two or more dimensional Euclidean spaces are quite commonly used in mathematics. Euclidean space is may also be defined as a real space utilized to denote vector field.

It is actually a flat and non-curvy space. In Euclidean geometry, the surface is always assumed to be flat. If it became curved or spherical, then it comes under non-Euclidean geometry.

An example of two-dimensional Euclidean space is a piece of paper, It does have two dimensions: length and breadth along a pair of horizontal and vertical lines, as shown in the figure below:

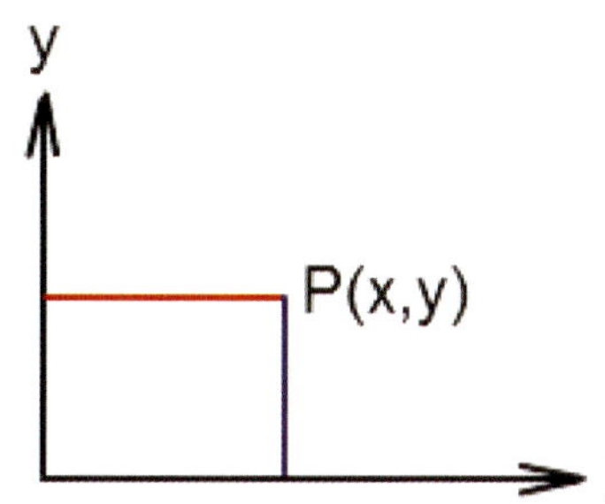

In three-dimensional Euclidean space, there is one additional dimension known as height perpendicular to both horizontal and vertical lines:

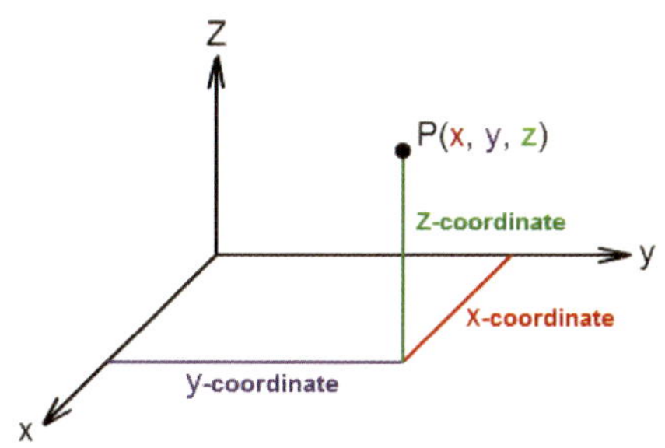

A point that is defined in a two-dimensional space, is represented by (x,y) and in three-dimensions by (x,y,z); where, x, y and z are the distances from X, Y and Z axes respectively.

Euclidean n-Space

Euclidean spaces do have finite number of dimensions. The number of dimensions in Euclidean space is denoted by n. Mathematically, the n-dimensional space denoting real coordinates is notified by R^n or E^n.

One-dimensional Euclidean space is represented by a real number line; while Euclidean space of two dimensions is composed of Cartesianplane with two mutually perpendicular lines. It is denoted by R^2 or $E^{2.}$

ln the same way, Euclidean space may possess any finite number of finite dimensions. An n-dimensional space is known as Euclidean n-space and is denoted by R^n or $E^{n.}$

The *n*-space is the system of n ordered tuples of real numbers, such as $(x_{1,} x_{2,} x_{3,...,} x_n)$. In other words Euclidean n-space is a vector space consisting of n vectors or n number of elements of real space.

Euclidean Topology

Euclidean topology is said to the topology on Euclidean *n*-space R^n,

where, *n* belongs to the set of natural numbers.

Euclidean topology is also termed as general topology or usual topology or ordinary topology.

Examples

1. Draw the following points in Euclidean two-dimensional space: (2, 1), (-1, -3), (-0.5, -1.5) and (-4, 6).

Solution

The figure showing given points is:

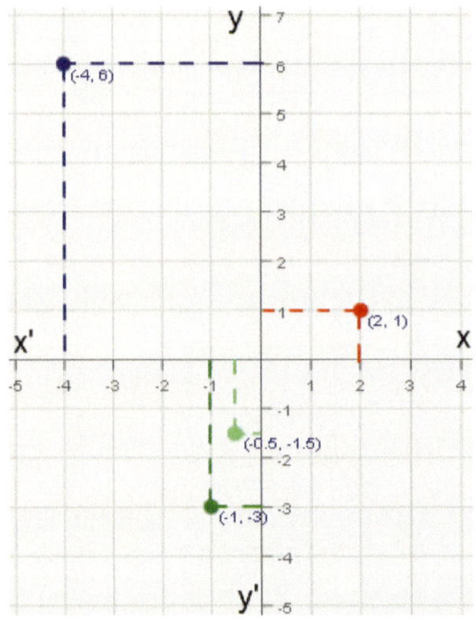

2. Show point (1, 2, 3) in R^3.

Solution

Point (1, 2, 3) is shown in the diagram below:

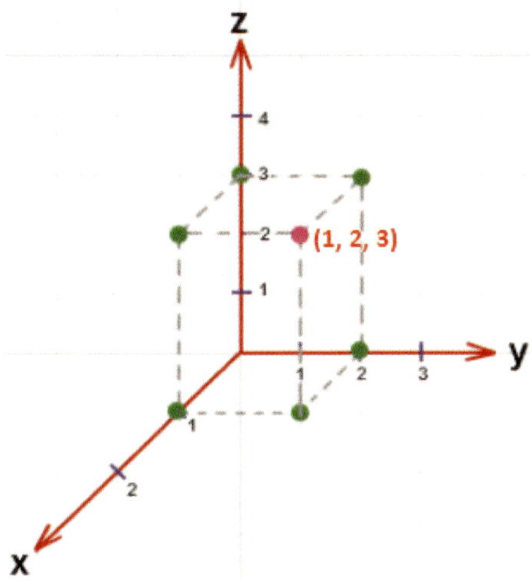

Euclidean Structure

These are distances between points and the angles between lines or vectors, which satisfy certain conditions, which makes a set of points a Euclidean space. The natural way to obtain these quantities is by introducing and using the standard inner product (also known as the dot product) on R^n. The inner product of any two real n-vectors x and y is defined by,

$$\mathrm{x}\cdot\mathrm{y}=\sum_{i=1}^{n}x_iy_i=x_1y_1+x_2y_2+\cdots+x_ny_n,$$

where,

x_i and y_i are i^{th} coordinates of vectors x and y respectively. The result is always a real number.

Distance

The inner product of x with itself is always non-negative. This product allows us to define the "length" of a vector x through square root:

$$\|\mathrm{x}\|=\sqrt{\mathrm{x}\cdot\mathrm{x}}=\sqrt{\sum_{i=1}^{n}(x_i)^2}.$$

This length function satisfies the required properties of a norm and is called the Euclidean norm on R^n.

Finally, one can use the norm to define a metric (or distance function) on R^n by,

$$d(\mathrm{x},\mathrm{y})=\|\mathrm{x}-\mathrm{y}\|=\sqrt{\sum_{i=1}^{n}(x_i-y_i)^2}.$$

This distance function is called the Euclidean metric. This formula expresses a special case of the Pythagorean Theorem.

This distance function (which makes a metric space) is sufficient to define all Euclidean geometry, including the dot product. Thus, a real coordinate space together with this Euclidean structure is called Euclidean space. Its vectors form an inner product space (in fact a Hilbert space), and a normed vector space.

The metric space structure is the main reason behind the use of real numbers R, not some other ordered field, as the mathematical foundation of Euclidean (and many other) spaces. Euclidean space is a complete metric space, a property which is impossible to achieve operating over rational numbers.

Angle

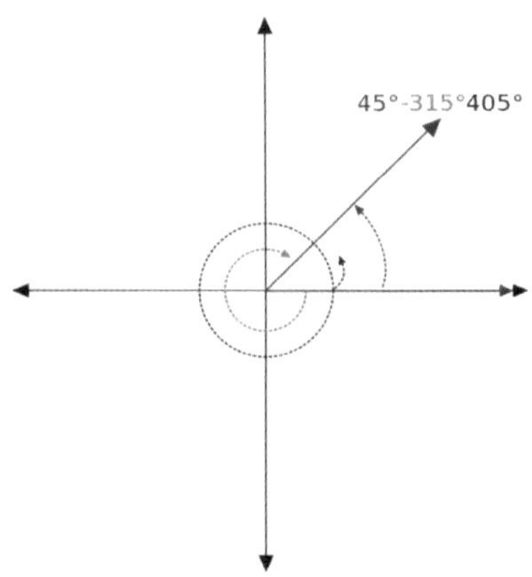

Positive and negative angles on the oriented plane

The (non-reflex) angle θ ($0° \le \theta \le 180°$) between vectors x and y is then given by,

$$\theta = arccos\left(\frac{\mathbf{x} \cdot \mathbf{y}}{\|\mathbf{x}\| \ \|\mathbf{y}\|}\right)$$

Where, Arccos is the arccosine function. It is useful only for $n > 1$, and the case $n = 2$ is somewhat special. Namely, on an oriented Euclidean plane one can define an angle between two vectors as a number defined modulo 1 turn (usually denoted as either 2π or $360°$), such that $\angle yx = -\angle xy$. This oriented angle is equal either to the angle θ from the formula above or to $-\theta$. If one non-zero vector is fixed (such as the first basis vector), then each non-zero vector is uniquely defined by its magnitude and angle.

The angle does not change if vectors x and y are multiplied by positive numbers.

Unlike the aforementioned situation with distance, the scale of angles is the same in pure mathematics, physics, and computing. It does not depend on the scale of distances; all distances may be multiplied by some fixed factor, and all angles will be preserved. Usually, the angle is considered a dimensionless quantity, but there are different units of measurement, such as radian (preferred in pure mathematics and theoretical physics) and degree (°) (preferred in most applications).

Rotations and Reflections

Symmetries of a Euclidean space are transformations which preserve the Euclidean metric (called *isometries*). Although aforementioned translations are most obvious of them, they have the same structure for any affine space and do not show a distinctive character of Euclidean geometry. Another family of symmetries leave one point fixed, which may be seen as the origin without loss of generality. All transformations, which preserves the origin and the Euclidean metric, are linear maps. Such transformations Q must, for any x and y, satisfy:

$$Q\mathrm{x} \cdot Q\mathrm{y} = \mathrm{x} \cdot \mathrm{y} \text{ (explain the notation)},$$
$$|Q\mathrm{x}| = |\mathrm{x}|.$$

Such transforms constitute a group called the *orthogonal group* O(n). Its elements Q are exactly solutions of a matrix equation:

$$Q^{\mathrm{T}}Q = QQ^{\mathrm{T}} = I,$$

Where, Q^{T} is the transpose of Q and I is the identity matrix.

But a Euclidean space is orientable. Each of these transformations either preserves or reverses orientation depending on whether its determinant is +1 or −1 respectively. Only transformations which preserve orientation, which form the *special orthogonal* group SO(n), are considered (proper) rotations. This group has, as a Lie group, the same dimension $n(n-1)/2$ and is the identity component of O(n).

Group	Diffeo-morphic to	Isomorphic to
SO(1)	{1}	
SO(2)	S^1	U(1)
SO(3)	$\mathbf{RP}^3$	SU(2)/{±1}
SO(4)	$(S^3 \times S^3)/\{\pm 1\}$	(SU(2) × SU(2))/{±1}
Note: elements of SU(2) are also known as versors.		

Groups SO(n) are well-studied for $n \leq 4$. There are no non-trivial rotations in 0- and 1-spaces. Rotations of a Euclidean plane ($n = 2$) are parameterised by the angle (modulo 1 turn). Rotations of a 3-space are parameterised with axis and angle, whereas a rotation of a 4-space is a superposition of two 2-dimensional rotations around perpendicular planes.

Among linear transforms in O(n) which reverse the orientation are hyper plane reflections. This is the only possible case for $n \leq 2$, but starting from three dimensions, such isometry in the general position is a rotoreflection.

Euclidean Group

The Euclidean group $E(n)$, also referred to as the group of all isometries ISO(n), treats translations, rotations, and reflections in a uniform way, considering them as group actions in the context of group theory, and especially in Lie group theory. These group actions preserve the Euclidean structure.

As the group of all isometries, ISO(n), the Euclidean group is important because it makes Euclidean geometry a case of Klein geometry, a theoretical framework including many alternative geometries.

The structure of Euclidean spaces – distances, lines, vectors, angles (up to sign), and so on – is invariant under the transformations of their associated Euclidean group. For instance, translations form a commutative subgroup that acts freely and transitively on E^n, while the stabilizer of any point there is the aforementioned O(n).

Along with translations, rotations, reflections, as well as the identity transformation, Euclidean motions comprise also glide reflections (for $n \geq 2$), screw operations and rotoreflections (for $n \geq 3$), and even more complex combinations of primitive transformations for $n \geq 4$.

The group structure determines which conditions a metric space needs to satisfy to be a Euclidean space:

1. Firstly, a metric space must be translationally invariant with respect to some (finite-dimensional) real vector space. This means that the space itself is an affine space, that the space is *flat*, not curved, and points do not have different properties, and so any point can be translated to any other point.
2. Secondly, the metric must correspond in the aforementioned way to some positive-defined quadratic form on this vector space, because point stabilizers have to be isomorphic to O(n).

Non-cartesian Coordinates

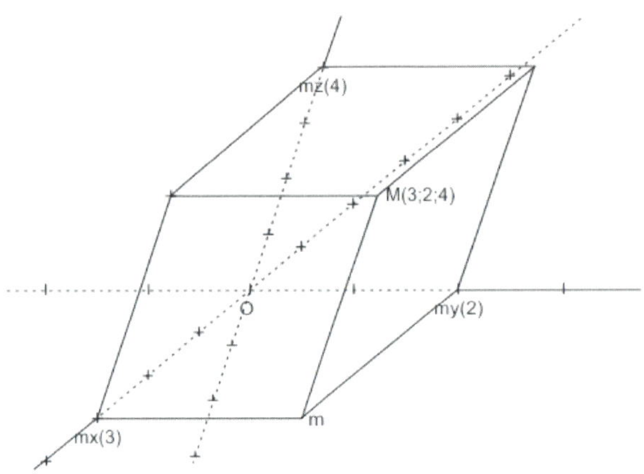

3-dimensional skew coordinates

Cartesian coordinates are arguably the standard, but not the only possible option for a Euclidean space. Skew coordinates are compatible with the affine structure of E^n, but make formulae for angles and distances more complicated.

Another approach, which goes in line with ideas of differential geometry and conformal geometry, is orthogonal coordinates, where coordinate hyper surfaces of different coordinates are orthogonal, although curved. Examples include the polar coordinate system on Euclidean plane, the second important plane coordinate system.

Geometric Shapes

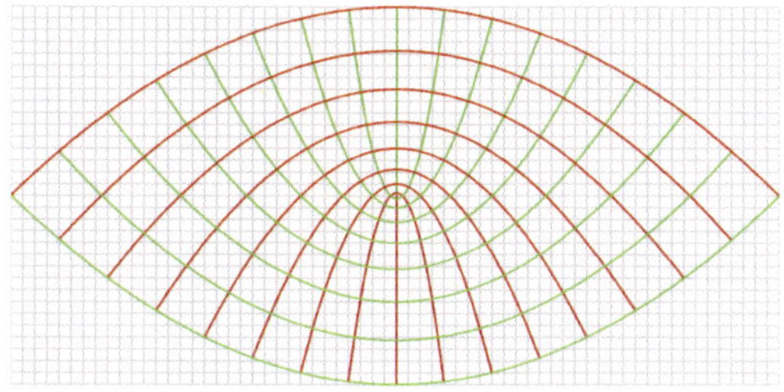

Parabolic coordinates.

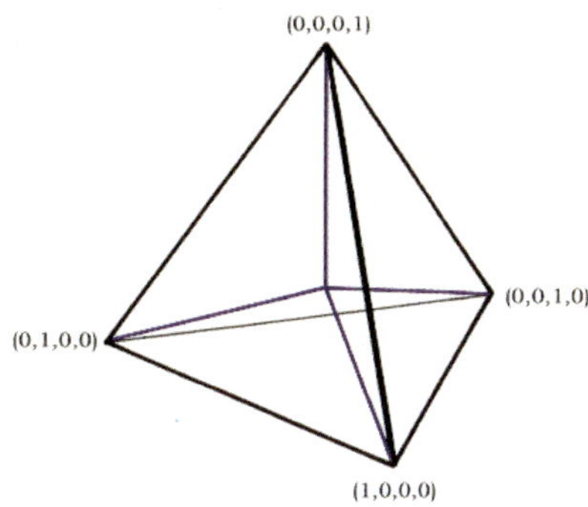

Barycentric coordinates in 3-dimensional space: four coordinates are related with one linear equation.

Polar coordinates: see Angle above	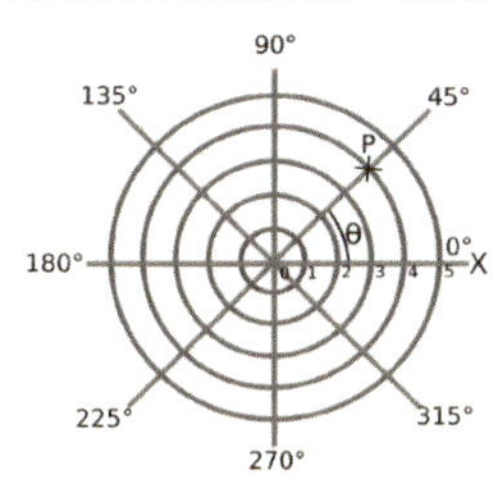

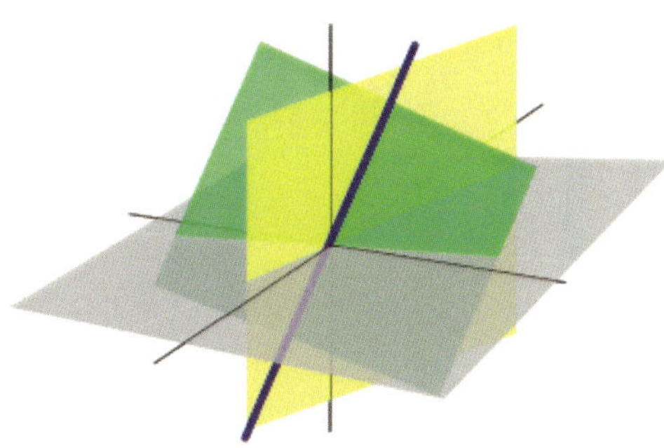

Three mutually transversal planes in the 3-dimensional space and their intersections, three lines.

Lines, Planes and Other Subspaces

The simplest (after points) objects in Euclidean space are flats, or Euclidean *subspaces* of lesser dimension. Points are 0-dimensional flats, 1-dimensional flats are called *(straight) lines*, and 2-dimensional flats are *planes*. (n – 1)-dimensional flats are called *hyper planes*.

Any two distinct points lie on exactly one line. Any line and a point outside it lie on exactly one plane. More generally, the properties of flats and their incidence of Euclidean space are shared with affine geometry, whereas the affine geometry is devoid of distances and angles.

Line Segments and Triangles

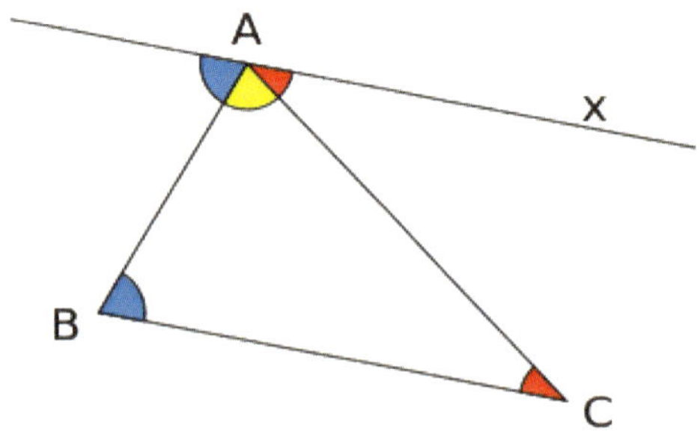

The sum of angles of a triangle is an important problem.
In a Euclidean space it invariably equals to 180°, or a half-turn.

This is not only a line which a pair (A, B) of distinct points defines. Points on the line which lie between A and B, together with A and B themselves, constitute a line segment AB. Any line segment has the length, which equals to distance between A and B. If $A = B$, then the segment is degenerate and its length equals to 0, otherwise the length is positive.

A (non-degenerate) triangle is defined by three points not lying on the same line. Any triangle lies on one plane. The concept of triangle is not specific to Euclidean spaces, but Euclidean triangles have numerous special properties and define many derived objects.

A triangle can be thought of as a 3-gon on a plane, a special (and the first meaningful in Euclidean geometry) case of a polygon.

Polytopes and Root Systems

The Platonic solids are the five polyhedra that are most regular in a combinatoric sense, but also, their symmetry groups are embedded into O(3)	
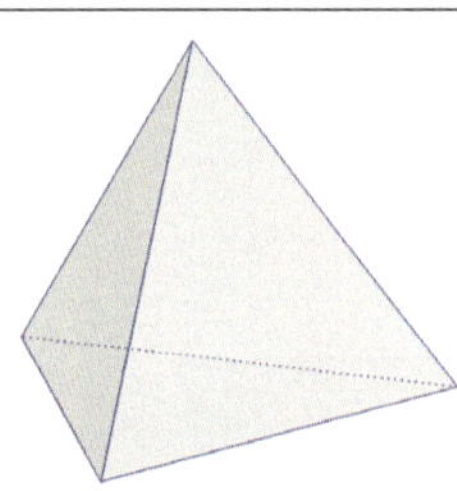 Tetrahedron	 Cube (green) and octahedron (cyan)
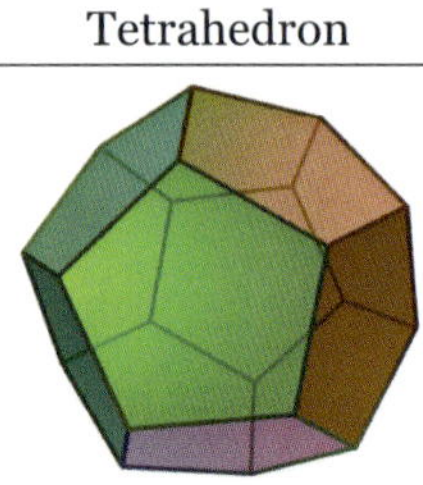 Dodecahedron	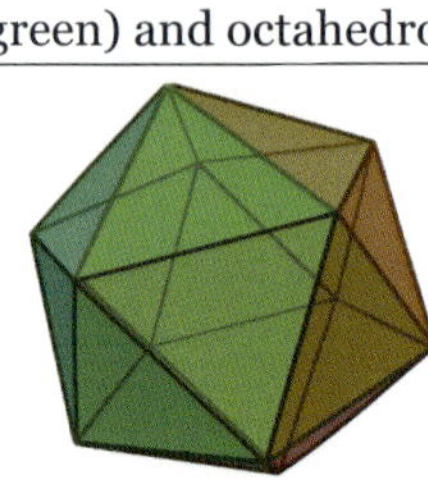 Icosahedron

Polytope is a concept that generalizes polygons on a plane and polyhedra in 3-dimensional space (which are among the earliest studied geometrical objects). A simplex is a generalization of a line segment (1-simplex) and a triangle (2-simplex). A tetrahedron is a 3-simplex.

The concept of a polytope belongs to affine geometry, which is more general than Euclidean. But Euclidean geometry distinguish *regular polytopes*. For example, affine geometry does not see the difference between an equilateral triangle and a right triangle, but in Euclidean space the former is regular and the latter is not.

Root systems are special sets of Euclidean vectors. A root system is often identical to the set of vertices of a regular polytope.

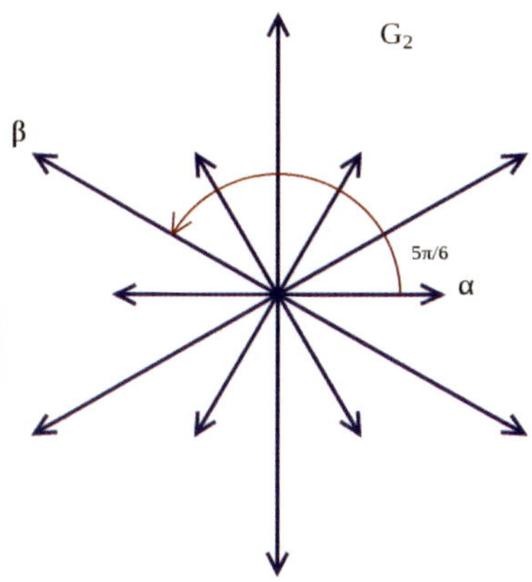

The root system G_2

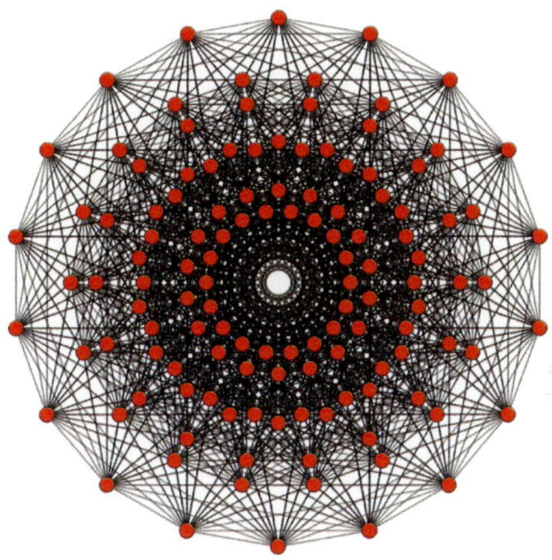

An orthogonal projection of the 2_{31} polytope, whose vertices are elements of the E_7 root system

Curve

A curve is a 1-dimensional entity which can be a straight line or a curved entity. In other words, a curve is a one-dimensional mathematical entity that may be straight or not. Hence, every line – straight or not can be a curve. Curves may be open – when the endpoints of a curve don't meet or closed – when the endpoints of the curve meet.

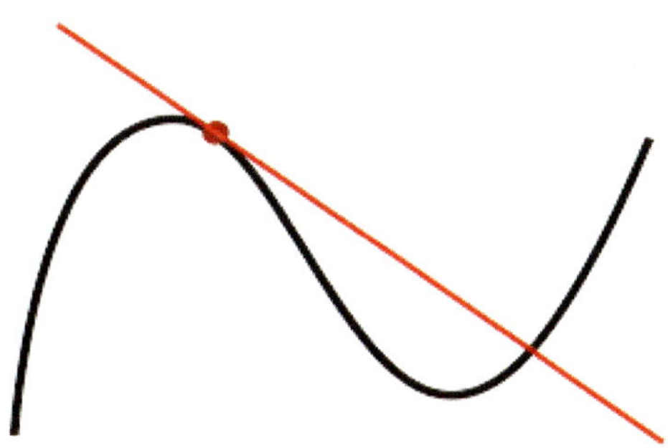

Sphere

A sphere is a round solid object such that it has only one curved surface and every point on this surface is equidistant from the centre of the sphere. The distance from the surface to the centre is called the 'radius' of the sphere.

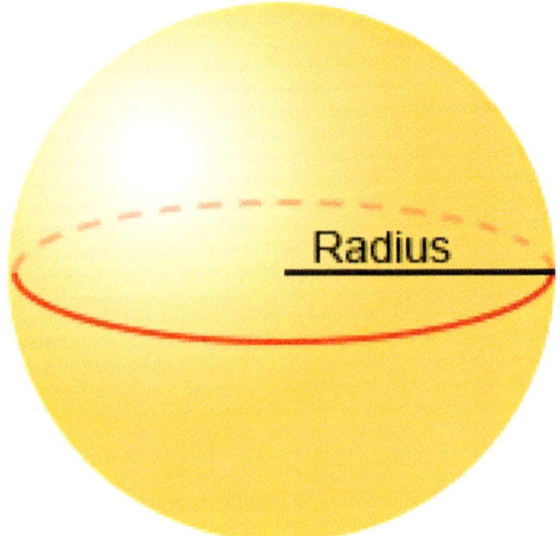

The properties of a sphere are as follows:

1. It is a perfectly symmetrical object. It is symmetrical about the plane containing any of the greater circles of the sphere. In other words, if we cut a sphere along any of its greater circle, we get two hemispheres that would be identical.

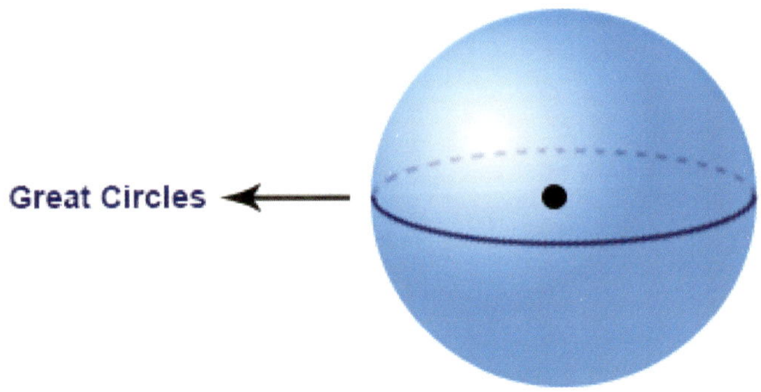

2. The surface of a sphere is uniformly curved. It never has any edges, vertices or faces. Just one single curved surface.
3. It is obtained as a solid of revolution by revolving a semicircle about its diameter.

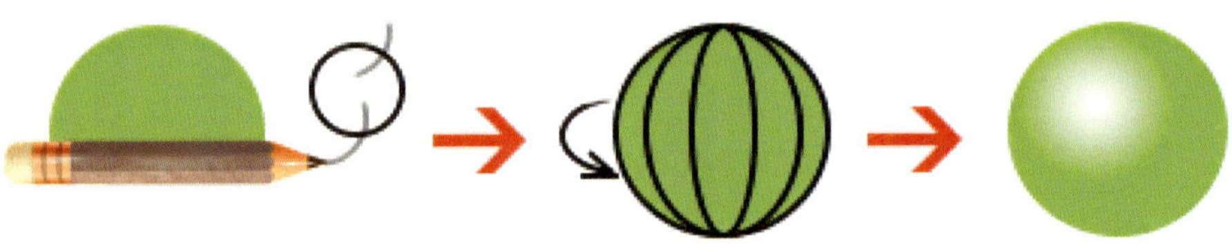

4. For a given volume, a sphere will have the smallest surface area as compared to other solids like prisms, pyramids, cylinders or cones. In other words, if the volume is fixed, say V cm^3 then of all the solids having the same volume, the sphere will have the smallest surface area.
5. For a given surface are, a sphere will have the largest volume as compared to other solids like prisms, pyramids, cylinders or cones. That means, that if we have a specific fixed surface area, say A cm^2, then of all the solids having this same surface area, the sphere will have the largest volume.

Applications

Aside from countless uses in fundamental mathematics, a Euclidean model of the physical space can be used to solve many practical problems with sufficient precision. Two usual approaches are a fixed, or *stationary* reference frame (i.e. the description of a motion of objects as their positions that change continuously with time), and the use of Galilean space-time symmetry (such as in Newtonian mechanics). To both of them the modern Euclidean geometry provides a convenient formalism; for example, the space of Galilean velocities is itself a Euclidean space.

Topographical maps and technical drawings are planar Euclidean. An idea behind them is the scale invariance of Euclidean geometry that permits to represent large objects in a small sheet of paper, or a screen.

Alternatives and Generalizations

Although Euclidean spaces are no longer considered to be the only possible setting for a geometry, they act as prototypes for other geometric objects. Ideas and terminology from Euclidean geometry (both traditional and analytic) are pervasive in modern mathematics, where other geometric objects share many similarities with Euclidean spaces, share part of their structure, or embed Euclidean spaces.

Curved Spaces

A smooth manifold is a Hausdorff topological space that is locally diffeomorphic to Euclidean space. Diffeomorphism does not respect distance and angle, but if one additionally prescribes a smoothly varying inner product on the manifold's tangent spaces, then the result is what is called a Riemannian manifold. Put differently, a Riemannian manifold is a space constructed by deforming and patching together Euclidean spaces. Such a space enjoys notions of distance and angle, but they behave in a curved, non-Euclidean manner. The simplest Riemannian manifold, consisting of R^n with a constant inner product, is essentially identical to Euclidean n-space itself. Less trivial examples are n-sphere and hyperbolic spaces. Discovery of the latter in the 19th century was branded as the non-Euclidean geometry.

Also, the concept of a Riemannian manifold permits an expression of the Euclidean structure in any smooth coordinate system, via metric tensor. From this tensor one can compute the Riemann curvature tensor. Where the latter equals to zero, the metric structure is locally Euclidean (it means that at least some open set in the coordinate space is isometric to a piece of Euclidean space), no matter whether coordinates are affine or curvilinear.

Indefinite Quadratic Form

If one replaces the inner product of a Euclidean space with an indefinite quadratic form, the result is a pseudo-Euclidean space. Smooth manifolds built from such spaces are called pseudo-Riemannian manifolds. Perhaps their most famous application is the theory of relativity, where flat space time is a pseudo-Euclidean space called Minkowski space, where rotations correspond to motions of hyperbolic spaces mentioned above. Further generalization to curved space times form pseudo-Riemannian manifolds, such as in general relativity.

Other Number Fields

Another line of generalization is to consider other number fields than one of real numbers. Over complex numbers, a Hilbert space can be seen as a generalization of Euclidean dot product structure, although the definition of the inner product becomes a sesquilinear form for compatibility with metric structure.

Dimensional Space

In mathematics, the dimensional space plays a very important role. It is defined as a mathematical space which refers to the position of some object. Geometrically, every object is actually defined in terms of minimum number of coordinates that are required to represent a point within it. In fact, the dimension of an object is said to be an intrinsic property that is independent of the space in which the object is located.

For Instance: On a circle situated in a plane, a point can be represented either by two-dimensional Cartesian coordinates or even by a single angle (the polar coordinate). Hence, a circle is a one-dimensional object even though it is located in a two-dimensional plane.

One-dimensional Space

The one-dimensional space is defined as a line which has only one dimension, namely the length. An object that is represented in one-dimensional space, is known as a one-dimensional object, such as a point since it requires only one coordinate in order to specify it, example - a point at 4 (or -7) on number line.

The number line is shown in the following figure:

Another common one dimensional coordinate system is an angle. A point can also be specified only by an angle in one dimension. The inclination of a point from a line is denoted by an angle which is called one dimension.

One-dimensional space in the form of an angle is shown in the following diagram:

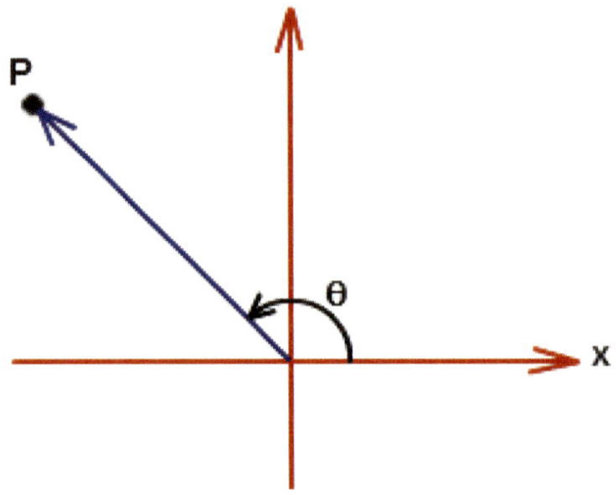

Two-dimensional Space

The two-dimensional space is defined as a system in which there are two dimensions. An object is known as two-dimensional object when it is denoted by the two dimensions - length and breadth only. There must not be a thickness in such shape. The two-dimensional space can be assumed to be situated in the plane of paper. A shape that only has two dimensions (such as width and height) and no thickness. The term two-dimensional space is abbreviated as “2D space”. Sometimes, it is also known as bi-dimensional space.

The most commonly used two-dimensional space is Cartesian coordinate system which is represented by two mutually perpendicular lines, known as X axis (horizontal line) and Y axis (vertical line) as shown in the following diagram:

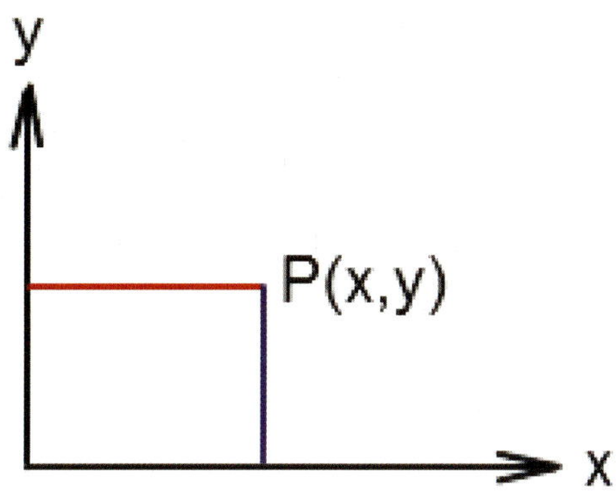

The coordinates of a point in two dimensional space are denoted by (x, y). Few common two-dimensional geometrical figures are circle, square, triangle, rectangle, polygon etc.

A polar coordinate system is also a two-dimensional space that denotes a point in terms of radius ρ and angle θ, as shown in the following diagram:

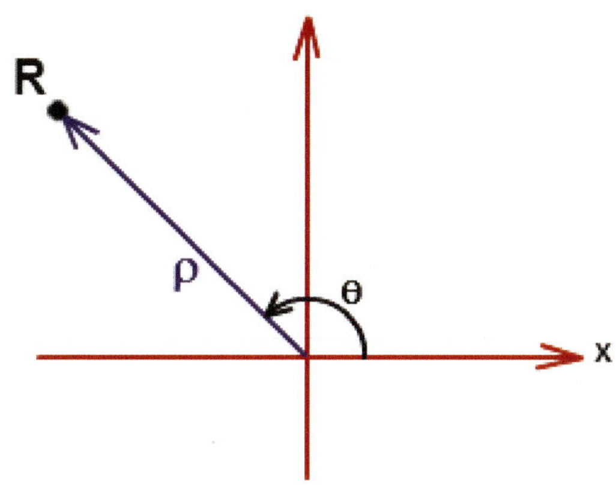

In this system, the coordinates of a point are (r, θ).

Two-dimensional shapes include triangles, squares or quadrilaterals, circles and many other shapes.

Polygons

Most shapes are polygons. Polygons are closed shapes with straight sides that do not cross. A regular polygon is a polygon with all sides of the same length and all angles the same size.

Any number of shapes can be made using straight lines that meet at a point.

The word 'polygon' comes from two greek words, *poly*, meaning 'many', and *gonia*, meaning 'angle'.

Types of Polygons

A polygon is a geometrical plane figure formed by joining three or more straight sides. In geometry, a plane is a flat surface. Some types of polygons include:

Number of sides	General name of polygon
3	triangle
4	quadrilateral
5	pentagon
6	hexagon
7	heptagon
8	octagon
9	nonagon
10	decagon
12	dodecagon

Polygons may be classified as either regular or irregular, depending upon the lengths of their sides.

Regular Polygons

A regular polygon has sides that are all the same length and angles that are all the same size. Equal sides and equal angles are sometimes drawn with small markings ('\\') so you can tell they are equal. Equilateral triangles and squares are regular polygons. Shapes that have equal sides but different angles cannot be called regular polygons.

Irregular Polygons

Irregular polygons have sides which are not all equal in length and angles which may be different in size. Isosceles and scalene triangles are irregular polygons, as are rectangles, kites and trapeziums.

Three-dimensional Space

The three-dimensional space is defined as a three parameter geometrical model for representing the physical universe consisting all the matter. These three dimensions include length, breadth and height; provided that any of the three dimensions must not lie in one plane. The third dimension height may also be referred as depth. The three-dimensional objects are called so since they require three coordinates in order to represent any point inside them.

For example: cube, cuboid, sphere, cylinder, cone are some common three-dimensional geometrical figures.

There are three types of three-dimensional spaces that are most-often used in mathematics - Cartesian coordinate system, spherical coordinate system and cylindrical coordinate system.

The Cartesian coordinate system consists of three mutually perpendicular lines, known as X, Y and Z axis as shown below:

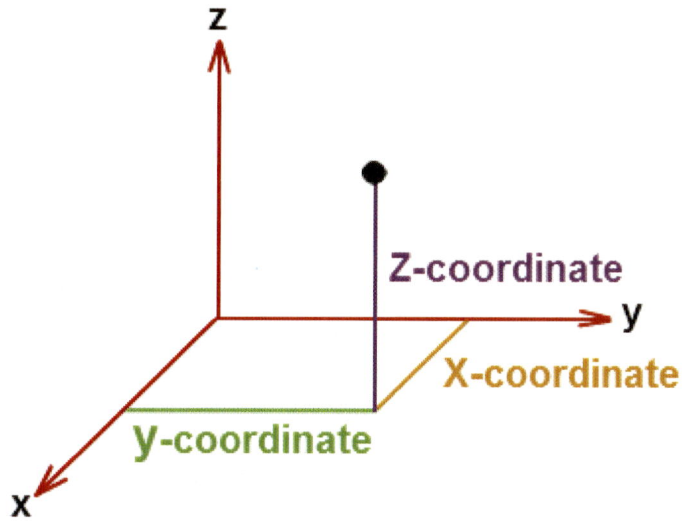

The coordinates of a point are denoted by (x, y, z) in this system.

The coordinates of a point is spherical coordinate system are represented by (ρ, θ, ψ) and in cylindrical coordinate system, they are (ρ, ψ, z).

Coordinate Geometry

A plane is a flat, two-dimensional surface that extends infinitely far. A plane is the two-dimensional analog of a point (zero dimensions), a line (one dimension), and three-dimensional space. A plane in three-dimensional space has the equation

$$ax + by + cz + d = 0,$$

Where, at least one of the numbers a, b and c must be non-zero. A plane in 3D coordinate space is determined by a point and a vector that is perpendicular to the plane.

A plane in 3D coordinate space is determined by a point and a vector that is perpendicular to the plane. Let $P_0 = (x_0, y_0, z_0)$ be the point given, and $\vec{n}$ the orthogonal vector. Also, let $P = (x, y, z)$ be any point in the 3D space, and r and r_0 the position vectors of points P and P_0 respectively. Now, if we let $\vec{n} = (a, b, c)$ then since $\overrightarrow{P_0P}$ is perpendicular to $\vec{n}$, we have

$$\begin{aligned}\overrightarrow{P_0P}.\vec{n} &= (\vec{r} - \vec{r}_0)\cdot\vec{n}\\ &= (x - x_0, y - y_0, z - z_0)\cdot(a, b, c)\\ &= a(x - x_0) + b(y - y_0) + c(z - z_0)\\ &= 0.\end{aligned}$$

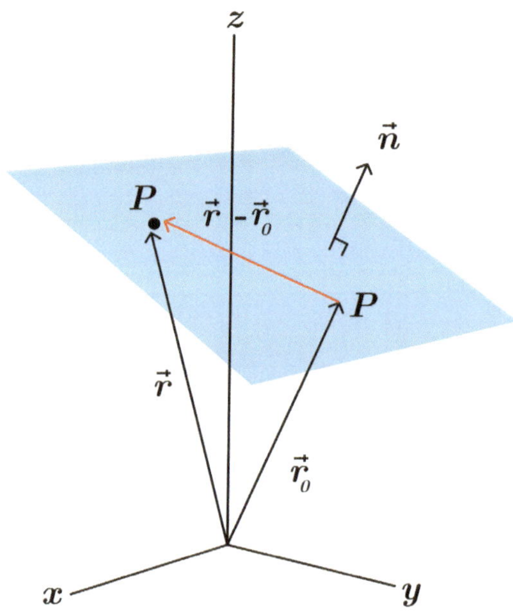

We can also write the above equation of the plane as,

$$ax+by+cz+d=0,$$

Where, $d=-(ax_0+by_0+cz_0)$.

Another way to think of the equation of the plane is as a flattened parallelepiped. A flattened parallelepiped, made of three vectors $\vec{a}=\langle x_1,y_1,z_1\rangle,\vec{b}=\langle x_2,y_2,z_2\rangle,\vec{c}=\langle x_3,y_3,z_3\rangle$, has volume 0. We can use the scalar triple product to compute this volume:

$$0=\vec{a}\cdot(\vec{b}\times\vec{c}),$$

Where, $(\vec{b}\times\vec{c})$ gives the vector that is normal to the plane.

Let's say that the endpoints of $(\vec{b}\times\vec{c})$ are (x,y,z) and (x_0,y_0,z_0) and the components of $\vec{a}$ are $\langle a,b,c\rangle$. Then by taking the dot product, we get the equation of a plane, which is

$$0=a(x-x_0)+b(y-y_0)+c(z-z_0).$$

Here is a problem to try:

Parallel to the Coordinate Planes

The equation of a plane which is parallel to each of the xy -, yz -, and zx - planes and going through a point $A=(a,b,c)$ is determined as follows:

1. The equation of the plane which is parallel to the xy -plane is $z=c$.
2. The equation of the plane which is parallel to the yz -plane is $x=a$.
3. The equation of the plane which is parallel to the zx -plane is $y=b$.

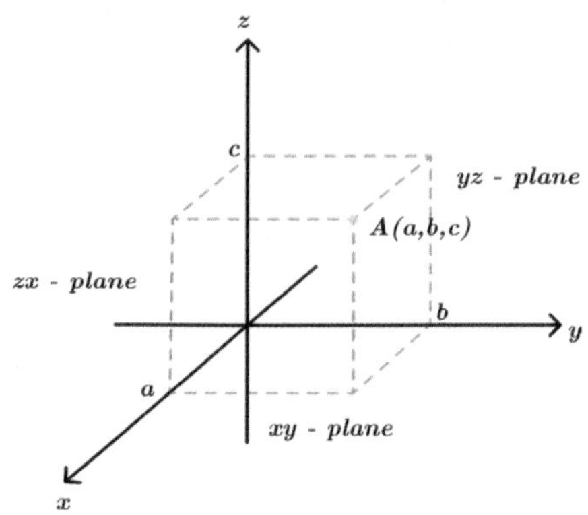

Here is an example based on the above:

What is the equation of the plane which passes through the point $B=(4,1,0)$ and is parallel to the yz -plane?

Since the x-coordinate of B is 4, the equation of the plane passing through B parallel to the yz-plane is:

$$x=4.$$

Normal Vector and a Point

If we know the normal vector of a plane and a point passing through the plane, the equation of the plane is established.

Thus, the equation of a plane through a point $A=(x_1, y_1, z_1)$ whose normal vector $\vec{n}=(a,b,c)$ is:

$$a(x-x_1)+b(y-y_1)+c(z-z_1)=0.$$

Check out the following examples:

If a plane is passing through the point $A=(1,3,2)$ and has normal vector $\vec{n}=(3,2,5)$, then what is the equation of the plane?

The equation of the plane which passes through $A=(1,3,2)$ and has normal vector $\vec{n}=(3,2,5)$ is

$$3(x-1)+2(y-3)+5(z-2)=0$$
$$3x-3+2y-6+5z-10=0$$
$$3x+2y+5z-19=0.$$

If a plane is passing through the point $A=(5,6,2)$ and has normal vector $\vec{n}=(-1,3,-7)$, then what is the equation of the plane?

The equation of the plane which passes through the point $A=(5,6,2)$ and has normal vector $\vec{n}=(-1,3,-7)$ is:

$$-1(x-5)+3(y-6)-7(z-2)=0$$
$$-x+5+3y-18-7z+14=0$$
$$-x+3y-7z+1=0.$$

Passing through Three Points

When we know three points on a plane, we can find the equation of the plane by solving simultaneous equations.

Let $ax+by+cz+d=0$ be the equation of a plane on which there are the following three points: $A=(1,0,2), B=(2,1,1)$ and $C=(-1,2,1)$. Then the equation of the plane is established as follows:

We already have the equation of the plane with 4 unknown constants:

$$ax+by+cz+d=0.$$

We also get the following 3 equations by substituting the coordinates of $A, B,$ and C into just above equation:

$$a\cdot 1+b\cdot 0+c\cdot 2+d=0$$
$$a\cdot 2+b\cdot 1+c\cdot 1+d=0$$
$$a\cdot(-1)+b\cdot 2+c\cdot 1+d=0,$$

Which gives $b=3a, c=4a, d=-9a.$

Now substitute equation $b=3a, c=4a, d=-9a$ in equation $ax+by+cz+d=0$.

Then we have,

$$ax+3ay+4az-9a=0$$
$$x+3y+4z-9=0.$$

Hence, the equation of the plane passing through the three points $A=(1,0,2,\ ), B(2,1,1)$ and $C=(-1,2,1)$ is:

$$x+3y+4z-9=0.$$

Using this method, we can find the equation of a plane if we know three points. Here are a couple of examples:

If a plane is passing through the three points $A=(0,0,2), B=(1,0,1)$, and $C=(3,1,1)$, then what is equation of the plane?

Let the equation of the plane be $ax+by+cz+d=0.$

Then since this plane includes the three points $A=(0,0,2), B=(1,0,1)$, and $C=(3,1,1)$, we have,

$$a\cdot 0+b\cdot 0+c\cdot 2+d=0$$
$$a\cdot 1+b\cdot 0+c\cdot 1+d=0$$
$$a\cdot 3+b\cdot 1+c\cdot 1+d=0$$

Which gives,

$$b = -2a, c = a, d = -2a.$$

Substituting $(b = 3a, c = 4a, d = -9a)$ into $(ax + by + cz + d = 0)$ we have,

$$ax + -2ay + az - 2a = 0$$
$$x - 2y + z - 2 = 0.$$

Hence, the equation of the plane passing through the three points $A = (0, 0, 2), B = (1, 0, 1)$ and $C = (3, 1, 1)$ is

$$x - 2y + z - 2 = 0.$$

If a plane is passing through the three points $A = (3, 1, 2), B = (6, 1, 2)$, and $C = (0, 2, 0)$, then what is the equation of the plane?

Let the equation of the plane be $ax + by + cz + d = 0$.

Then since this plane includes the three points $A = (0, 0, 2), B = (1, 0, 1)$, and $C = (3, 1, 1)$, we have,

$$a \cdot 3 + b \cdot 1 + c \cdot 2 + d = 0$$
$$a \cdot 6 + b \cdot 1 + c \cdot 2 + d = 0$$
$$a \cdot 0 + b \cdot 2 + c \cdot 0 + d = 0$$

Which gives $a = 0, c = \frac{1}{2}b, d = -2b.$

Substituting $(a = 0, c = \frac{1}{2}b, d = -2b)$ into $(ax + by + cz + d = 0)$ we have

$$0x + -by + \frac{1}{2}bz - 2b = 0$$
$$x - y + \frac{1}{2}z - 2 = 0$$
$$2x - 2y + z - 4 = 0.$$

Hence, the equation of the plane passing through the three points $A = (0, 0, 2), B = (1, 0, 1)$, and $C = (3, 1, 1)$ is $2x - 2y + z - 4 = 0$.

Properties

In three-dimensional Euclidean space, we may exploit the following facts that do not hold in higher dimensions:

1. Two planes are either parallel or they intersect in a line.
2. A line is either parallel to a plane, intersects it at a single point, or is contained in the plane.
3. Two lines perpendicular to the same plane must be parallel to each other.
4. Two planes perpendicular to the same line must be parallel to each other.

Euclidean Distance

The euclidean distance is the distance between two points in euclidean space. The two points P and Q in two dimensional euclidean spaces and P with the coordinates (p1, p2), Q with the coordinates (q1, q2). The line segment with the endpoints of P and Q will form the hypotenuse of a right angled triangle. The distance between two points p and q is defined as the square root of the sum of the squares of the differences between the corresponding coordinates of the points. The two-dimensional euclidean geometry, the euclidean distance between two points a = (ax, ay) and b = (bx, by) is defined as:

$$d(a,b)=\sqrt{(bx-ax)^2+(by-ay)^2}$$

Euclidean Distance Algorithm

Euclidean distance algorithm computes the minimum distance between a column vector x and a collection of column vectors in the code book matrix cb. The algorithm computes the minimum distance to x and finds the column vector in cb that is closest to x.

The steps of Euclidean distance algorithm shows:

$$d(a,b)=|p-q|$$

$$\sqrt{(p_1-q_1)^2+(p_2-q_2)^2+\ldots+(p_n-q_n)^2}$$
$$=\sqrt{\sum_{i=1}^{n}(p_i-q_i)^2}$$

In one dimension, the distance between two points, x_1 and x_2, on a line is simply the absolute value of the difference between the two points as:

$$\sqrt{(X_2-X_1)^2}=|X_2-X_1|$$

In two dimensions, the distance between P = (p1, p2) and q = (q1, q2) as:

$$\sqrt{(p_1-q_1)^2+(p_2-q_2)^2}$$

Steps of algorithm:

Step 1: Load the column vector x;

Step 2: Load the code book;

Step 3: Minimum distance is initially set to the first element of cb;

Step 4: i.e. set idx=1;

Step 5: Compute distance by normalized values of (x-cb) for all cb;

Step 6: If d is less than distance set distance is equal to d;

Step 7: Set idx=index;

Step 8: End.

Euclidean Distance in Three Dimension

Let $P(x_1, y_1, z_1)$ and $Q(x_2, y_2, z_2)$ be two points in $\mathbb{R}^3$. Let M be the foot of the perpendicular from Q to the plane through P that is parallel to the (x, y)-plane; then M has coordinates (x_2, y_2, z_1). Next, let N be the point in this plane with coordinates (x_1, y_2, z_1).

Then the triangles PQM and PMN are both right-angled triangles, with right angles at M and N, respectively, as shown below:

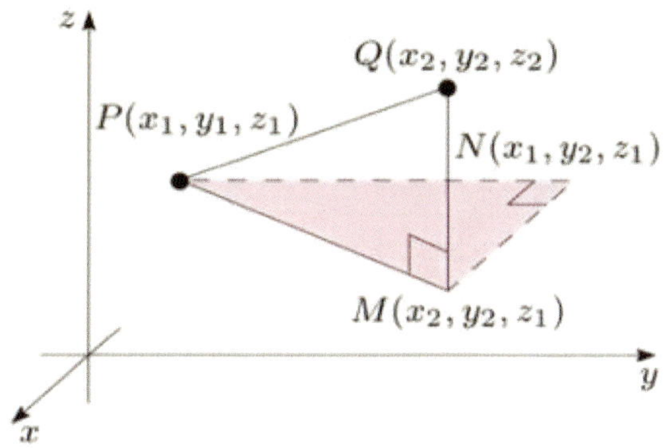

The length of PN is $|y_2 - y_1|$, the length of NM is $|x_2 - x_1|$, and the length of MQ is $|z_2 - z_1|$. It follows from Pythagoras' Theorem that:

$PM^2 = NM^2 + PN^2$,

Then,

$PM^2 = (x_2 - x_1)^2 + (y_2 - y_1)^2$.

Using Pythagoras' Theorem again, we obtain:

$PQ^2 = PM^2 + MQ^2$

So,

$$PQ^2 = (X_2 - X_1)^2 + (y_2 - y_1)^2 + (Z_2 - Z_1)^2,$$

That is,

$$PQ = \sqrt{(X_2 - X_1)^2 + (y_2 - y_1)^2 + (Z_2 - Z_1)^2}.$$

Distance Formula in Three-dimensional Euclidean Space

The distance between two points (x_1, y_1, z_1) and (x_2, y_2, z_2) in $\mathbb{R}^3$ is:

$$\sqrt{(X_2 - X_1)^2 + (y_2 - y_1)^2 + (Z_2 - Z_1)^2}.$$

For example, it follows from this formula that the distance between the points (1, 2, 3) and (4, −2, 15) is:

$$\sqrt{(4-1)^2 + (-2-2)^2 + (15-3)^2} = \sqrt{169} = 13.$$

Euclidean Function

The input source data is a feature class; it will first be converted internally to a raster before the euclidean analysis is performed. The resolution will be smaller of the height or width of the extent of the feature class, divided by 250 and the resolution can be set with the output cell size parameter.

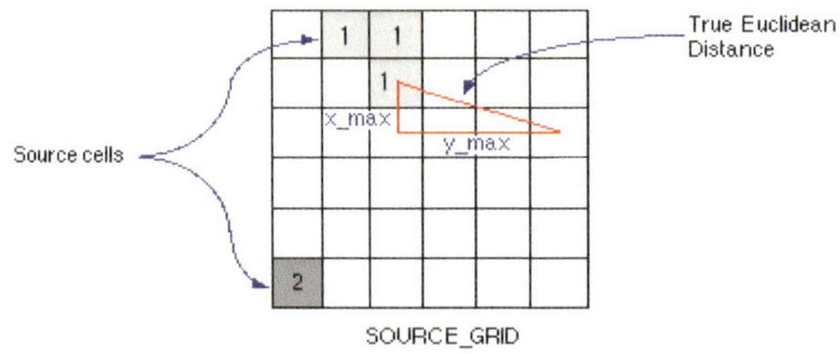

The euclidean distance is calculated from the center of the source cells to the center of each of the surrounding cells and true distance is calculated to each cell in the distance functions. The euclidean algorithm works as follows:

For each cell, the distance is calculated to each source cell by calculating the hypotenuse, with the x-max and y-max.

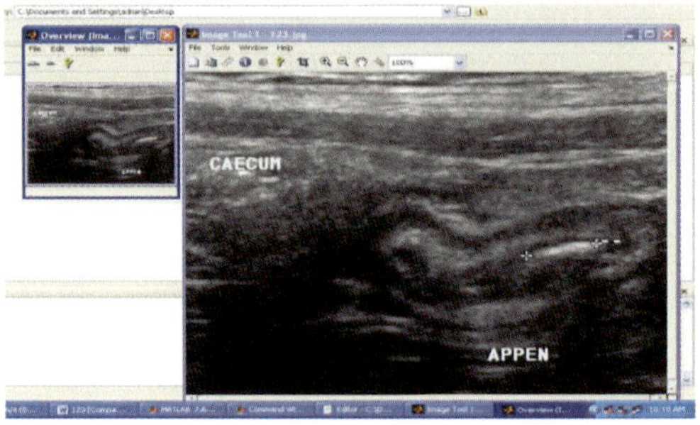

Figure: Sonographic image of Appendicitis

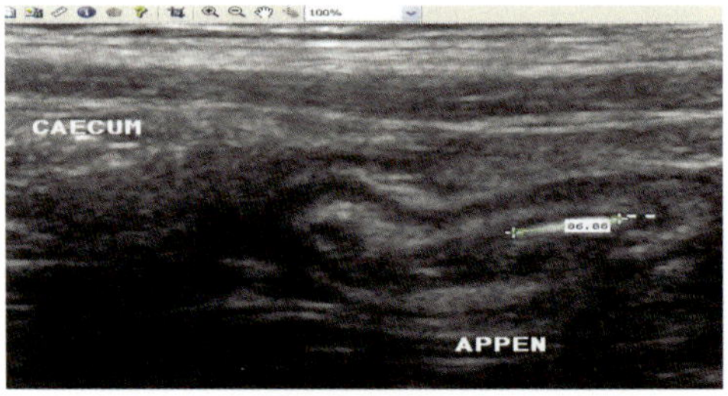

Figure: Euclidean Distance Measure

The shortest distance to a source is determined if it is less than the specified maximum distance and the value is assigned to the cell location on the output raster. The output values for the euclidean distance raster are floating-point distance values and the cell is at an equal distance to two or more sources, which assigned to the source as encountered first in the scanning process. The euclidean distance raster tells how close each cell is to the nearest source and raster defines which source zone and cell value is the closest. The euclidean direction identifies the direction to the closest source cell. Figure shows euclidean distance measure.

Image and Euclidean Distance

M X N images can be easily discussed in an MN dimensional euclidean space, called image space. It is natural to adopt the base $e_1, e_2, \ldots, e_{MN}$ to form a coordinate system of the image space, where e_{kN+1} corresponds to an ideal point source with unit intensity at location (k, l). Thus an image x = ($x^1, x^2, \ldots, x^{MN}$), where x^{kN+l} is the gray level at the (k, l)th pixel, is represented as a point in the image space, and x^{kN+l} with respect to e_{kN+1}. The origin of the image space is an image whose gray levels are zero everywhere.

Although the algebra of the image space can be easily formulated as above, the euclidean distance of images could not be determined until the metric coefficients of the basis are given. The metric coefficients g_{ij} i, j =1, 2,....., MN, are defined as:

$$g_{ij} = < e_i, e_j > = \sqrt{< e_i, e_i >}\sqrt{< e_j, e_j >}.\cos \theta_{ij}.$$

Where,

<,> is the scalar product, and θ_{ij} is the angle between e_i and e_j.

Note that, if $< e_i\ e_i > = < e_j\ e_j > = \ldots$, i.e. all the base vectors have the same length, then g_{ij} depends completely on the angle θ_{ij}. Given the metric coefficients, the euclidean distance of two images x, y is written by,

$$d_E^2(x, y) = \sum_{i,j=1}^{MN} g_{ij}\left(x^i - y^i\right)\left(x^j - y^j\right) = (x-y)^T G(x-y).$$

Where, the symmetric matrix $G = (g_{ij})_{MNxMN}$ will be referred to as metric matrix. For images of fixed size M by N, every MNth order and positive definite matrix G induces a euclidean distance. But most of them are not appropriate for measuring image distances. For example, suppose any two base vectors e_i, e_j ($i \neq j$), no matter which pixels they correspond to are mutually perpendicular, the basis then forms a Cartesian coordinate system. Accordingly, G is the identity matrix, and it induces the traditional euclidean distance given by,

$$d_E^2(x, y) = \sum_{k=1}^{MN}\left(x^k - y^k\right)^2$$

Geometrically, this defect is due to the orthogonality of the base vectors $e_1, e_2, \ldots, e_{MN}$ which correspond to pixels. Clearly, the information about the spatial relationship, i.e. the distances between the pixels, cannot be reflected by all mutually perpendicular base vectors. Such information, however, often appears in intuitive image distance as in the following statement: A slightly deformed image is very similar to the original one. Here, slightly deformed means that pixels in the deformed image are close to the corresponding pixels in the original image. This implies that a good euclidean distance for images should contain the information of pixel distances. Accordingly, the metric coefficients, which define the euclidean distance, have to be related to the pixel distances.

If the metric coefficients depend properly on the pixel distances, the obtained euclidean distance is insensitive to small deformation. The two distances that are considered are: one is the image distance measured in the high dimensional image space, the other is the pixel distance. Let P_iP_j, i, j = 1, 2,....., MN, be pixels. The pixel distance, written as $|P_i - P_j|$, is the distance between P_i and P_j on the image lattice. For example, if P_i is at location (k, l), and P_j is at (k′, l′), $|P_i - P_j|$ may be $\sqrt{(k-k')^2+(l-l')^2}$. Let euclidean distance, set $\theta_{ij} = \pi/2$. Then $g_{ij} = \cos\theta_{ij} = 0$.

Hence,

$$d_E(x,y)=\sqrt{g_{ii}\left(x^i-y^i\right)^2+g_{jj}\left(x^j-y^j\right)^2+2g_{ij}\left(x^i-y^i\right)\left(x^j-y^j\right)}=\sqrt{g_{ii}+g_{jj}}=\sqrt{2}$$

$$g_{ij}=\cos\theta_{ij}\approx 1,$$

Thus, obtained euclidean distance is,

$$d_E(x,y)=\sqrt{g_{ii}+g_{jj}-2g_{ij}}\approx 0\,.$$

Butterfly Theorem

The butterfly theorem is a well-known result from Euclidean geometry.

Statement

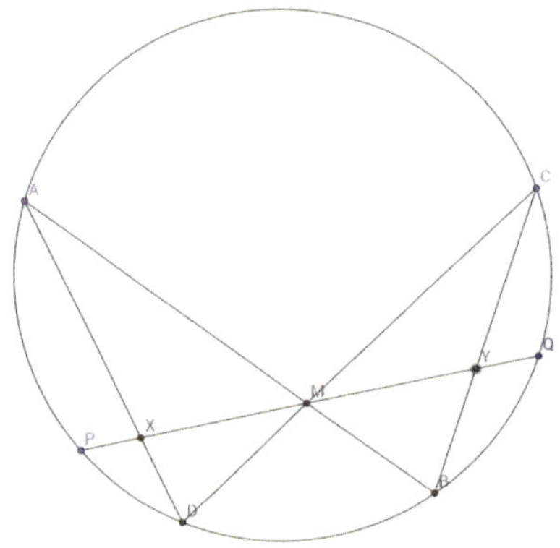

Let M be the midpoint of chord PQ of a circle.

Draw two other chords AB and CD through M.

If AD and BC intersect chord PQ at X and Y, respectively, M is the midpoint of XY.

Looking at the diagram, you can probably tell how the butterfly theorem got its name.

Proof

There are various proofs for the butterfly theorem. We're going to prove it by using the properties of similar triangles.

We're going to start by dropping perpendiculars x_1 and x_2 from X to AB and CD, respectively, and likewise y_2 and y_1 from Y to AB and CD, respectively.

Here's what the diagram looks like after we do that:

Let, $MQ = PM = a$, $XM = x$, and $MY = y$.

Now we have a lot of similar triangles we can work with, and from these similar triangles we have a lot of ratios that are equal to one another:

$$\frac{x}{y}=\frac{x_1}{y_2},\frac{x}{y}=\frac{x_2}{y_1},\frac{x_1}{y_1}=\frac{AX}{CY},\frac{x_2}{y_2}=\frac{XD}{YB}.$$

So, we have $\frac{x^2}{y^2}=\frac{x_1}{y_2}.\frac{x_2}{y_1}=\frac{x_1}{y_1}.\frac{x^2}{y^2}=\frac{AX\cdot XD}{CY\cdot YB}$.

Now we're going to use the intersecting chords theorem:

From that theorem, we have, $\frac{AX\cdot XD}{CY\cdot YB}=\frac{PX\cdot XQ}{PY\cdot YQ}$.

So, we can write $\frac{x^2}{y^2}=\frac{PX\cdot XQ}{PY\cdot YQ}=\frac{(a-x)(a+x)}{(a+y)(a-y)}=\frac{a^2-x^2}{a^2-y^2}$.

In other words, $\frac{x^2}{y^2}=\frac{a^2-x^2}{a^2-y^2}$. But that can only happen if $x^2=y^2$.

This means $x=y$, which implies M is the midpoint of XY.

Ceva's Theorem

Ceva's Theorem is a criterion for the concurrence of cevians in a triangle.

Statement

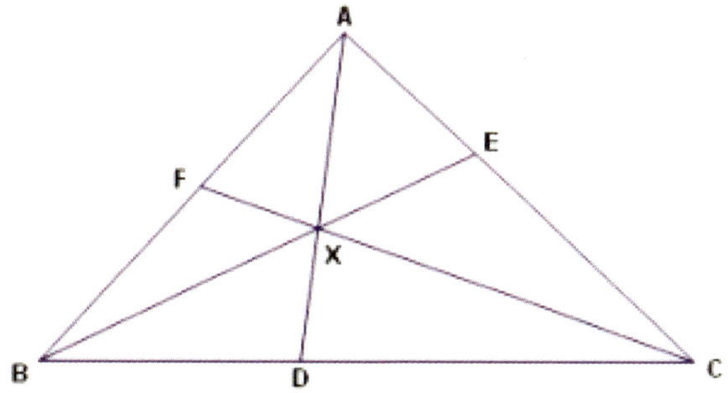

Let ABC be a triangle, and let D, E, F be points on lines BC, CA, AB, respectively. Lines AD, BE, CF are concurrent if and only if:

$$\frac{BD}{DC}\cdot\frac{CE}{EA}\cdot\frac{AF}{FB}=1.$$

Where, lengths are directed. This also works for the reciprocal of each of the ratios, as the reciprocal of 1 is 1.

The proof using Routh's Theorem is extremely trivial, so we will not include it.

Proof

We will use the notation $[ABC]$ to denote the area of a triangle with vertices A, B, C.

First, suppose AD, BE, CF meet at a point X. We note that triangles ABD, ADC have the same altitude to line BC, but bases BD and DC. It follows that $\frac{BD}{DC}=\frac{[ABD]}{[ADC]}$. The same is true for triangles XBD, XDC, so

$$\frac{BD}{DC}=\frac{[ABD]}{[ADC]}=\frac{[XBD]}{[XDC]}=\frac{[ABD][XBD]}{[ADC][XDC]}=\frac{[ABX]}{[AXC]}.$$

Similarly, $\frac{CE}{EA}=\frac{[BCX]}{[BXA]}$ and $\frac{AF}{FB}=\frac{[CAX]}{[CXB]}$, so

$$\frac{BD}{DC}\cdot\frac{CE}{EA}\cdot\frac{AF}{FB}=\frac{[ABX]}{[AXC]}\cdot\frac{[BCX]}{[BXA]}\cdot\frac{[CAX]}{[CXB]}=1.$$

Now, suppose D, E, F satisfy Ceva's criterion, and suppose AD, BE intersect at X. Suppose the line CX intersects line AB at F'. We have proven that F' must satisfy Ceva's criterion. This means that,

$$\frac{AF'}{F'B}=\frac{AF}{FB},$$

So,

$$F'=F,$$

And line CF concurrs with AD and BE.

Proof by Barycentric Coordinates

Since, $D\in BC$, we can write its coordinates as $(0, d, 1-d)$

The equation of line AD is then, $z=\frac{1-d}{d}y$.

Similarly,

Since, $E=(1-e,0,e)$, and $F=(f,1-f,0)$, we can see that the equations of BE and CF respectively are $x=\frac{1-e}{e}z$ and $y=\frac{1-f}{f}x$.

Multiplying the three together yields the solution to the equation:

$$xyz=\frac{1-e}{e}.z.\frac{1-f}{f}.x.\frac{1-d}{d}y$$

Dividing by xyz yields:

$1=\frac{1-e}{e}.\frac{1-f}{f}.\frac{1-d}{d}$, which is equivalent to Ceva's theorem.

Trigonometric Form

The trigonometric form of Ceva's Theorem (Trig Ceva) states that cevians AD, BE, CF concur if and only if,

$$\frac{\sin BAD}{\sin DAC}.\frac{\sin CBE}{\sin EBA}.\frac{\sin ACF}{\sin FCB}=1$$

Proof

First, suppose AD, BE, CF concur at a point X.

$$\frac{[BAX]}{[XAC]}=\frac{\frac{1}{2}AB\cdot AX\cdot\sin BAX}{\frac{1}{2}AX\cdot AC\cdot\sin XAC}=\frac{AB\cdot\sin BAD}{AC\cdot\sin DAC},$$

And similarly,

$$\frac{[CBX]}{[XBA]}=\frac{BC\cdot\sin CBE}{BA\cdot\sin EBA};\quad\frac{[ACX]}{[XCB]}=\frac{CA\cdot\sin ACF}{CB\cdot\sin FCB}.$$

It follows that:

$$\frac{\sin BAD}{\sin DAC}.\frac{\sin CBE}{\sin EBA}.\frac{\sin ACF}{\sin FCB}=\frac{AB\cdot\sin BAD}{AC\cdot\sin DAC}.\frac{BC\cdot\sin CBE}{BA\cdot\sin EBA}.\frac{CA\cdot\sin ACF}{CB\cdot\sin FCB}$$

$$=\frac{[BAX]}{[XAC]}.\frac{[CBX]}{[XBA]}.\frac{[ACX]}{[XCB]}=1.$$

Here, sign is irrelevant, as we may interpret the sines of directed angles mod π to be either positive or negative.

The converse follows by an argument almost identical to that used for the first form of Ceva's Theorem.

References

- Ball, W.W. Rouse (1960). A Short Account of the History of Mathematics (4th ed. [Reprint. Original publication: London: Macmillan & Co., 1908] ed.). New York: Dover Publications. pp. 50–62. ISBN 0-486-20630-0
- Geometry, introduction-to-euclids-geometry: toppr.com, Retrieved 22 May 2018
- E.D. Solomentsev (7 February 2011). "Euclidean space". Encyclopedia of Mathematics. Springer. Retrieved 1 May 2014
- Properties-of-sphere, geometry: tutorvista.com, Retrieved 12 June 2018
- 3d-coordinate-geometry-equation-of-a-plane: brilliant.org, Retrieved 29 April 2018

3

Differential Geometry

Differential geometry is a branch of geometry that applies the techniques and principles of differential and integral calculus along with linear and multilinear algebra, to study geometry. The aim of this chapter is to explore the central aspects of differential geometry, such as Riemannian geometry, Finsler geometry, Symplectic geometry, Contact geometry, Euler's theorem, Cohn-Vossen's inequality, Darboux's theorem, etc.

Differential geometry is the branch of mathematics that studies the geometry of curves, surfaces, and manifolds (the higher-dimensional analogs of surfaces). The discipline owes its name to its use of ideas and techniques from differential calculus, though the modern subject often uses algebraic and purely geometric techniques instead. Although basic definitions, notations, and analytic descriptions vary widely, the following geometric questions prevail: How does one measure the curvature of a curve within a surface (intrinsic) versus within the encompassing space (extrinsic)? How can the curvature of a surface be measured? What is the shortest path within a surface between two points on the surface? How is the shortest path on a surface related to the concept of a straight line?

While curves had been studied since antiquity, the discovery of calculus in the 17th century opened up the study of more complicated plane curves such as those produced by the French mathematician René Descartes (1596–1650) with his "compass". In particular, integral calculus led to general solutions of the ancient problems of finding the arc length of plane curves and the area of plane figures. This in turn opened the stage to the investigation of curves and surfaces in space—an investigation that was the start of differential geometry.

Some of the fundamental ideas of differential geometry can be illustrated by the strake, a spiraling strip often designed by engineers to give structural support to large metal cylinders such as smokestacks. A strake can be formed by cutting an annular strip (the region between two concentric circles) from a flat sheet of steel and then bending it into a helix that spirals around the cylinder, as illustrated in the figure. What should the radius r of the annulus be to produce the best fit? Differential geometry supplies the solution to this problem by defining a precise measurement for the curvature of a curve; then r can be adjusted until the curvature of the inside edge of the annulus matches the curvature of the helix.

An important question remains: Can the annular strip be bent, without stretching, so that it forms a strake around the cylinder? In particular, this means that distances measured along the surface (intrinsic) are unchanged. Two surfaces are said to be isometric if one can be bent (or transformed) into the other without changing intrinsic distances. (For example, because a sheet of paper can be rolled into a tube without stretching, the sheet and tube are "locally" isometric—only locally because new, and possibly shorter, routes are created by connecting the two edges of the paper.) Thus, the second question becomes: Are the annular strip and the strake isometric?

To answer this and similar questions, differential geometry developed the notion of the curvature of a surface.

Differential Geometry of Curves

The differential geometry of curves and surfaces has two aspects. One, which may be called classical differential geometry, started with the beginnings of calculus. Roughly speaking, classical differential geometry is the study of local properties of curves and surfaces. By local properties we mean those properties which depend only on the behavior of the curve or surface in the neighborhood of a point. The methods which have shown themselves to be adequate in the study of such properties are the methods of differential calculus. Because of this, the curves and surfaces considered in differential geometry will be defined by functions which can be differentiated a certain number of times.

The other aspect is the so-called global differential geometry. Here one studies the influence of the local properties on the behavior of the entire curve or surface.

Parametrized Curves

We denote by R^3 by the set of triples (x, y, z) of real numbers. Our goal is to characterize certain subsets of R^3 (to be called curves) that are, in a certain sense, one-dimensional and to which the methods of differential calculus can be applied. A natural way of defining such subsets is through differentiable functions. We say that a real function of a real variable is differentiable (or smooth) if it has, at all points, derivatives of all orders (which are automatically continuous).

Definition: A parametrized differentiable curve is a differentiable $map\ \alpha : I \to R^3$ of an open interval I = (a, b) of the real line R into R^3.

The word differentiable in this definition means that   is a correspondence which maps each $t \in I$ into a point $\alpha(t) = (x(t), y(t), z(t)) \in R^3$ in such a way that the functions x(t), yet), z(t) are differentiable. The variable *t* is called the parameter of the curve. The word interval is taken in a generalized sense, so that we do not exclude the cases $a = -\infty, b = +\infty$.

If we denote by x'(t) the first derivative of x at the point t and use similar notations for the functions y and z, the vector $(x'(t), y'(t), z'(t)) = \alpha'(t) \in R^3$ is called the tangent vector (or velocity vector) of the curve α, at t. The image set $\alpha(I) \subset R^3$ is called the trace of α.

Example: The parametrized differentiable curve given by

$$\alpha(t) = (a \cos t, a \sin t, bt), \qquad t \in R,$$

has as its trace in R^3 a helix of pitch $21\pi b$ on the cylinder $x^2 + y^2 = a^2$. The parameter *t* here measures the angle which the x axis makes with the line joining the origin o to the projection of the point α(t) over the xy plane.

Example: The map $\alpha : R \to R^2$ given by $\alpha(t) = (t^3, t^2)$, $t \in R$, is a parameterized differentiable curve. Notice that $\alpha'(0) = (0, 0)$; that is, the velocity vector is zero for $t = 0$.

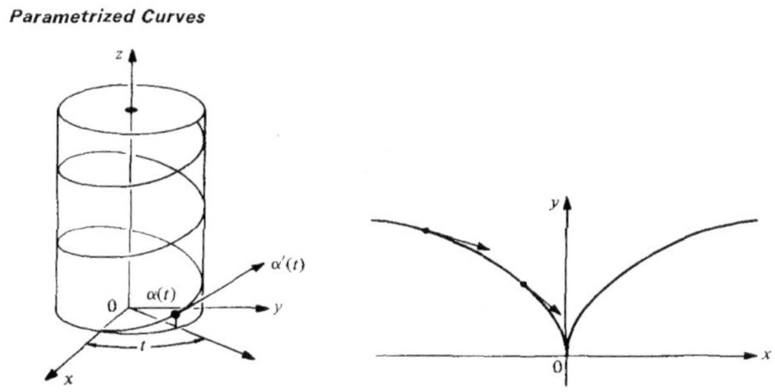

Example: The map $\alpha : R \to R^2$ given by $\alpha(t) = (t^3 - 4t, t^2 - 4)$, $t \in R$, is a parametrized differentiable curve. Notice that $\alpha(2) = \alpha(-2) = (0, 0)$; that is, the map α is not one-to-one.

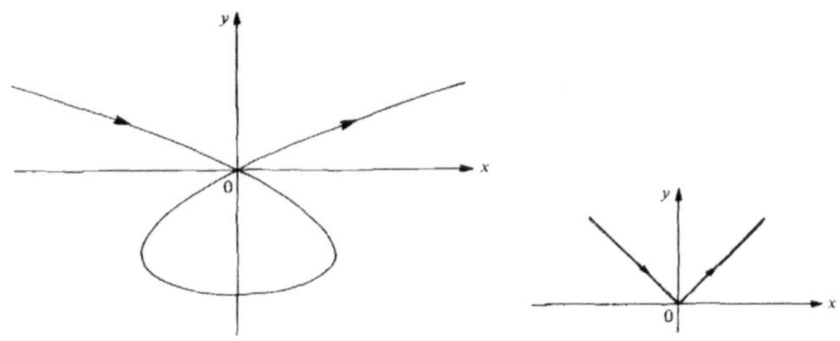

Example: The map $\alpha : R \to R^2$ given by $\alpha(t) = (t, |t|)$, $t \in R$, is not a parametrized differentiable curve, since $|t|$ is not differentiable at $t = 0$.

Example: The two distinct parametrized curves

$$\alpha(t) = (\cos t, \sin t),$$

$$\beta(t) = (\cos 2t, \sin 2t),$$

where $t \in (0 - \varepsilon, 2\pi + \varepsilon)$, $\varepsilon > 0$, have the same trace, namely, the circle $x^2 + y^2 = 1$. Notice that the velocity vector of the second curve is the double of the first one.

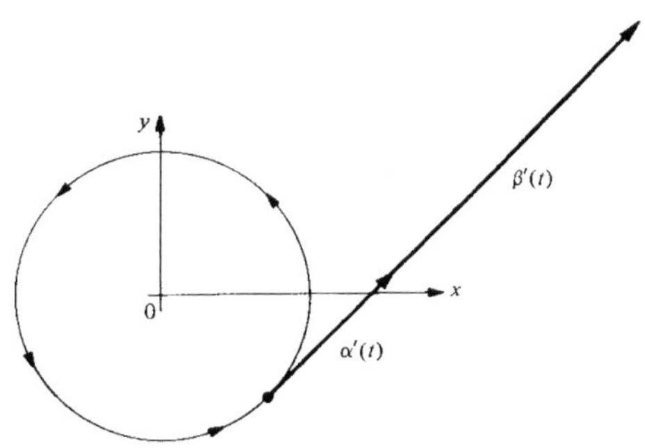

We shall now recall briefly some properties of the inner (or dot) product of vectors in R^3. Let u $= (u_1, u_2, u_3) \in R^3$ and define its norm (or length) by

$$|u| = \sqrt{u_1^2 + u_2^2 + u_3^2}$$

Geometrically, $|u|$ is the distance from the point (u_1, u_2, u_3) to the origin $0 = (0, 0, 0)$. Now, let

$u=(u_1,u_2,u_3)$ and $v=(v_1,v_2,v_3)$ belong to R³, and let $\theta, 0\le\theta\le\pi$, be the angle formed by the segments $0u$ and $0v$. The inner product u· v is defined by

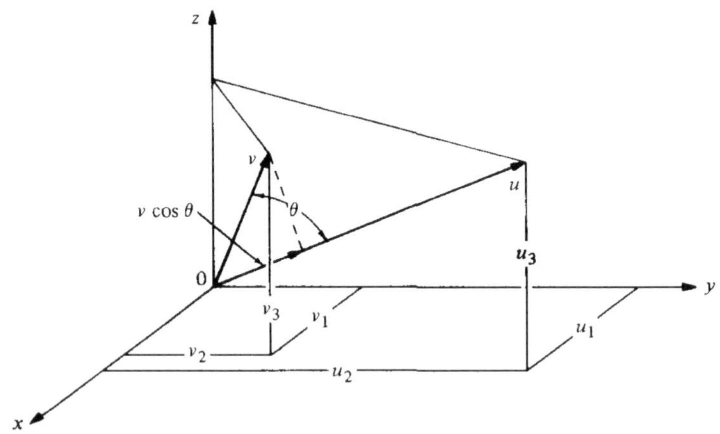

The following properties hold:

1. Assume that u and v are nonzero vectors. Then $u\cdot v=0$ if and only if u is orthogonal to v.

2. $u\cdot v=v\cdot u$.

3. $\lambda(u\cdot v)=\lambda u\cdot v=u\cdot\lambda v.$

4. $u\cdot(v+w)=u\cdot v+u\cdot w.$

A useful expression for the inner product can be obtained as follows. Let e_1 = (1, 0, 0), e_2 = (0, 1, 0), and e_3 = (0, 0, 1). It is easily checked that $e_i\cdot e_j=1$ if $i=j$ and that $e_i\cdot e_j$ if $i\ne j$, where i, j = 1, 2, 3. Thus, by writing

$$u=u_1e_1+u_2e_2+u_3e_3,\qquad v=v_1e_1+v_2e_2+v_3e_3,$$

and using properties 3 and 4, we obtain

$$u\cdot v=u_1v_1+u_2v_2+u_3v_3.$$

From the above expression it follows that if $u(t)$ and $v(t)$, $t\in I$, are differentiable curves, then $u(t)\cdot v(t)$ is a differentiable function, and

$$\frac{d}{dt}(u(t)\cdot v(t))=u'(t)\cdot v(t)+u(t)\cdot v'(t).$$

Regular Curves - Arc Length

Let $\alpha: I\to R^3$ be a parametrized differentiable curve. For each $t\in I$ where $\alpha'(t)\ne 0$, there is a well-defined straight line, which contains the point α (t) and the vector $\alpha'(t)$. This line is called the tangent line to α at t. For the study of the differential geometry of a curve it is essential that there exists such a tangent line at every point. Therefore, we call any point t where α (V) = 0 a singular point of α and restrict our attention to curves without singular points.

Definition: A parametrized differentiable curve $\alpha : I \to R^3$ is said to be regular if $\alpha'(t) \neq 0$ for all $t \in I$.

From now on we shall consider only regular parametrized differentiable curves (and, for convenience, shall usually omit the word differentiable).

Given: $t \in I$, the arc length of a regular parametrized curve $\alpha : I \to R^3$, from the point t_0, is by definition

$$s(t) = \int_{t_0}^{t} |\alpha'(t)| dt,$$

Where

$$|\alpha'(t)| = \sqrt{(x'(t))^2 + (y'(t))^2 + (z'(t))^2}$$

is the length of the vector $\alpha'(t)$. Since $\alpha'(t) \neq 0$, the arc. length s is a differentiable function of t and $ds/dt = |\alpha'(t)|$.

It can happen that the parameter t is already the arc length measured from some point. In this case, $ds/dt = 1 = |\alpha'(t)|$; that is, the velocity vector has constant length equal to 1. Conversely, if $|\alpha'(t)| \equiv 1$, then

$$s = \int_{t_0}^{t} dt = t - t_0;$$

i.e., t is the arc length of α measured from some point.

To simplify our exposition, we shall restrict ourselves to curves parametrized by arc length. In general, it is not necessary to mention the origin of the arc length s, since most concepts are defined only in terms of the derivatives of $\alpha(s)$.

It is convenient to set still another convention. Given the curve α para-metrized by arc length $S \in (a, b)$, we may consider the curve β defined in (–b, –a) by β (–s) = α (S), which has the same trace as the first one but is described in the opposite direction. We say, then, that these two curves differ by a change of orientation.

Differential Geometry of Surfaces

Differential geometry of a 2D manifold or surface embedded in 3D is the study of the intrinsic properties of the surface as well as the effects of a given parameterization on the surface. For the discussion below, we will consider the local behavior of a surface S at a single point p with a given local parameterization (u, v). We would like to calculate the principle curvature values (κ_1, κ_2), the maximum and minimum curvature values at p, as well as the local orthogonal arc length parameterization (s, t) which is aligned with the directions of maximum and minimum curvature.

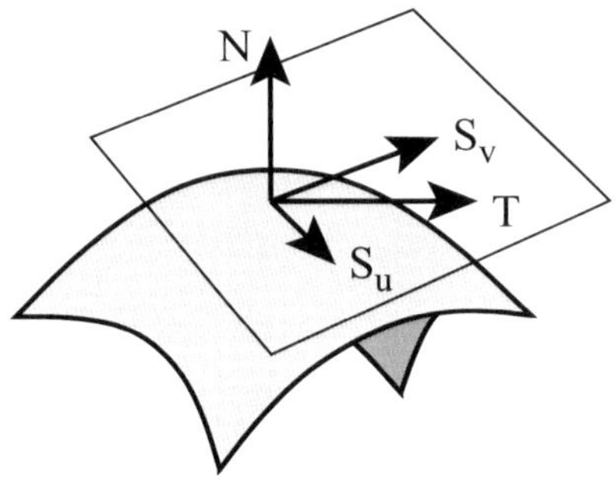

Figure: Tangent plane of S

The First Fundamental Form

With the given parameterization, we can compute a pair of tangent vectors $\begin{bmatrix} S_u & S_v \end{bmatrix} = \begin{bmatrix} \frac{\partial S}{\partial u} & \frac{\partial S}{\partial v} \end{bmatrix}$ assuming the u and v directions are distinct. These two vectors locally parameterize the tangent plane to S at p. A general tangent vector T can be constructed as a linear combination of $\begin{bmatrix} S_u & S_v \end{bmatrix}$ scaled by infinitesimal coefficients $U = \begin{bmatrix} \partial u & \partial v \end{bmatrix}^t$.

$$T = \partial u S_u + \partial v S_v = \begin{bmatrix} S_u & S_v \end{bmatrix} \begin{bmatrix} \partial u \\ \partial v \end{bmatrix} \qquad (1)$$

Consider rotating the direction of the tangent T around p by setting $\begin{bmatrix} \partial u & \partial v \end{bmatrix} = \begin{bmatrix} \cos\theta & \sin\theta \end{bmatrix}$. In general, the magnitude and direction of T will vary based on the skew and scale of the basis vectors $\begin{bmatrix} S_u & S_v \end{bmatrix}$. The distortion of the local parameterization is described by the metric tensor or the first fundamental form I_S. This measures the length of a tangent vector on a given parametric basis by examining the quantity $T \cdot T$.

$$\begin{aligned} T.T &= (S_u \cdot S_v)\partial u^2 + (S_u \cdot S_v)\partial u \partial v + (S_v \cdot S_v)\partial v^2 && (2) \\ &= E\partial u^2 + 2F\partial u \partial v + G\partial v^2 && (3) \\ &= \begin{bmatrix} \partial u & \partial v \end{bmatrix} \begin{bmatrix} S_u \\ S_v \end{bmatrix} \cdot \begin{bmatrix} S_u & S_v \end{bmatrix} \begin{bmatrix} \partial u \\ \partial v \end{bmatrix} && (4) \\ &= \begin{bmatrix} \partial u & \partial v \end{bmatrix} \begin{bmatrix} S_u \cdot S_u & S_u \cdot S_v \\ S_u \cdot S_v & S_v \cdot S_v \end{bmatrix} \begin{bmatrix} \partial u \\ \partial v \end{bmatrix} && (5) \\ &= \begin{bmatrix} \partial u & \partial v \end{bmatrix} \begin{bmatrix} E & F \\ F & G \end{bmatrix} \begin{bmatrix} \partial u \\ \partial v \end{bmatrix} && (6) \\ &= U^t I_S U && (7) \end{aligned}$$

Equation 7 defines I_S.

$$\begin{aligned} I_S = g_{ij} &= \begin{bmatrix} S_u \\ S_v \end{bmatrix} \cdot \begin{bmatrix} S_u & S_v \end{bmatrix} && (8) \\ &= \begin{bmatrix} S_u \cdot S_u & S_u \cdot S_v \\ S_u \cdot S_v & S_v \cdot S_v \end{bmatrix} && (9) \end{aligned}$$

$$= \begin{bmatrix} E & F \\ F & G \end{bmatrix} \qquad (10)$$

If an orthonormal arc length parameterization (s,t) is chosen then there will be no local metric distortion and I_S will reduce to the identity matrix.

The Second Fundamental Form

The unit normal *N* of a surface *S* at *p* is the vector perpendicular to *S*, i.e. the tangent plane of *S*, at *p*. *N* can be calculated given a general nondegenerate parametrization (*u*, *v*).

$$N = \frac{S_u \times S_v}{\|S_u \times S_v\|} \qquad (11)$$

The curvature of a surface *S* at a point *p* is defined by the rate that *N* rotates in response to a unit tangential displacement as in figure. A normal section curve at p is constructed by intersecting *S* with a plane normal to it, i.e a plane that contains *N* and a tangent direction *T*. The curvature of this curve is the curvature of *S* in the direction *T*. The curvature κ of a curve is the reciprocal of the radius *ρ* of the best fitting osculating circle, i.e. $k = \frac{1}{\rho}$. The directional derivative N_T of *N* in the direction *T* is a linear combination of the partial derivative vectors $\begin{bmatrix} N_u & N_v \end{bmatrix} = \begin{bmatrix} \frac{\partial N}{\partial u} & \frac{\partial N}{\partial v} \end{bmatrix}$ and is parallel to the tangent plane.

$$N_u = \frac{(S_{uu} \times S_v) + (S_u \times S_{vu})}{\|S_u \times S_v\|} + (S_u \times S_v)\frac{\partial \|S_u \times S_v\|^{-1}}{\partial u} \qquad (12)$$

$$N_v = \frac{(S_{uv} \times S_v) + (S_u \times S_{vv})}{\|S_u \times S_v\|} + (S_u \times S_v)\frac{\partial \|S_u \times S_v\|^{-1}}{\partial u} \qquad (13)$$

$$N_T = \partial u N_u + \partial v N_v = \begin{bmatrix} N_u & N_v \end{bmatrix} \begin{bmatrix} \partial u \\ \partial v \end{bmatrix} \qquad (14)$$

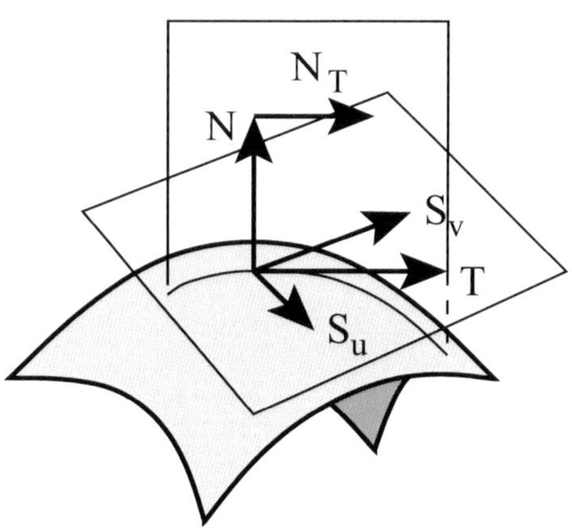

Figure: Normal section of S

As with the tangent vector *T*, N_T is subject to scaling by the metric of the parameter space. In an arc length parameter direction *s*, $-S_s \cdot N_s$ is the curvature of the normal section. Given a general parameterization (u, v), we can compute $-T \cdot N_T$. This quantity defines the second fundamental form II_S which describes curvature information distorted by the local metric.

$$-T \cdot N_T = -(S_u \cdot N_u)\partial u^2 - (S_u \cdot N_v + S_v \cdot N_u)\partial u \partial v - (S_v \cdot N_v)\partial v^2 \qquad (15)$$

$$= \begin{bmatrix} \partial u & \partial v \end{bmatrix} \begin{bmatrix} -S_u \\ -S_v \end{bmatrix} \cdot \begin{bmatrix} N_u & N_v \end{bmatrix} \begin{bmatrix} \partial u \\ \partial v \end{bmatrix} \qquad (16)$$

$$= \begin{bmatrix} \partial u & \partial v \end{bmatrix} \begin{bmatrix} -S_u \cdot N_u & -S_u \cdot N_v \\ -S_v \cdot N_u & -S_v \cdot N_v \end{bmatrix} \begin{bmatrix} \partial u \\ \partial v \end{bmatrix} \qquad (17)$$

$$= U^t II_S U \qquad (18)$$

Equation 18 defines II_S. II_S can be simplified to a set of quanties which are easier to compute by substituting the expressions in Equations [12, 13] for N_u and N_v respectively. Two of the terms of in of the entries of the matrix in Equation 21 vanish due to the dot product of orthogonal vectors. Further simplification using properties of the box product of vectors yields Equation 22. Equation 23 is derived by subsituting N from Equation 11 and shows that II_S can be computed simply by the dot product of the second partial derivatives of S with N. Equation 23 also demonstrates that II_S is a symmetric matrix.

$$II_S = h_{ij} = \begin{bmatrix} S_u \\ S_v \end{bmatrix} \cdot \begin{bmatrix} -N_u & -N_v \end{bmatrix} \qquad (19)$$

$$= \begin{bmatrix} -S_u \cdot N_u & -S_u \cdot N_v \\ -S_v \cdot N_u & -S_v \cdot N_v \end{bmatrix} \qquad (20)$$

$$= \begin{bmatrix} -\dfrac{S_u \cdot (S_{uu} \times S_v)}{\|S_u \times S_v\|} & -\dfrac{S_u \cdot (S_{uv} \times S_v)}{\|S_u \times S_v\|} \\ -\dfrac{S_v \cdot (S_u \times S_{vu})}{\|S_u \times S_v\|} & -\dfrac{S_v \cdot (S_u \times S_{vv})}{\|S_u \times S_v\|} \end{bmatrix} \qquad (21)$$

$$= \begin{bmatrix} \dfrac{S_{uu} \cdot (S_u \times S_v)}{\|S_u \times S_v\|} & \dfrac{S_{uv} \cdot (S_u \times S_v)}{\|S_u \times S_v\|} \\ \dfrac{S_{vu} \cdot (S_u \times S_v)}{\|S_u \times S_v\|} & \dfrac{S_{vv} \cdot (S_u \times S_v)}{\|S_u \times S_v\|} \end{bmatrix} \qquad (22)$$

$$= \begin{bmatrix} S_{uu} \cdot N & S_{uv} \cdot N \\ S_{uv} \cdot N & S_{vv} \cdot N \end{bmatrix} \qquad (23)$$

$$= \begin{bmatrix} L & M \\ M & N \end{bmatrix} \qquad (24)$$

Equations [23, 24] can be substituted for II_S to yield alternate formulas for $-T \cdot N_T$.

$$-T \cdot N_T = (S_{uu} \cdot N)\partial u^2 + 2(S_{uv} \cdot N)\partial u \partial v + (S_{vv} \cdot N)\partial v^2 \qquad (25)$$

$$= L\partial u^2 + 2M\partial u \partial v + N\partial v^2 \qquad (26)$$

$$= \begin{bmatrix} \partial u & \partial v \end{bmatrix} \begin{bmatrix} S_{uu} \cdot N & S_{uv} \cdot N \\ S_{vu} \cdot N & S_{vv} \cdot N \end{bmatrix} \begin{bmatrix} \partial u \\ \partial v \end{bmatrix} \quad (27)$$

$$= \begin{bmatrix} \partial u & \partial v \end{bmatrix} \begin{bmatrix} L & M \\ M & N \end{bmatrix} \begin{bmatrix} \partial u \\ \partial v \end{bmatrix} \quad (28)$$

II_S contains information about the curvature of S, but it will be distorted by the metric if the given parameterization is not an arc length parameterization. It is still necessary to counter act this distortion in order to compute the curvature of the S.

Coordinate Transformations

Our purpose in studying differential geometry and the fundamental forms has been to compute the principle curvature values (k_1, k_2) and directions (S_{k1}, S_{k2}) of a parametric surface S at a point p. Thus far we have derived IIS which contains curvature information but can be distorted by the metric of the parameterization. I_S describes this metric information. We must combine II_S and I_S to find the arc length curvature values.

The given parameterization (u, v) defines a set of basis tangent vectors $\begin{bmatrix} S_u & S_v \end{bmatrix}$. We can construct an orthonormal basis $\begin{bmatrix} S_s & S_t \end{bmatrix}$ due to an arc length parameterization (s, t) where S_s and S_u are aligned. A point T can be expressed in either coordinate system.

$$T = \begin{bmatrix} \partial u & \partial v \end{bmatrix} \begin{bmatrix} S_u \\ S_v \end{bmatrix} = \begin{bmatrix} \partial s & \partial t \end{bmatrix} \begin{bmatrix} S_s \\ S_t \end{bmatrix} \quad (29)$$

The coordinates $\begin{bmatrix} \partial s & \partial t \end{bmatrix}$ measure the length of T as opposed to the coordinates $\begin{bmatrix} \partial u & \partial v \end{bmatrix}$.

$$T \cdot T = \partial s^2 + \partial t^2 \neq \partial u^2 + \partial v^2 \quad (30)$$

We would like to work in the $\begin{bmatrix} S_s & S_t \end{bmatrix}$, so we must find the transformations to and from $\begin{bmatrix} S_u & S_v \end{bmatrix}$. The quantities a, b, and θ in figure (a) are defined by $\begin{bmatrix} S_u & S_v \end{bmatrix}$.

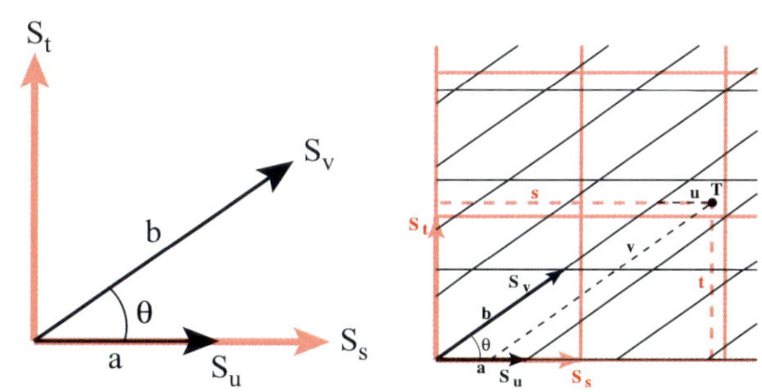

(a) *Constructing an orthonormal basis on top of the given parametric basis*

(b) *A point T in (u, v) and (s, t) coordinates*

Figure: Effects of the metric

$$\begin{bmatrix} S_u \\ S_v \end{bmatrix} = \begin{bmatrix} a & 0 \\ b\cos\theta & b\sin\theta \end{bmatrix} \begin{bmatrix} S_s \\ S_t \end{bmatrix} \quad (31)$$

$$= A\begin{bmatrix} S_s \\ S_t \end{bmatrix} \quad (32)$$

$$\begin{bmatrix} S_s \\ S_t \end{bmatrix} = \frac{1}{ab\sin\theta}\begin{bmatrix} b\sin\theta & 0 \\ -b\cos\theta & a \end{bmatrix}\begin{bmatrix} S_u \\ S_v \end{bmatrix} \quad (33)$$

$$= A^{-1}\begin{bmatrix} S_u \\ S_v \end{bmatrix} \quad (34)$$

The first fundamental form I_S can be represented by these coordinate transformations.

$$I_S = \begin{bmatrix} S_u \\ S_v \end{bmatrix} \cdot \begin{bmatrix} S_u & S_v \end{bmatrix} \quad (35)$$

$$= A\begin{bmatrix} S_s \\ S_t \end{bmatrix} \cdot \begin{bmatrix} S_s & S_t \end{bmatrix} A^t \quad (36)$$

$$= A\begin{bmatrix} 1 & 0 \\ 0 & 1 \end{bmatrix} A^t \quad (37)$$

$$= AA^t \quad (38)$$

$$= \begin{bmatrix} a^2 & ab\cos\theta \\ ab\cos\theta & b^2 \end{bmatrix} \quad (39)$$

We can also derive transformations between coordinates of the different sets of basis vectors by substituting Equations [32, 34] respectively into Equation 29.

$$\begin{bmatrix} \partial s & \partial t \end{bmatrix} = \begin{bmatrix} \partial u & \partial v \end{bmatrix} A \quad (40)$$

$$\begin{bmatrix} \partial u & \partial v \end{bmatrix} = \begin{bmatrix} \partial s & \partial t \end{bmatrix} A^{-1} \quad (41)$$

To verify that this makes sense, consider $T \cdot T$ again and transform the coordinates using Equations [40, 41].

$$T \cdot T = \begin{bmatrix} \partial u & \partial v \end{bmatrix}\begin{bmatrix} S_u \\ S_v \end{bmatrix} \cdot \begin{bmatrix} S_u & S_v \end{bmatrix}\begin{bmatrix} \partial u \\ \partial v \end{bmatrix} \quad (42)$$

$$= \begin{bmatrix} \partial u & \partial v \end{bmatrix} I_S \begin{bmatrix} \partial u \\ \partial v \end{bmatrix} \quad (43)$$

$$= \begin{bmatrix} \partial s & \partial t \end{bmatrix} A^{-1} I_S \left(A^{-1}\right)^t \begin{bmatrix} \partial s \\ \partial t \end{bmatrix} \quad (44)$$

$$= \begin{bmatrix} \partial s & \partial t \end{bmatrix} A^{-1}\left(AA^t\right)\left(A^{-1}\right)^t \begin{bmatrix} \partial s \\ \partial t \end{bmatrix} \quad (45)$$

$$= \begin{bmatrix} \partial s & \partial t \end{bmatrix}\begin{bmatrix} \partial s \\ \partial t \end{bmatrix} \quad (46)$$

$$= \partial s^2 + \partial t^2 \quad (47)$$

It is now possible to transform our given parameters to coordinates on an orthonormal basis where we can measure lengths. We can also transform coordinates on the orthonormal basis back to coordinates on our given parameterization.

Curvature

The curvature at a point p on the surface S in any tangential direction T is $-T \cdot N_T$.

$$-T\cdot N_T = \begin{bmatrix}\partial u & \partial v\end{bmatrix}\begin{bmatrix}S_u \\ S_v\end{bmatrix}\cdot\begin{bmatrix}-N_u & -N_v\end{bmatrix}\begin{bmatrix}\partial u \\ \partial v\end{bmatrix} \quad (48)$$

$$= \begin{bmatrix}\partial u & \partial v\end{bmatrix} II_S \begin{bmatrix}\partial u \\ \partial v\end{bmatrix} \quad (49)$$

This function is parameterized by the given parameterization. We would like to study the function that sweeps a unit length tangent around p, so it is more convenient to use arc length coordinates. We must transform coordinates so that setting the arc length coordinates $\begin{bmatrix}\partial s & \partial t\end{bmatrix} = \begin{bmatrix}\cos\phi & \sin\phi\end{bmatrix}$ sweeps out the curvature value.

$$-T\cdot N_T = \begin{bmatrix}\partial s & \partial t\end{bmatrix} A^{-1} II_S \left(A^{-1}\right)^t \begin{bmatrix}\partial s \\ \partial t\end{bmatrix} \quad (50)$$

$$= \begin{bmatrix}\partial s & \partial t\end{bmatrix} II_S \begin{bmatrix}\partial s \\ \partial t\end{bmatrix} \quad (51)$$

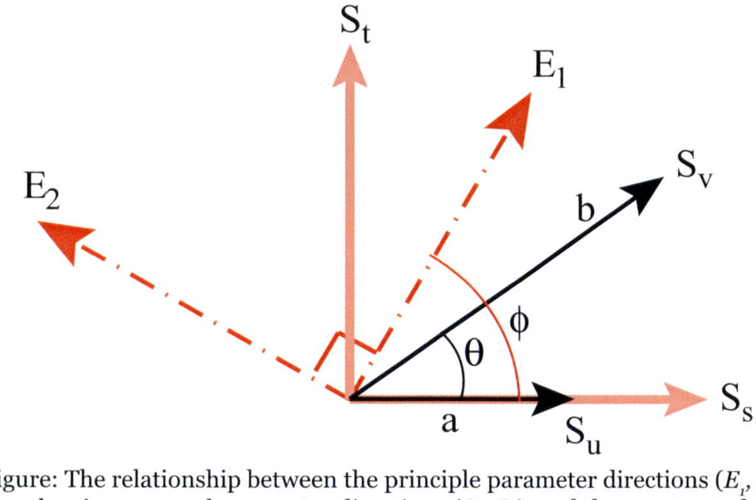

Figure: The relationship between the principle parameter directions (E_1, E_2), the given general parameter directions (S_u, S_v), and the constructed orthonormal basis (S_s, S_t)in the tangent plane to S at p.

The curvature function will have maximum and minimum values that occur at directions that are orthogonal to each other. The eigenvalues of $II_{\hat{S}}$ are the principle curvature values (κ_1, κ_2). The eigenvectors of $II_{\hat{S}}$ are the coordinates of the principle curvature directions in the arc length coordinates. These coordinates will not be directly useful because we only know the $\begin{bmatrix}S_u & S_v\end{bmatrix}$ vectors, so we would actually prefer the eigenvectors in coordinates on this basis.

$$II_{\hat{S}} = A^{-1} II_S \left(A^{-1}\right)^t \tag{52}$$

$$= \frac{1}{a^2 b^2 \sin^2\theta} \begin{bmatrix} b\sin\theta & 0 \\ -b\cos\theta & a \end{bmatrix} \begin{bmatrix} L & M \\ M & N \end{bmatrix} \begin{bmatrix} b\sin\theta & -b\cos\theta \\ 0 & a \end{bmatrix} \tag{53}$$

$$= \frac{1}{a^2 b^2 \sin^2\theta} \cdot \begin{bmatrix} b^2 \sin^2\theta & b\sin\theta(-b\cos\theta L + aM) \\ b\sin\theta(-b\cos\theta L + aM) & b^2\cos^2\theta L - 2ab\cos\theta M + a^2 N \end{bmatrix} \tag{54}$$

Solving for the eigenvalues, we find the characteristic equation. The principle curvature values are the eigenvalues.

$$0 = \left(a^2 b^2 - (ab\cos\theta)^2\right)\lambda^2 - \left(b^2 L - 2ab\cos\theta M + a^2 N\right)\lambda + \left(LN - M^2\right) \tag{55}$$

$$0 = \left(EG - F^2\right)\lambda^2 - (GL - 2FM + EN)\lambda + \left(LN - M^2\right) \tag{56}$$

Riemannian Geometry

Riemannian geometry considers manifolds with the additional structure of a Riemannian metric, a type (0, 2) positive definite symmetric tensor field. To a first order approximation this means that a Riemannian manifold is a Euclidean space: we can measure lengths of vectors and angles between them. Immediately we can define the length of curves by the usual integral, and then the distance between points comes from the glb of lengths of curves. Thus, Riemannian manifolds become metric spaces in the topological sense.

A Riemannian structure is also frequently used as a tool for the study of other properties of manifolds. For example, if we are given a second order linear elliptic partial differential operator, then the second-order coefficients of that operator form a Riemannian metric and it is natural to ask what the general properties of that metric tell us about the partial differential equations involving that operator. Conversely, on a Riemannian manifold there is singled out naturally such an operator (the Laplace-Beltrami operator), so that it makes sense, for example, to talk about solving the heat equation on a Riemannian manifold. The Riemannian metrics have nice properties not shared by just any topological metrics, so that in topological studies they are also used as a tool for the study of manifolds.

The generalization of Riemannian geometry to the case where the metric is not assumed to be positive definite, but merely nondegenerate, forms the basis for general relativity theory. Some of the major theorems and concepts are identical in the generalization.

Theorem: (Existence of Riemannian metrics). On any smooth manifold there exists a Riemannian metric.

The key idea of the proof is that locally we always have Riemannian metrics carried over from the standard one on Cartesian space by coordinate mappings, and we can glue them together smoothly with a partition of unity. In the gluing process the property of being positive definite is preserved due to the convexity of the set of positive definite symmetric matrices. What happens for indefinite metrics? The set of nonsingular symmetric matrices of a given index is not convex, so that the existence proof breaks down. In fact, there is a condition on the manifold, which can be reduced to topological invariants, in order that a semi-Riemannian metric of index ν exist: there must be a sub bundle of the tangent bundle of rank ν. When $\nu = 1$ the structure is called a Lorentz structure; that is the case of interest in general relativity theory; the topological condition for a compact manifold to have a Lorentz structure is easily understood: the Euler characteristic must be 0.

The proof in the Lorentz case can be done by using the fact that for any simple curve there is a diffeomorphism which is the identity outside any given neighborhood of the curve and which moves one end of the curve to the other. In the compact case start with a vector field having discrete singularities. Then by choosing disjoint simple curves from these singularities to inside a fixed ball, we can obtain a diffeomorphism which moves all of them inside that ball. If the Euler characteristic is 0, then by the Hopf index theorem, the index of the vector field on the boundary of the ball is 0, so the vector field can be extended to a nonsingular vector field inside the ball.

Conversely, by the following theorem, a Lorentz metric would give a nonsingular rank 1 subbundle. If that field is nonorientable, pass to the double covering for which the lift of it is orientable. Then there is a nonsingular vector field which is a global basis of the line field, so the Euler characteristic is 0.

In the noncompact case, take a countable exhaustion of the manifold by an increasing family of compact sets. Then the singularities of a vector field can be pushed outside each of the compact sets sequentially, leaving a nonsingular vector field on the whole in the limit. Thus, every noncompact (separable) manifold has a Lorentz structure.

Theorem: (Existence of semi-Riemannian metrics). A smooth manifold has a semi-Riemannian metric of index ν if and only if there is a subbundle of the tangent bundle of rank ν.

The idea of the proof is the subbundle will be the directions in which the semi-Riemannian metrics will be negative definite. If we change the sign on the subbundle and leave it unchanged on the orthogonal complement, we will get a Riemannian metric. The construction goes both ways.

Although the idea for the proof of theorem is correct, there are some nontrivial technical difficulties to entertain us. One direction is relatively easy.

Proof of "*if*" part if M has a smooth tangent subbundle V of rank ν, then M has a semi-Riemannian metric of index ν.

Definition: A frame at a point p of a semi-Riemannian manifold is a basis of the tangent space M_p with respect to which the component matrix of the metric tensor is diagonal with -1's followed by 1's on the diagonal. A local frame field is a local basis of vector fields which is a frame at each point of its domain.

Lemma: (Technical Lemma 1). Local frames exist in a neighborhood of every point.

TL 1 is important for other purposes than the proof of the theorem at hand. For the proof of TL 1 one modifies the Gramm-Schmidt procedure, starting with a smooth local basis and shrinking the domain at each step if necessary to divide by the length for normalization.

If we write $g = g_{ij}\omega^i\omega^j$, then a local frame is exactly one for which the coframe of 1-forms (ε^i) satisfies

$$g = -(\epsilon^1)^2 - \ldots - (\epsilon^v)^2 + \ldots + (\epsilon^n)^2 .$$

The Gramm-Schmidt procedure amounts to iterated completion of squares, viewing g as a homogeneous quadratic polynomial in the ω^i. The modifications needed to handle the negative signs are probably easier in this form.

Lemma: (Technical Lemma 2). If V is a smooth tangent subbundle of rank v and g is *a* Riemannian metric, then the (1, 1) tensor fields A and the semi-Riemannian metric h given as follows are smooth:

$$A = \begin{cases} -1 & on\ V \\ 1 & on\ V^{\perp}. \end{cases}$$

$$h(v, w) = g(Av, w).$$

(Their expressions in terms of smooth local frames adapted to V for g are constant, hence smooth.)

Now the converse:

Proof of "only if " part if there is a semi-Riemannian metric h of index v, then there is a tangent subbundle V of rank v.

The outline of the proof goes as follows. Take a Riemannian metric g. Then h and g are related by a (1, 1) tensor field A as above. We know that A is symmetric with respect to g-frames, and has ν negative eigenvalues (counting multiplicities) at each point. Thus, the subspace spanned by the eigenvectors of these negative eigenvalues at each point is ν-dimensional. The claim is that those subspaces form a smooth subbundle, even though it may be impossible to choose the individual eigenvectors to form smooth vector fields.

Lemma: (Technical Lemma 3). If $A : V \to V$ is a symmetric linear operator, smoothly dependent on coordinates $x^1, \ldots, x^n$, and λ_0 is a simple eigenvalue at the origin, then λ_0 extends to a smooth simple eigenvalue function in a neighborhood of the origin having a smooth eigenvector field.

Let $P(X) = \det(XI - A)$. Use the implicit function theorem to solve $P(X) = 0$, getting $X = \lambda(x^1, \ldots, x^n)$. Then we can write $P(X) = (X - \lambda)Q(X)$, and any nonzero column of $Q(A)$ is an eigenvector, by the Cayley-Hamilton theorem.

Lemma: (Technical Lemma 4). Suppose that $W : R^n \to \wedge^v V$ is a smooth function with decomposable, nonzero values. Then locally there are smooth vector fields having wedge product equal to W.

For $\omega \in \wedge^{v-1} V^*$, the interior product $i(\omega)W$ is always in the subspace carried by W. Choose v of these ω's which give linearly independent interior products with W at one point; then they do so locally.

Lemma: (Technical Lemma 5). If $A : V \rightarrow V$ is a symmetric linear operator of index v, smoothly dependent on coordinates x^i, then the extension of A to a derivation of the Grassmann algebra $\wedge^* V \rightarrow \wedge^* V$ has a unique minimum simple eigenvalue $\lambda_1 + \cdots + \lambda_v$ on $\wedge^v V$. The eigenvectors $W: R^n \rightarrow \wedge^v V$ are decomposable.

Finsler Geometry

A metric generalization of Riemannian geometry, where the general definition of the length of a vector is not necessarily given in the form of the square root of a quadratic form as in the Riemannian case. Such a generalization was first developed in the paper by P. Finsler.

The object studied in Finsler geometry is a real N-dimensional differentiable manifold M (of class at least C^3) with a system of local coordinates x^i, on which a real non-negative scalar function $F(x, y)$ in $2N$ independent variables x^i and y^i is given, where y^i are the components of the contravariant vectors tangent to M at the point x^i. Suppose that $F(x, y)$ belongs to the class in x^i, and that in each tangent space M_χ to M there is a domain M_χ^* such that, first, it is conical (in the sense that if any vector y^i tangent at some point x^i belongs to M_χ^*, then every other tangent vector that is collinear with y^i and tangent at the same point x^i also belongs to M_χ^*), and secondly, $F(x, y)$ belongs to the class C^5 in $y^i \in M_\chi^*$. Non-zero vectors $y^i \in M_\chi^*$ are called admissible. Suppose further that for every admissible y^i and every point x^i:

$$F(x, y) > 0, \quad \det \frac{\partial^2 F^2(x, y)}{\partial y^i \partial y^j} \neq 0$$

and also that $F(x, y)$ is positively homogeneous of degree one in y^i, that is, $F(x, ky) = kF(x, y)$ for every $k > 0$ and all x^i and admissible y^i. Under these conditions the triple $(M, M_\chi^*, F(x, y))$ is called a Finsler space, and F a Finsler metric. The value of $F(x, y)$ is interpreted as the length of the vector y^i tangent at x^i.

If a Finsler space admits a coordinate system x^i such that F does not depend on these x, then it is called a Minkowski space. The latter is related to a Finsler space in the same way as a Euclidean space is related to a Riemannian space. A Finsler space is called positive definite if one imposes a condition on F that ensures that the quadratic form $z^i z^j \{\partial^2 F^2(x, y) / \partial y^i \partial y^j\}$ is positive definite for all x^i and non-zero y^j.

Imposing the condition of homogeneity in y^i on has a clear geometrical meaning from the point of view of invariant concepts in centro-affine spaces, the tangent spaces M_χ being such spaces. Namely, the ratio of the lengths of any two collinear vectors y_1^i and $y_2^i = k y_1^i$ in M_χ^* can be invariantly defined in the following way: $y_1^1 / y_2^1 = y_1^2 / y_2^2 = \ldots = k$, which does not include any metric functions. Thus, the homogeneity condition imposed on F is a condition that the Finslerian definition of length is consistent with the particular centro-affine definition; the Finsler metric is needed to compare the lengths of non-collinear vectors.

The Tensor

$$g_{ij} = \frac{1}{2}\frac{\partial^2 F^2(x,y)}{\partial y^i \partial y^j}$$

is called the Finsler metric tensor. By Euler's theorem on homogeneous functions,

$$F^2(x,y) = g_{ij}(x,y)y^i y^j, \qquad yi = \frac{1}{2}\frac{\partial F^2(x,y)}{\partial y^i},$$

where, by definition, $y_i = g_{ij}(x,y)y^j$. These formulas are an immediate generalization of their Riemannian analogues, and follow from just the homogeneity condition. Finsler geometry reduces to Riemannian geometry in the case when the metric tensor $g_{ij}(x,y)$ is assumed to be independent of y^n. The last condition can be written in the form $C_{ijk} = 0$, where

$$C_{ijk}(x,y) = \frac{1}{2}\frac{\partial g_{ij}(x,y)}{\partial y^k} \equiv \frac{1}{4}\frac{\partial^3 F^2(x,y)}{\partial y^i \partial y^j \partial y^k}$$

is called the Cartan torsion tensor. It satisfies the identity $y^j C_{ijk} = 0$. All Finsler relations can be turned into their Riemannian analogues by setting $C_{ijk} = 0$. The Christoffel symbols $\gamma_{ij}^k(x,y)$, which are constructed from the Finsler metric tensor by the same formula as in Riemannian geometry, do not obey the transformation law of the coefficients of a connection. Nevertheless, one can construct the coefficients of a connection from the first derivatives of the Finsler metric tensor so that (as also in Riemannian geometry) the covariant derivative of the metric tensor vanishes. They are called the Cartan connection coefficients and have the form

$$\Gamma_{ij}^k(x,y) = \gamma_{ij}^k - C_{in}^k G_j^n - C_{jn}^k G_i^n + C_{ijn} G^{kn},$$

where

$$G_i^n = -2C_{jm}^n G^m + y^m \gamma_{mj}^n, \quad 2G^m = y^n y^k \gamma_{nk}^m.$$

From the commutators of various covariant derivatives one can find expressions for the Finsler curvature tensors.

In each tangent space M_χ the Finsler metric defines an $(N-1)$-dimensional hypersurface $F(x,y)=1$ (where the x^i are regarded as fixed and the y^i as varying), called the indicatrix. The indicatrix is formed by the ends of the unit tangent vectors $l^i = y^i / F(x,y)$ tangent at the point x^i. The fundamental significance of the concept of the indicatrix is already evident from the fact that, because the Finsler metric is homogeneous, the indicatrix at x^i uniquely determines the form of $F(x,y)$ at this point x^i. In the Riemannian case the indicatrix is a sphere. Generally speaking, the indicatrix of a Finsler space can be a surface of a rather general form. The Finsler metric tensor induces a Riemannian metric on the indicatrix, converting it into a Riemannian space. For each fixed χ the Finsler metric tensor is Riemannian in the variables y. The pair $(M_\chi^*, g_{ij}(x,y))$, where the χ^n are fixed and the y^n are variable, is called the tangent Riemannian space at χ (a Euclidean space in the case of Riemannian geometry); the Riemannian curvature tensor of this space reduces to the expression $C^j_{mh}C^m_{ik} - C^j_{mk}C^m_{ih}$. The indicatrix is a hypersurface that is imbedded in the tangent Riemannian space. The most immediate example of a Finsler metric function is the f-th root of a form of order f.

Let $f(x)$ and $r^A(x)$ be real scalar functions of class C^3 satisfying at each point x the conditions $f \neq 0,1$ or 2, and $r^A \neq 0$, and let $S_i^A(\chi)$ be N linearly independent real covariant vector fields of class C^3, C^3, $A=1,\dots,N$. Then for

$$F_1(x,y) = \left[\sum_{A=1}^{N} r^A(\chi).\left(S_m^A(\chi)y^m\right)^{f(\chi)}\right]^{1/f(x)}$$

the curvature of the indicatrix is constant and equal to $f^2/4(f-1)$, and for

$$F_2(x,y) = \prod_{A=1}^{N}\left(S^A(\chi)y^m\right)^{r^A(x)}, \quad \sum_{A=1}^{N} r^A(\chi) = 1,$$

the curvature tensor of the indicatrix is zero. The determinant of the Finsler metric tensor is independent of y^i if and only if $C_i = 0$, where $C_i = C^n_{in}$. If a Finsler space is positive definite and the indicatrix is a convex surface, then $C_i \neq 0$. The function F_2 is the only known example (1984) of a Finsler metric for which $C_i = 0$ (not counting the proper Riemannian case).

One can select special types of Finsler spaces by postulating some special form of the characteristic Finsler tensors. If the base manifold M admits a field of frames $S_j^A(x)$ globally, and $F^*(y^A)$ is the metric function of some Minkowski space, then one can introduce a Finsler metric on M:

$$F(x^n, y^i) = F^*\left(S_j^A(x^n)y^i\right).$$

In this case the Finsler space and the metric are called 1-form. The functions F_1 and F_2 are 1-form when f and r^A are constants. 1-form spaces may be reckoned to be the simplest from the point of view of the way the variables χ^n enter in the metric. A Finsler space is called C-reducible if it is not Riemannian, if $N > 2$ and if the Cartan torsion tensor can be represented in the form

$$C_{ijm} = \frac{1}{N+1}\left(h_{ij}C_m + h_{jm}C_i + h_{mi}C_j\right),$$

where $h_{ij} = g_{ij} - l_i l_j$. C-reducible spaces can have metrics of only two types: either the Kropina metric $F_3 = \alpha^2/\beta$, or the Randers metric $F_4 = \alpha + \beta$, where $\beta = b_i(x)y^i$, $\alpha^2 = a_{ij}y^i y^j, b_i(x)$ is a covariant vector field, and $a_{ij}(x)$ is a Riemannian metric tensor. For example, the Lagrange function of an electric test charge in a gravitational or electromagnetic field is a Randers metric. The Finsler metric tensor corresponding to F_2 has signature $(+--\dots)$, which makes it of interest in developing a Finslerian generalization of the general theory of relativity; this signature is also encountered in the case of the choice of metric tensors of the form F_1. Such a generalization can be based on the concept of an oscillating Riemannian space to a Finsler space, according to which the Finsler metric tensor associates with each vector field $y^i(x)$ the so-called osculating Riemannian metric tensor $g_{mn}(x, y(x))$. Choosing tensor fields $z^A(x)$ depending only on the x_i, from which one constructs the Finsler metric according to $F(x,y)=v(z^A(x),y)$, where v is a scalar function, one can regard the z^A as genuine gravitational field variables. The Finslerian geometrization of space-time also makes it possible to develop a theory of physical fields with various internal symmetries, relying on the concept of the group of transformations of the tangent vectors y^i that leave the Finsler metric invariant.

Symplectic Geometry

Symplectic geometry is an even dimensional geometry. It lives on even dimensional spaces, and measures the sizes of 2-dimensional objects rather than the 1-dimensional lengths and angles that are familiar from Euclidean and Riemannian geometry. It is naturally associated with the field of complex rather than real numbers. However, it is not as rigid as complex geometry: one of its most intriguing aspects is its curious mixture of rigidity (structure) and flabbiness (lack of structure).

A differential 2-form ω on (real) manifold M is a gadget that at any point $p \in M$ eats two tangent vectors and spits out a real number in a skew-symmetric, bilinear way. That is, it gives a family of skew-symmetric bilinear functions

$$\omega_p : T_p M \times T_p M \to \mathbb{R}$$

depending smoothly on the point $p \in M$. A 2-form $\omega \in \Omega^2(M)$ is symplectic if it is both closed (its exterior derivative satisfies $d\omega = 0$) and nondegenerate (each function ω_p is nondegenerate). Nondegeneracy is equivalent to the statement that for each nonzero tangent vector $v \in T_pM$, there is a symplectic buddy: a vector $w \in T_pM$ so that $(v, w) = 1$. A symplectic manifold is a (real) manifold M equipped with a symplectic form ω.

Nondegeneracy has important consequences. Purely in terms of linear algebra, at any point $p \in M$ we may choose a basis of T_pM that is compatible with ω_p, using a skewsymmetric analogue of the Gram-Schmidt procedure. We start by choosing any nonzero vector v_1 and then finding a symplectic buddy w_1. These must be linearly independent by skew-symmetry. We then peel off the two-dimensional subspace that v_1 and w_1 span and continue recursively, eventually arriving at a basis

$$v_1, w_1, \dots, v_d, w_d,$$

which contains an even number of basis vectors. So symplectic manifolds are even-dimensional. This also allows us to think of each tangent space as a complex vector space where each v_i and w_i span a complex coordinate subspace. Moreover, the top wedge power, $\omega^d \in \Omega^2(M)$, is nowhere-vanishing, since at each tangent space,

$$\omega\ (v\ , \dots, w\ ) \neq\ .$$

In other words, ω^d is a volume form, and M is necessarily orientable.

Symplectic geometry enjoys connections to algebraic combinatorics, algebraic geometry, dynamics, mathematical physics, and representation theory. The key examples underlying these connections include:

1. $M = S^2 = \mathbb{C}P^1$ with (v, w) = signed area of the parallelogram spanned by v and w;
2. M any Riemann surface as in figure below with the area for ω as in (1);

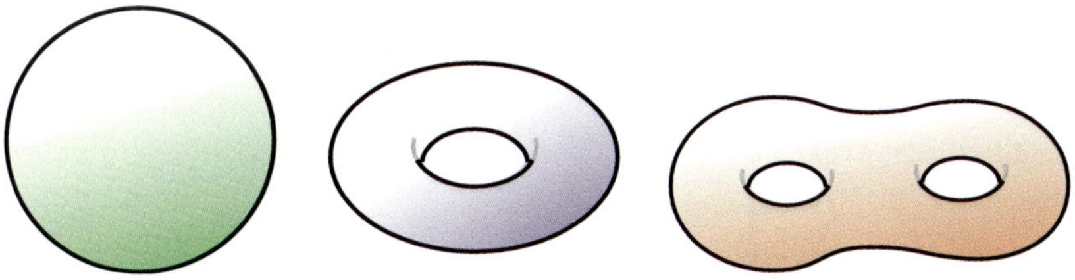

Figure: The area form on a Riemannian surface defines its symplectic geometry.

3. $M = \mathbb{R}^{2d}$ with $\omega_{std} = \sum dx_i \wedge dy_i$;
4. $M = T{*}X$, the cotangent bundle of any manifold X, thought of as a phase space, with p-coordinates coming from X being positions, q-coordinates in the cotangent directions representing momentum directions, and $\omega = \sum dp_i \wedge dq_i$;
5. M any smooth complex projective variety with ω induced from the Fubini-Study form (this includes smooth normal toric varieties, for example);
6. $M = \mathcal{O}_\lambda$ a coadjoint orbit of a compact connected semisimple Lie group G, equipped with the Kostant Kirillov-Souriau form ω. For the group $G = (n)$, this class of examples includes complex projective space $\mathbb{C}P^{n-1}$, Grassmannians $\mathcal{G}r(\mathbb{C}^n)$, the full flag variety $\mathcal{F}\ell(\mathbb{C}^n)$, and all other partial flag varieties.

Plenty of orientable manifolds do not admit a symplectic structure. For example, even-dimensional spheres that are at least four-dimensional are not symplectic. The reason is that on a compact manifold, Stokes' theorem guarantees that $[\omega] \neq 0 \in H^2(M; \mathbb{R})$. In other words, compact symplectic manifolds must exhibit nonzero topology in degree 2 cohomology. The only sphere with this property is S^2.

Comparison with Riemannian Geometry

Symplectic geometry has a number of similarities with and differences from Riemannian geometry, which is the study of differentiable manifolds equipped with nondegenerate, symmetric 2-tensors (called metric tensors). Unlike in the Riemannian case, symplectic manifolds have no local invariants such as curvature. This is a consequence of Darboux's theorem which states that a neighborhood of any point of a $2n$-dimensional symplectic manifold is isomorphic to the standard symplectic structure on an open set of $\mathbb{R}^{2n}$. Another difference with Riemannian geometry is that not every differentiable manifold need admit a symplectic form; there are certain topological restrictions. For example, every symplectic manifold is even-dimensional and orientable. Additionally, if M is a closed symplectic manifold, then the 2nd de Rham cohomology group $H^2(M)$ is nontrivial; this implies, for example, that the only n-sphere that admits a symplectic form is the

2-sphere. A parallel that one can draw between the two subjects is the analogy between geodesics in Riemannian geometry and pseudoholomorphic curves in symplectic geometry: Geodesics are curves of shortest length (locally), while pseudoholomorphic curves are surfaces of minimal area. Both concepts play a fundamental role in their respective disciplines.

Examples and Structures

Every Kähler manifold is also a symplectic manifold. Well into the 1970s, symplectic experts were unsure whether any compact non-Kähler symplectic manifolds existed, but since then many examples have been constructed (the first was due to William Thurston); in particular, Robert Gompf has shown that every finitely presented group occurs as the fundamental group of some symplectic 4-manifold, in marked contrast with the Kähler case.

Most symplectic manifolds, one can say, are not Kähler; and so do not have an integrable complex structure compatible with the symplectic form. Mikhail Gromov, however, made the important observation that symplectic manifolds do admit an abundance of compatible almost complex structures, so that they satisfy all the axioms for a Kähler manifold *except* the requirement that the transition maps be holomorphic.

Gromov used the existence of almost complex structures on symplectic manifolds to develop a theory of pseudoholomorphic curves, which has led to a number of advancements in symplectic topology, including a class of symplectic invariants now known as Gromov–Witten invariants. These invariants also play a key role in string theory.

Contact Geometry

Contact geometry forms the mathematical basis of geometrical optics, in the same sense in which symplectic geometry forms the basis of classical mechanics. The 'optical-mechanical analogy' of Hamilton allows one to translate problems and results from symplectic geometry to the language of contact geometry and vice versa. However, a direct approach in terms of contact geometry is in many cases preferable, at least from the point of view of geometrical intuition: it shows the geometrical content of formulas from the symplectic theory. The relation between symplectic and contact geometry is similar to the relation between the geometries of linear spaces and projective geometry: in order to obtain a contact counterpart of a symplectic piece of theory one has to replace functions by hyp er surf aces, aißne spaces by projective spaces, etc.

A contact structure on a manifold is a field of tangent hyperplanes (contact hyperplanes) that is non degenerate at any point.

Locally such a field is defined as the field of zeros of a 1-form a, called a contact form. The non degeneracy condition is: d α is non degenerate on the hyperplanes on which a vanishes; equivalently, in (2n + l)-space:

$$\alpha \Lambda (d\alpha)^n \neq 0.$$

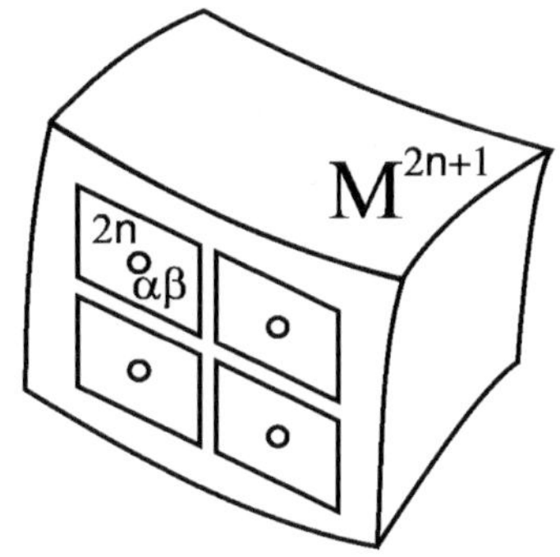

Figure: A contact structure: a maximally nonintegrable field of tangent hyperplanes

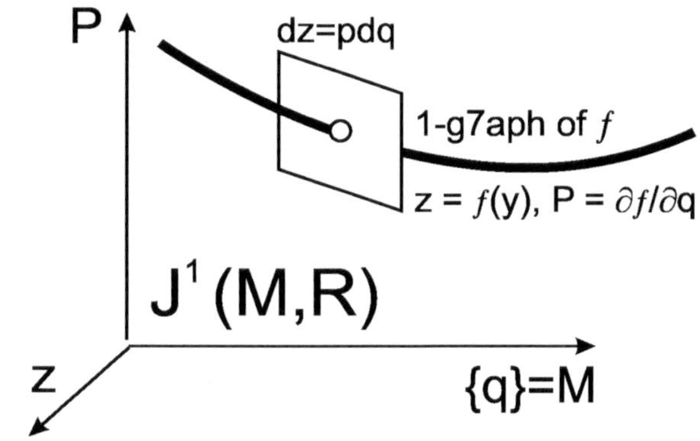

Figure: The contact structure of the manifold of I-jets of functions

This condition can be regarded as a condition for maximal nonintegrability, since Frobenius's integrability condition for the field of hyperplanes $\alpha = 0$ is $\alpha \wedge d\alpha = 0$.

Example 1: R^{2n+1}, $\alpha = dz - P_1 dq_1 - ... - P_n dq_n$. All contact structures on a manifold of fixed dimension are locally contactomorphic, and all contact forms are locally diffeomorphic (and hence may be written in the above normal form, using suitable Darboux local coordinates).

Example 2: The manifold $J^1(M, R)$ of 1-jets of functions $M \to R$ has a natural contact structure, '$dz - pdq$' (a I-jet (z, p, q) of a function f is its Taylor polynomial $f(x) = z + p(x - q) + ...$ of degree 1).

Applications

Like symplectic geometry, contact geometry has broad applications in physics, e.g. geometrical optics, classical mechanics, thermodynamics, geometric quantization, integrable systems and to control theory. Contact geometry also has applications to low-dimensional topology; for example, it has been used by Kronheimer and Mrowka to prove the property P conjecture, by Michael Hutchings to define an invariant of smooth three-manifolds, and by Lenhard Ng to define invariants of knots. It was also used by Yakov Eliashberg to derive a topological characterization of Stein manifolds of dimension at least six.

Contact Forms and Structures

A contact structure on an odd dimensional manifold is a smoothly varying family of codimension one subspaces of each tangent space of the manifold, satisfying a non-integrability condition. The family may be described as a section of a bundle as follows:

Given an n-dimensional smooth manifold M, and a point $p \in M$, a contact element of M with contact point p is an $(n - 1)$-dimensional linear subspace of the tangent space to M at p. A contact element can be given by the kernel of a linear function on the tangent space to M at p. However, if a subspace is given by the kernel of a linear function ω, then it will also be given by the zeros of λω where $\lambda \neq 0$ is any nonzero real number. Thus, the kernels of $\{ \lambda\omega : \lambda \neq 0 \}$ all give the same contact element. It follows that the space of all contact elements of M can be identified with a quotient of the cotangent bundle T^*M (with the zero section 0_M removed), namely:

$$PT^*M = (T^*M - 0_M)/\sim \text{ where, for } \omega_i \in T_p^*M,\ \omega_1 \sim \omega_2 \Leftrightarrow \exists\, \lambda \neq 0 : \omega_1 = \lambda\omega_2 .$$

A contact structure on an odd dimensional manifold M, of dimension $2k+1$, is a smooth distribution of contact elements, denoted by ξ, which is generic at each point. The genericity condition is that ξ is non-integrable.

Assume that we have a smooth distribution of contact elements ξ, given locally by a differential 1-form α; i.e. a smooth section of the cotangent bundle. The non-integrability condition can be given explicitly as:

$$\alpha \wedge (\mathrm{d}\alpha)^k \neq 0 \text{ where } (\mathrm{d}\alpha)^k = \underbrace{\mathrm{d}\alpha \wedge \ldots \wedge \mathrm{d}\alpha}_{k-\text{times}} .$$

Notice that if ξ is given by the differential 1-form α, then the same distribution is given locally by $\beta = f \cdot \alpha$, where f is a non-zero smooth function. If ξ is co-orientable then α is defined globally.

Properties

It follows from the Frobenius theorem on integrability that the contact field ξ is *completely non-integrable*. This property of the contact field is roughly the opposite of being a field formed by the tangent planes to a family of nonoverlapping hypersurfaces in M. In particular, you cannot find a hypersurface in M whose tangent spaces agree with ξ, even locally. In fact, there is no submanifold of dimension greater than n whose tangent spaces lie in ξ.

Relation with Symplectic Structures

A consequence of the definition is that the restriction of the 2-form $\omega = d\alpha$ to a hyperplane in ξ is a nondegenerate 2-form. This construction provides any contact manifold M with a natural symplectic bundle of rank one smaller than the dimension of M. Note that a symplectic vector space is always even-dimensional, while contact manifolds need to be odd-dimensional.

The cotangent bundle T^*N of any n-dimensional manifold N is itself a manifold (of dimension $2n$) and supports naturally an exact symplectic structure $\omega = d\lambda$. (This 1-form λ is sometimes called the Liouville form). There are several ways to construct an associated contact manifold, one of dimension $2n - 1$, one of dimension $2n + 1$.

Projectivization

Let M be the projectivization of the cotangent bundle of N: thus M is fiber bundle over a M whose fiber at a point x is the space of lines in T^*N, or, equivalently, the space of hyperplanes in TN.

The 1-form λ does not descend to a genuine 1-form on *M*. However, it is homogeneous of degree 1, and so it defines a 1-form with values in the line bundle O(1), which is the dual of the fibrewise tautological line bundle of *M*. The kernel of this 1-form defines a contact distribution.

Energy Surfaces

Suppose that *H* is a smooth function on T*N, that *E* is a regular value for *H*, so that the level set $L=\{(q,p)\in T^*N\mid H(q,p)=E\}$ is a smooth submanifold of codimension 1. A vector field *Y* is called an Euler (or Liouville) vector field if it is transverse to *L* and conformally symplectic, meaning that the Lie derivative of $d\lambda$ with respect to *Y* is a multiple of $d\lambda$ in a neighborhood of *L*.

Then the restriction of $i_Y d\lambda$ to *L* is a contact form on *L*.

This construction originates in Hamiltonian mechanics, where *H* is a Hamiltonian of a mechanical system with the configuration space *N* and the phase space *T*N*, and *E* is the value of the energy.

The Unit Cotangent Bundle

Choose a Riemannian metric on the manifold *N* and let *H* be the associated kinetic energy. Then the level set $H = 1/2$ is the *unit cotangent bundle* of *N*, a smooth manifold of dimension $2n$-1 fibering over *N* with fibers being spheres. Then the Liouville form restricted to the unit cotangent bundle is a contact structure. This corresponds to a special case of the second construction, where the flow of the Euler vector field *Y* corresponds to linear scaling of momenta p's, leaving the q's fixed. The vector field *R*, defined by the equalities

$\lambda(R) = 1$ and $d\lambda(R, A) = 0$ for all vector fields *A*,

is called the Reeb vector field, and it generates the geodesic flow of the Riemannian metric. More precisely, using the Riemannian metric, one can identify each point of the cotangent bundle of *N* with a point of the tangent bundle of *N*, and then the value of *R* at that point of the (unit) cotangent bundle is the corresponding (unit) vector parallel to *N*.

First Jet Bundle

On the other hand, one can build a contact manifold *M* of dimension $2n + 1$ by considering the first jet bundle of the real valued functions on *N*. This bundle is isomorphic to $T^*N\times$Rusing the exterior derivative of a function. With coordinates (*x*, *t*), *M* has a contact structure

$$\alpha = dt + \lambda.$$

Conversely, given any contact manifold *M*, the product $M\times$R has a natural structure of a symplectic manifold. If α is a contact form on *M*, then

$$\omega = d(e^t\alpha)$$

is a symplectic form on $M\times$R, where *t* denotes the variable in the R-direction. This new manifold is called the symplectization (sometimes symplectification in the literature) of the contact manifold *M*.

Examples

As a prime example, consider R^3, endowed with coordinates (x, y, z) and the one-form $dz - y\,dx$. The contact plane ξ at a point (x, y, z) is spanned by the vectors $X_1 = \partial_y$ and $X_2 = \partial_x + y\,\partial_z$.

By replacing the single variables x and y with the multivariables $x_1, \ldots, x_n, y_1, \ldots, y_n$, one can generalize this example to any R^{2n+1}. By a theorem of Darboux, every contact structure on a manifold looks locally like this particular contact structure on the $(2n + 1)$-dimensional vector space.

An important class of contact manifolds is formed by Sasakian manifolds.

Legendrian Submanifolds and Knots

The most interesting subspaces of a contact manifold are its Legendrian submanifolds. The non-integrability of the contact hyperplane field on a $(2n + 1)$-dimensional manifold means that no $2n$-dimensional submanifold has it as its tangent bundle, even locally. However, it is in general possible to find n-dimensional (embedded or immersed) submanifolds whose tangent spaces lie inside the contact field. Legendrian submanifolds are analogous to Lagrangian submanifolds of symplectic manifolds. There is a precise relation: the lift of a Legendrian submanifold in a symplectization of a contact manifold is a Lagrangian submanifold. The simplest example of Legendrian submanifolds are Legendrian knots inside a contact three-manifold. Inequivalent Legendrian knots may be equivalent as smooth knots; that is, there are knots which are smoothly isotopic where the isotopy cannot be chosen to be a path of Legendrian knots.

Legendrian submanifolds are very rigid objects; typically there are infinitely many Legendrian isotopy classes of embeddings which are all smoothly isotopic. Symplectic field theoryprovides invariants of Legendrian submanifolds called relative contact homology that can sometimes distinguish distinct Legendrian submanifolds that are topologically identical (i.e. smoothly isotopic).

Reeb Vector Field

If α is a contact form for a given contact structure, the Reeb vector field R can be defined as the unique element of the (one-dimensional) kernel of dα such that $\alpha(R) = 1$. If a contact manifold arises as a constant-energy hypersurface inside a symplectic manifold, then the Reeb vector field is the restriction to the submanifold of the Hamiltonian vector field associated to the energy function. (The restriction yields a vector field on the contact hypersurface because the Hamiltonian vector field preserves energy levels.)

The dynamics of the Reeb field can be used to study the structure of the contact manifold or even the underlying manifold using techniques of Floer homology such as symplectic field theory and, in three dimensions, embedded contact homology. Different contact forms whose kernels give the same contact structure will yield different Reeb vector fields, whose dynamics are in general very different. The various flavors of contact homology depend a priori on the choice of a contact form, and construct algebraic structures the closed trajectories of their Reeb vector fields; however, these algebraic structures turn out to be independent of the contact form, i.e. they are invariants of the underlying contact structure, so that in the end, the contact form may be seen as an auxiliary choice. In the case of embedded contact homology, one obtains an invariant of the underlying three-manifold, i.e. the embedded contact homology is independent of contact structure; this allows one to obtain results that hold for any Reeb vector field on the manifold.

The Reeb field is named after Georges Reeb.

Theorems of Differential Geometry

Euler's Theorem

The normal curvatures of a surface in an arbitrary direction (in the tangent plane) at point P can be expressed in terms of principal curvatures k_1 and k_2 at point P and the angle Φ between the arbitrary direction and the principal direction corresponding to k_1, namely,

$$k_n = k_1 \cos^2 \Phi + k_2 \sin^2 \Phi.$$

This is known as Euler's theorem. For simplicity, we assume that the iso-parametric curves of a surface are lines of curvature, which leads to $F=M=0$

$$k_n = \frac{Ldu^2 + Ndv^2}{Edu^2 + Gdv^2}$$

For $v = const$ iso-parametric lines $dv = 0$ and for $u = const$ iso-parametric lines $du = 0$, thus the principal curvatures k_1 and k_2 are given by:

$$k_1 = \frac{L}{E},\ k_2 = \frac{N}{G}.$$

The angle Φ between the direction $\frac{dv}{du}$ and the principal direction corresponding to (arbitrary) is evaluated as

$$\cos\Phi = E\frac{du}{ds}\frac{du_1}{ds_1}.$$

Since $ds_1 = \sqrt{Edu\frac{2}{1}}$ and $ds = \sqrt{Edu^2 + Gdv^2}$ we deduce following equations

$$\cos\Phi = \sqrt{E}\frac{du}{ds}, \quad \sin\Phi = \sqrt{G}\frac{dv}{ds}.$$

As a consequence, we have the Euler's theorem.

Next we explain Euler's theorem in a more simple way. Let us consider a section of the surface cut by a plane parallel to the tangent plane at the point P, and at an infinitesimal distance $h>0$ from it. We also consider a plane through P containing the normal vector. If we denote the intersection points of the surface and the two planes by Q and Q', the signed radius of curvature of this normal section by ϱ, and the length of QQ' by 2R.

$$(|\varrho| - h)^2 + R^2 = |\varrho|^2,$$

Thus

$$R^2 = 2h\,|\varrho|$$

to the first order. If Φ is the inclination of this normal section to the principal direction corresponding to k_1, Euler's theorem provides

$$k_1 \cos^2 \Phi + k_2 \sin^2 \Phi = \frac{1}{\varrho} = \pm\frac{2h}{R^2}$$

If we set

$$\xi = R\cos\Phi, \qquad \eta = R\sin\Phi,$$

we obtain

$$\frac{\xi^2}{2h\varrho_1} + \frac{\eta^2}{2h\varrho_2} = \pm 1$$

where ϱ_1 and ϱ_2 are principal radius of curvatures.

Hilbert's Theorem

In differential geometry, Hilbert's theorem (1901) states that there exists no complete regular surface S of constant negative gaussian curvature K immersed in $\mathbb{R}^3$. This theorem answers the question for the negative case of which surfaces in $\mathbb{R}^3$ can be obtained by isometrically immersing complete manifolds with constant curvature.

Proof

The proof of Hilbert's theorem is elaborate and requires several lemmas. The idea is to show the nonexistence of an isometric immersion

$$\varphi = \psi \circ exp_p : S' \to \mathbb{R}^3$$

of a plane S' to the real space $\mathbb{R}^3$. This proof is basically the same as in Hilbert's paper, although based in the books of Do Carmo and Spivak.

Observations: In order to have a more manageable treatment, but without loss of generality, the curvature may be considered equal to minus one, $K = -1$. There is no loss of generality, since it is being dealt with constant curvatures, and similarities of $\mathbb{R}^3$ multiply K by a constant. The exponential map is a local diffeomorphism (in fact a covering map, by Cartan-Hadamard $exp_p : T_p(S) \to S$ theorem), therefore, it induces an inner product in the tangent space of S at p: $T_p(S)$. Furthermore, S' denotes the geometric surface $T_p(S)$ with this inner product. If $\psi : S \to \mathbb{R}^3$ is an isometric immersion, the same holds for

$$\varphi = \psi^\circ exp_o : S' \to \mathbb{R}^3.$$

The first lemma is independent from the other ones, and will be used at the end as the counter statement to reject the results from the other lemmas.

Lemma: The area of S' is infinite.

Proof's Sketch:

The idea of the proof is to create a global isometry between H and S'. Then, since H has an infinite area, S' will have it too.

The fact that the hyperbolic plane H has an infinite area comes by computing the surface integral with the corresponding coefficients of the First fundamental form. To obtain these ones, the hyperbolic plane can be defined as the plane with the following inner product around a point $q \in \mathbb{R}^2$ with coordinates (u,v).

$$E = \left\langle \frac{\partial}{\partial u}, \frac{\partial}{\partial u} \right\rangle = 1 \qquad F = \left\langle \frac{\partial}{\partial u}, \frac{\partial}{\partial v} \right\rangle = \left\langle \frac{\partial}{\partial v}, \frac{\partial}{\partial u} \right\rangle = 0 \qquad G = \left\langle \frac{\partial}{\partial v}, \frac{\partial}{\partial v} \right\rangle = e^u$$

Since the hyperbolic plane is unbounded, the limits of the integral are infinite, and the area can be calculated through

$$\int_{-\infty}^{\infty} \int_{-\infty}^{\infty} e^u du dv = \infty$$

Next it is needed to create a map, which will show that the global information from the hyperbolic plane can be transfer to the surface S', i.e. a global isometry. $\varphi : H \to S'$ will be the map, whose domain is the hyperbolic plane and image the 2-dimensional manifold S', which carries the inner product from the surface S with negative curvature. φ will be defined via the exponential map, its inverse, and a linear isometry between their tangent spaces,

$$\psi : T_p(H) \to T_{p'}(S').$$

That is

$$\varphi = exp_{p'} \circ \psi \circ exp_p^{-1},$$

where $p \in H, p' \in S'$. That is to say, the starting point $p \in H$ goes to the tangent plane from H through the inverse of the exponential map. Then travels from one tangent plane to the other through the isometry ψ, and then down to the surface $'$ with another exponential map.

The following step involves te use of polar coordinates, (ρ,θ) and (ρ',θ'), around p and p' respectively. The requirement will be that the axis are mapped to each other, that is $\theta = 0$ goes to $\theta' = 0$. Then preserves the first fundamental form.

In a geodesic polar system, the Gaussian curvature K can be expressed as

$$K = -\frac{(\sqrt{G})_{\rho\rho}}{\sqrt{G}}.$$

In addition K is constant and fulfills the following differential equation

$$(\sqrt{G})_{\rho\rho} + K \cdot \sqrt{G} = 0$$

Since H and S' have the same constant Gaussian curvature, then they are locally isometric (Minding's Theorem). That means that φ is a local isometry between H and S'. Furthermore, from the Hadamard's theorem it follows that φ is also a covering map. Since S' is simply connected, φ is a homeomorphism, and hence, a (global) isometry. Therefore, H and S' are globally isometric, and because H has an infinite area, then $S' = T_p(S)$ has an infinite area, as well.

Lemma: For each $p \in S'$ exists a parametrization $x : U \subset \mathbb{R}^2 \to S'$, $p \in x(U)$, such that the coordinate curves of x are asymptotic curves of $x(U) = V'$ and form a Tchebyshef net.

Lemma: Let $V' \subset S'$ be a coordinate neighborhood of S' such that the coordinate curves are asymptotic curves in V'. Then the area A of any quadrilateral formed by the coordinate curves is smaller than 2π.

The next goal is to show that x is a parametrization of S'.

Lemma: For a fixed t, the curve $x(s,t), -\infty < s < +\infty$, is an asymptotic curve with s as arc length.

The following 2 lemmas will demonstrate the existence of a parametrization $x : \mathbb{R}^2 \to S'$.

Lemma: x is a local diffeomorphism.

Lemma: x is surjective.

Lemma: On S' there are two differentiable linearly independent vector fields which are tangent to the asymptotic curves of.

Lemma: x is injective.

Proof of Hilbert's Theorem

First, it will be assumed that an isometric immersion from a complete surface with negative curvature S exists: $\psi : S \to \mathbb{R}^3$.

As stated in the observations, the tangent plane $T_p(S)$ is endowed with the metric induced by the exponential map $exp_p : T_p(S) \to S$. Moreover, $\varphi = \psi \circ exp_p : S' \to \mathbb{R}^3$ is an isometric immersion and Lemmas show the existence of a parametrization $x : \mathbb{R}^2 \to S'$ of the whole S', such that the coordinate curves of x are the asymptotic curves of S'. This result was provided by Lemma. Therefore, S' can be covered by a union of "coordinate" quadrilaterals Q_n with $Q_n \subset Q_{n+1}$. By Lemma, the area of each quadrilateral is smaller than 2π. On the other hand, by Lemma, the area of S' is infinite, therefore has no bounds. This is a contradiction and the proof is concluded.

Slice Theorem

A theorem reducing the description of the action of a transformation group on some neighborhood U of a given orbit to that of the stabilizer H of a point χ of this orbit on some space S which is "normal" to the orbit at χ. Namely, this theorem claims that U is the homogeneous fibre space over

G/H with fibre S. Below, the precise setting and formulation are given, together with a counterpart of the theorem for algebraic transformation groups, called the étale slice theorem.

Slice Theorem for Topological Transformation Groups

Let G be a topological transformation group of a Hausdorff space X. A subspace S of X is called a slice at a point $x \in S$ if the following conditions hold:

1. S is invariant under the stabilizer G_x of x;
2. The union $G(S)$ of all orbits intersecting S is an open neighbourhood of the orbit $G(\chi)$ of x;
3. If $G \times G_x\ S$ is the homogeneous fibre space over G/G_x with fibre S, then the equivariant mapping $\varphi G \times G_x\ S \to X$, which is uniquely defined by the condition that its restriction to the fibre S over G_x is the identity mapping $S \to S$, is a homeomorphism of $G \times G_x\ S$ onto $G(S)$. Equivalent definitions are obtained by replacing iii) either by:
4. There is an equivariant mapping $\pi : G(S) \to G(x)$ that is the identity on $G(\chi)$ and is such that $\pi^{-1}(x) = S$; or by
5. S is closed in $G(S)$ and $g(S) \cap S \neq \phi$ implies $g \in G_x$.

The slice theorem claims that if certain conditions hold, then there is a slice at a point $x \in X$. The conditions depend on the case under consideration. The necessity of certain conditions is explained by the following observation.

If there is a slice S at x, then there is a neighbourhood U of x (namely, $G(S)$) such that the stabilizer of every point of U is conjugate to a subgroup of G_x. In general, this property fails (e.g., take $G = SL_2(C)$ acting on the space of binary forms of degree 3 in the variables t_1, t_2 by linear substitutions. Then the stabilizer of $x = t_1^2 t_2$ is trivial but every neighbourhood of x contains a point whose stabilizer has order 3).

The first case in which the validity of the slice theorem was investigated is that of a compact Lie group G. In this case, it has been proved that if X is a fully regular space, then there is a slice at every point $x \in X$. If, moreover, X is a differentiable manifold and G acts smoothly, then at every x there is a differentiable slice of a special kind. Namely, in this case there is an equivariant diffeomorphism φ, being the identity on $G(\chi)$, of the normal vector bundle of $G(\chi)$ onto an open neighbourhood of $G(\chi)$ in X. The image under φ of the fibre of this bundle over x is a slice at x which is a smooth submanifold of X diffeomorphic to a vector space.

The slice theorem is an indispensable tool in the theory of transformation groups, which frequently makes it possible to reduce an investigation to simple group actions like linear ones. In particular, the slice theorem is the key ingredient in the proofs of the following two basic facts of the theory:

1. If G is a compact Lie group, X a separable metrizable space and there are only finitely many conjugacy classes of stabilizers of points in X, then there is an equivariant embedding of X in a Euclidean vector space endowed with an orthogonal action of G.

2. Let G be a compact Lie group acting smoothly on a connected differentiable manifold X. Then there are a subgroup G^* of G and a dense open subset Ω of X such that $\operatorname{codim}(X \setminus \Omega) \geq 1$ and the stabilizer of every point $x \in X$ is conjugate to a subgroup of G^* which coincides with G^* if $x \in \Omega$.

There are versions of the slice theorem for non-compact groups. For instance, let H be an algebraic complex reductive group and $H \to GL(V)$ its finite-dimensional algebraic representation, both defined over the real numbers R. Let $H_{\mathbf{R}}$ be the Lie group of real points of H and G a subgroup of $H_{\mathbf{R}}$ containing the connected component of identity element. Let X be a closed H-invariant differentiable submanifold of $V_{\mathbf{R}}$, the space of real points of V. Then for every closed orbit $G(\chi)$ in X there is an equivariant diffeomorphism φ of a G-invariant neighbourhood $\tilde{U}$ of $G(\chi)$ in the normal vector bundle of $G(\chi)$ onto a G-invariant saturated neighbourhood U of ( ) in X (a neighbourhood U is saturated if the fact fact that the closure of an orbit intersects U implies that this orbit lies in U). In this case, the image under φ of the fibre of the natural projection $\tilde{U} \to G(x)$ over x is a slice at x for the action of G on X.

Slice Theorem for Algebraic Transformation Groups

Let G be an algebraic transformation group of an algebraic variety X, all defined over an algebraically closed field k. It happens very rarely that the literal analogue of the slice theorem above holds in this setting. Loosely speaking, the reason is that Zariski-open sets are "too big". One obtains an algebraic counterpart of a compact Lie group action on a differentiable manifold by taking G to be a reductive group and X an affine variety. In this setting, a counterpart of a slice is given by the notion of an étale slice, defined as follows.

Let x be a point of X such that the orbit $G(x)$ is closed. Let S be an affine G_x-invariant subvariety of X containing x. As above, one can consider the homogeneous fibre space $G \times G_x\ S$ over $G \times G_x$ with fibre S, and mapping $\varphi: G \times G_x\ S \to X$. In this situation, $G \times G_x\ S$ is an affine variety, φ is a morphism, there are the categorical quotients $\pi : X \to X /\!/ G$, $\pi G \times_{Gx} S : G \times G_x S \to (G \times G_x S) /\!/ G$ and an induced morphism $\varphi /\!/ G : (G \times G_x\ S) /\!/ G \to X /\!/ G$. The subvariety S is called an étale slice at x if:

1. $\pi G \times_{Gx} S$ is obtained from πX by means of the base change $\varphi /\!/ G$; and

2. $\varphi /\!/ G$ is an étale morphism.

The étale slice theorem, claims that there is an étale slice at every point $x \in X$ such that the orbit $G(x)$ is closed.

If k is C, the field of complex numbers, and x is a smooth point of X, then the étale slice theorem implies that there exists an analytic slice at x. More precisely, there is an invariant analytic neighbourhood of $G(x)$ in X which is analytically isomorphic to an invariant analytic neighborhood of $G(x)$ in the normal vector bundle of $G(x)$.

Like the slice theorem for topological transformation groups, the étale slice theorem is an indispensable result in the investigation of algebraic transformation groups.

Cohn-Vossen's Inequality

In differential geometry, Cohn-Vossen's inequality, named after Stephan Cohn-Vossen, relates the integral of Gaussian curvature of a non-compact surface to the Euler characteristic. It is akin to the Gauss–Bonnet theorem for a compact surface.

A divergent path within a Riemannian manifold is a smooth curve in the manifold that is not contained within any compact subset of the manifold. A complete manifold is one in which every divergent path has infinite length with respect to the Riemannian metric on the manifold. Cohn-Vossen's inequality states that in every complete Riemannian 2-manifold S with finite total curvature and finite Euler characteristic, we have

$$\iint_S K\,dA \le 2\pi\chi(S),$$

where K is the Gaussian curvature, dA is the element of area, and χ is the Euler characteristic.

Examples

- If S is a compact surface (without boundary), then the inequality is an equality by the usual Gauss–Bonnet theorem for compact manifolds.
- If S has a boundary, then the Gauss–Bonnet theorem gives

  $$\iint_S K\,dA = 2\pi\chi(S) - \int_{\partial S} k_g\,ds$$

  where k_g is the geodesic curvature of the boundary, and its integral the total curvature which is necessarily positive for a boundary curve, and the inequality is strict. (A similar result holds when the boundary of S is piecewise smooth.)
- If S is the plane R^2, then the curvature of S is zero, and $\chi(S) = 1$, so the inequality is strict: $0 < 2\pi$.

Darboux's Theorem

Darboux theorem may may refer to one of the following assertions:

- Darboux theorem on local canonical coordinates for symplectic structure;
- Darboux theorem on intermediate values of the derivative of a function of one variable.

Darboux Theorems for Symplectic Structure

A symplectic structure on an even-dimensional manifold M^{2n} is a closed nondegenerate C^∞-smooth differential 2-form ω:

$$\omega \in \Lambda^2(M), \qquad d\omega = 0, \qquad \forall v \in T_pM \quad \exists w \in T_pM : \omega_p(v, w) \neq 0.$$

The matrix $S(z)$ of a symplectic structure, $S_{ij}(z)=\omega\left(\frac{\partial}{\partial z_i},\frac{\partial}{\partial z_i}\right)=-S_{ji}(z)$ in any local coordinate system $(z_1,\ldots,z_{2n})$ is antisymmetric and nondegenerate: $\omega=\frac{1}{2}\sum_{i,j=1}^{2n}S_{ij}(z)dz_i\Delta dz_j$.

The *standard symplectic structure* on $\mathbb{R}^{2n}$ in the *standard canonical coordinates* $(x_1,\ldots, x_n, p_1,\ldots, p_n)$ is given by the form

$$\omega=\sum_{i=1}^{n}dx_i\wedge dp_i.$$

Local Equivalence

Theorem (*Darboux theorem*, sometimes also referred to as the *Darboux-Weinstein theorem*).

Any symplectic structure locally is C^∞-equivalent to the standard to the standard syplectic structure (1): for any point $a\in M$ there exists a neighborhood $M\supseteq U\ni a$ and "canonical" coordinate functions $(x, p):(U, a)\to(\mathbb{R}^{2n}, 0)$, such that in these coordinates ω takes the form $\sum dx_i\wedge dp_i$.

In particular, any two symplectic structures ω_1, ω_2 on M are locally equivalent near each point: there exists the germ of a diffeomorphism $h:(M, a)\to(M, a)$ such that $h^*\omega_1=\omega_2$.

Relative Versions

Together with the "absolute" version, one has a "relative" version of the Darboux theorem: if M is a smooth manifold with two symplectic structures ω_1, ω_2, and N is a submanifold on which the two 2-forms coincide, then near each point $a\in N\subseteq M$ one has a diffeomorphism h: $(M, a)\to(M, a)$ transforming ω_1 to ω_2 and identical on N:

$$\omega1=\omega2\,|_{TN}\Rightarrow\exists h\in \mathrm{Diff}(M,a):h^*\omega_1=\omega_2,h\,|_N\equiv id.$$

The assertion of the Darboux theorem on local normalization of antisymmetric 2-forms should be compared with a similar question about symmetric nondegenerate forms, which (if positive) define a Riemannian metric on MM. It is well known that, although at a given point aa the Riemannian metric can be brought to the canonical form $\langle v,v\rangle=\Sigma_{i=1}^{n}\ v_i^2$, such transformation is in general impossible in any open neighborhood of aa: the obstruction, among other things, is represented by the curvature of the metric (which is zero for the "constant" standard Euclidean metric).

In the same way the relative Darboux theorem means that submaniolds of the symplectic manifold have no "intrinsic" geometry: any two submanifolds N, N′ with equivalent (eventually, quite degenerate) restrictions of ω on TN, resp., TN′, can be transformed to each other by a diffeomorphism preserving the symplectic structure.

Darboux Theorem for Intermediate Values of Differentiable Functions

If f:$[a, b]\to\mathbb{R}$ is a function which is differentiable at all points of the segment $[a, b]\subseteq\mathbb{R}$ (the right and left derivatives are assumed at the endpoints a,b respectively), then its derivative assumes all intermediate values (i.e., the range of the derivative $f'=\frac{df}{dx}:[a,b]\to\mathbb{R}$ is a connected set).

For functions $f \in C^1[a, b]$ whose derivative is continuous, this is a simple consequence of the intermediate value theorem for the derivative. For functions whose derivative exists at all points but is discontinuous, e.g., $f(x) = x^2 \sin(1/x)$, $0 \neq x \in [-1, 1]$, $f(0)=0$, the assertion follows from the Fermat principle (the derivative of a differential function at an extremal point vanishes), applied to a suitable combination $f(x) - \alpha x$.

References

- Abraham, Ralph; Marsden, Jerrold E. (1978). Foundations of Mechanics. London: Benjamin-Cummings. ISBN 0-8053-0102-X
- Differential-geometry, science: britannica.com, Retrieved 19 June 2018
- Finsler-geometry: encyclopediaofmath.org, Retrieved 22 May 2018
- Weinstein, Alan (1981). "Symplectic Geometry" (PDF). Bulletin of the American Mathematical Society. 5 (1): 1–13. doi:10.1090/s0273-0979-1981-14911-9
- Darboux-theorem: encyclopediaofmath.org, Retrieved 13 March 2018

4

Algebraic Geometry

The branch of mathematics that studies the zeros of multivariate polynomials is known as algebraic geometry. Abstract algebraic techniques, such as commutative algebra, are used for the solution of geometrical problems. This chapter closely examines the key concepts of algebraic geometry such as algebraic curve, algebraic variety, affine variety, Zariski's main theorem, Bézout's theorem, etc. to provide an extensive understanding of the field.

Algebraic geometry is the study of the geometric properties of solutions to polynomial equations, including solutions in dimensions beyond three. (Solutions in two and three dimensions are first covered in plane and solid analytic geometry, respectively.)

Algebraic geometry emerged from analytic geometry after 1850 when topology, complex analysis, and algebra were used to study algebraic curves. An algebraic curve C is the graph of an equation f(x, y) = 0, with points at infinity added, where f(x, y) is a polynomial, in two complex variables, that cannot be factored. Curves are classified by a nonnegative integer—known as their genus, g—that can be calculated from their polynomial.

The equation f(x, y) = 0 determines y as a function of x at all but a finite number of points of C. Since x takes values in the complex numbers, which are two-dimensional over the real numbers, the curve C is two-dimensional over the real numbers near most of its points. C looks like a hollow sphere with g hollow handles attached and finitely many points pinched together—a sphere has genus 0, a torus has genus 1, and so forth. The Riemann-Roch theorem uses integrals along paths on C to characterize g analytically.

A birational transformation matches up the points on two curves via maps given in both directions by rational functions of the coordinates. Birational transformations preserve intrinsic properties of curves, such as their genus, but provide leeway for geometers to simplify and classify curves by eliminating singularities (problematic points).

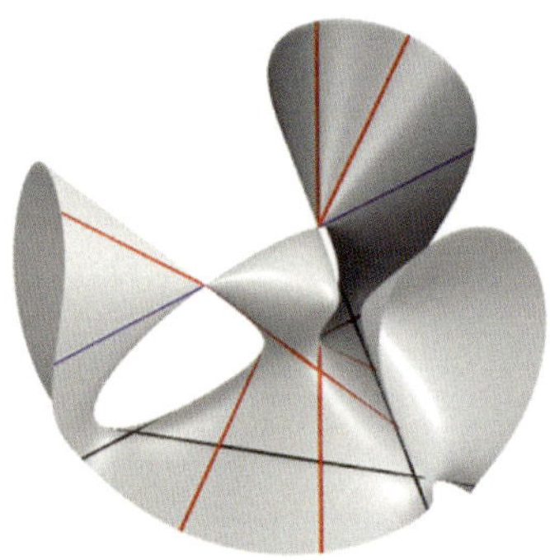

An algebraic curve generalizes to a variety, which is the solution set of r polynomial equations in n complex variables. In general, the difference n−r is the dimension of the variety—i.e., the number of independent complex parameters near most points. For example, curves have (complex)

dimension one and surfaces have (complex) dimension two. The French mathematician Alexandre Grothendieck revolutionized algebraic geometry in the 1950s by generalizing varieties to schemes and extending the Riemann-Roch theorem.

Arithmetic geometry combines algebraic geometry and number theory to study integer solutions of polynomial equations.

Basic Notions

Zeros of Simultaneous Polynomials

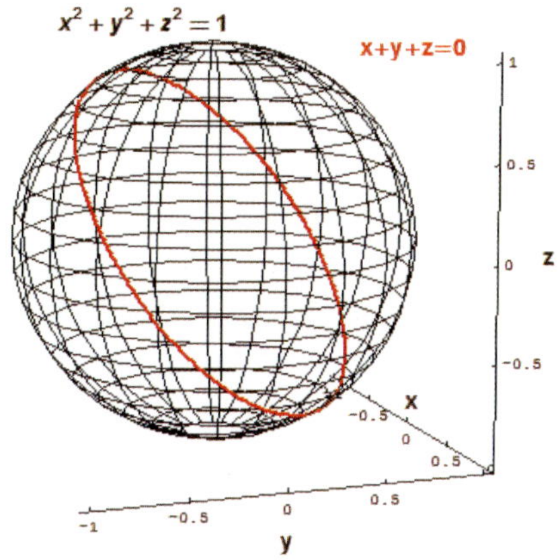

Sphere and slanted circle

In classical algebraic geometry, the main objects of interest are the vanishing sets of collections of polynomials, meaning the set of all points that simultaneously satisfy one or more polynomial equations. For instance, the two-dimensional sphere of radius 1 in three-dimensional Euclidean space R^3 could be defined as the set of all points (x, y, z) with

$$x^2+y^2+z^2-1=0.$$

A "slanted" circle in R^3 can be defined as the set of all points (x, y, z) which satisfy the two polynomial equations

$$x^2+y^2+z^2-1=0,$$

$$x+y+z=0.$$

Affine Varieties

First we start with a field k. In classical algebraic geometry, this field was always the complex numbers C, but many of the same results are true if we assume only that k is algebraically closed. We consider the affine space of dimension n over k, denoted $A^n(k)$ (or more simply A^n, when k is clear from the context). When one fixes a coordinate system, one may identify $A^n(k)$ with k^n. The purpose of not working with k^n is to emphasize that one "forgets" the vector space structure that k^n carries.

A function $f : A^n \to A^1$ is said to be *polynomial* (or *regular*) if it can be written as a polynomial, that is, if there is a polynomial p in $k[x_1,...,x_n]$ such that $f(M) = p(t_1,...,t_n)$ for every point M with coordinates $(t_1,...,t_n)$ in A^n. The property of a function to be polynomial (or regular) does not depend on the choice of a coordinate system in A^n.

When a coordinate system is chosen, the regular functions on the affine n-space may be identified with the ring of polynomial functions in n variables over k. Therefore, the set of the regular functions on A^n is a ring, which is denoted $k[A^n]$.

We say that a polynomial *vanishes* at a point if evaluating it at that point gives zero. Let S be a set of polynomials in $k[A^n]$. The *vanishing set of S* (or *vanishing locus* or *zero set*) is the set $V(S)$ of all points in A^n where every polynomial in S vanishes. Symbolically,

$$V(S) = \{(t_1,...,t_n) \mid \forall p \in S, p(t_1,...,t_n) = 0\}.$$

A subset of A^n which is $V(S)$, for some S, is called an *algebraic set*. The V stands for *variety* (a specific type of algebraic set to be defined below).

Given a subset U of A^n, can one recover the set of polynomials which generate it? If U is *any* subset of A^n, define $I(U)$ to be the set of all polynomials whose vanishing set contains U. The I stands for ideal: if two polynomials f and g both vanish on U, then $f+g$ vanishes on U, and if h is any polynomial, then hf vanishes on U, so $I(U)$ is always an ideal of the polynomial ring $k[A^n]$.

Two natural questions to ask are:

- Given a subset U of A^n, when is $U = V(I(U))$?
- Given a set S of polynomials, when is $S = I(V(S))$?

The answer to the first question is provided by introducing the Zariski topology, a topology on A^n whose closed sets are the algebraic sets, and which directly reflects the algebraic structure of $k[A^n]$. Then $U = V(I(U))$ if and only if U is an algebraic set or equivalently a Zariski-closed set. The answer to the second question is given by Hilbert's Nullstellensatz. In one of its forms, it says that $I(V(S))$ is the radical of the ideal generated by S. In more abstract language, there is a Galois connection, giving rise to two closure operators; they can be identified, and naturally play a basic role in the theory; the example is elaborated at Galois connection.

For various reasons we may not always want to work with the entire ideal corresponding to an algebraic set U. Hilbert's basis theorem implies that ideals in $k[A^n]$ are always finitely generated.

An algebraic set is called *irreducible* if it cannot be written as the union of two smaller algebraic sets. Any algebraic set is a finite union of irreducible algebraic sets and this decomposition is unique. Thus its elements are called the *irreducible components* of the algebraic set. An irreducible algebraic set is also called a *variety*. It turns out that an algebraic set is a variety if and only if it may be defined as the vanishing set of a prime ideal of the polynomial ring.

Some authors do not make a clear distinction between algebraic sets and varieties and use *irreducible variety* to make the distinction when needed.

Regular Functions

Just as continuous functions are the natural maps on topological spaces and smooth functions are the natural maps on differentiable manifolds, there is a natural class of functions on an algebraic set, called *regular functions* or *polynomial functions*. A regular function on an algebraic set V contained in A^n is the restriction to V of a regular function on A^n. For an algebraic set defined on the field of the complex numbers, the regular functions are smooth and even analytic.

It may seem unnaturally restrictive to require that a regular function always extend to the ambient space, but it is very similar to the situation in a normal topological space, where the Tietze extension theorem guarantees that a continuous function on a closed subset always extends to the ambient topological space.

Just as with the regular functions on affine space, the regular functions on V form a ring, which we denote by $k[V]$. This ring is called the *coordinate ring of V*.

Since regular functions on V come from regular functions on A^n, there is a relationship between the coordinate rings. Specifically, if a regular function on V is the restriction of two functions f and g in $k[A^n]$, then $f - g$ is a polynomial function which is null on V and thus belongs to $I(V)$. Thus $k[V]$ may be identified with $k[A^n]/I(V)$.

Morphism of Affine Varieties

Using regular functions from an affine variety to A^1, we can define regular maps from one affine variety to another. First we will define a regular map from a variety into affine space: Let V be a variety contained in A^n. Choose m regular functions on V, and call them $f_1, ..., f_m$. We define a *regular map f* from V to A^m by letting $f = (f_1, ..., f_m)$. In other words, each f_i determines one coordinate of the range of f.

If V' is a variety contained in A^m, we say that f is a *regular map* from V to V' if the range of f is contained in V'.

The definition of the regular maps apply also to algebraic sets. The regular maps are also called *morphisms*, as they make the collection of all affine algebraic sets into a category, where the objects are the affine algebraic sets and the morphisms are the regular maps. The affine varieties is a subcategory of the category of the algebraic sets.

Given a regular map g from V to V' and a regular function f of $k[V']$, then $f \circ g \in k[V]$. The map $f \rightarrow f \circ g$ is a ring homomorphism from $k[V']$ to $k[V]$. Conversely, every ring homomorphism from $k[V']$ to $k[V]$ defines a regular map from V to V'. This defines an equivalence of categories between the category of algebraic sets and the opposite category of the finitely generated reduced k-algebras. This equivalence is one of the starting points of scheme theory.

Rational Function and Birational Equivalence

In contrast to the preceding topics, this topic concerns only varieties and not algebraic sets. On the other hand, the definitions extend naturally to projective varieties, as an affine variety and its projective completion have the same field of functions.

If V is an affine variety, its coordinate ring is an integral domain and has thus a field of fractions which is denoted $k(V)$ and called the *field of the rational functions* on V or, shortly, the *function field* of V. Its elements are the restrictions to V of the rational functions over the affine space containing V. The domain of a rational function f is not V but the complement of the subvariety (a hypersurface) where the denominator of f vanishes.

As with regular maps, one may define a *rational map* from a variety V to a variety V''. As with the regular maps, the rational maps from V to V' may be identified to the field homomorphisms from $k(V')$ to $k(V)$.

Two affine varieties are *birationally equivalent* if there are two rational functions between them which are inverse one to the other in the regions where both are defined. Equivalently, they are birationally equivalent if their function fields are isomorphic.

An affine variety is a *rational variety* if it is birationally equivalent to an affine space. This means that the variety admits a rational parameterization. For example, the circle of equation $x^2 + y^2 - 1 = 0$ is a rational curve, as it has the parameterization.

$$x = \frac{2t}{1+t^2}$$

$$y = \frac{1-t^2}{1+t^2},$$

which may also be viewed as a rational map from the line to the circle.

The problem of resolution of singularities is to know if every algebraic variety is birationally equivalent to a variety whose projective completion is nonsingular. It was solved in the affirmative in characteristic 0 by Heisuke Hironaka in 1964 and is yet unsolved in finite characteristic.

Projective Variety

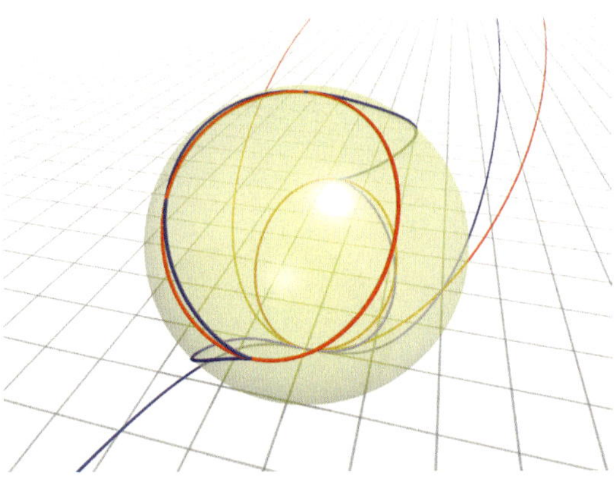

Parabola ($y = x^2$, red) and cubic ($y = x^3$, blue) in projective space

Just as the formulas for the roots of second, third, and fourth degree polynomials suggest extending real numbers to the more algebraically complete setting of the complex numbers, many properties of algebraic varieties suggest extending affine space to a more geometrically complete projective space. Whereas the complex numbers are obtained by adding the number i, a root of the polynomial $x^2 + 1$, projective space is obtained by adding in appropriate points "at infinity", points where parallel lines may meet.

To see how this might come about, consider the variety $V(y - x^2)$. If we draw it, we get a parabola. As x goes to positive infinity, the slope of the line from the origin to the point (x, x^2) also goes to positive infinity. As x goes to negative infinity, the slope of the same line goes to negative infinity.

Compare this to the variety $V(y - x^3)$. This is a cubic curve. As x goes to positive infinity, the slope of the line from the origin to the point (x, x^3) goes to positive infinity just as before. But unlike before, as x goes to negative infinity, the slope of the same line goes to positive infinity as well; the exact opposite of the parabola. So the behavior "at infinity" of $V(y - x^3)$ is different from the behavior "at infinity" of $V(y - x^2)$.

The consideration of the *projective completion* of the two curves, which is their prolongation "at infinity" in the projective plane, allows us to quantify this difference: the point at infinity of the parabola is a regular point, whose tangent is the line at infinity, while the point at infinity of the cubic curve is a cusp. Also, both curves are rational, as they are parameterized by x, and the Riemann-Roch theorem implies that the cubic curve must have a singularity, which must be at infinity, as all its points in the affine space are regular.

Thus many of the properties of algebraic varieties, including birational equivalence and all the topological properties, depend on the behavior "at infinity" and so it is natural to study the varieties in projective space. Furthermore, the introduction of projective techniques made many theorems in algebraic geometry simpler and sharper: For example, Bézout's theorem on the number of intersection points between two varieties can be stated in its sharpest form only in projective space. For these reasons, projective space plays a fundamental role in algebraic geometry.

Nowadays, the *projective space* P^n of dimension n is usually defined as the set of the lines passing through a point, considered as the origin, in the affine space of dimension $n + 1$, or equivalently to the set of the vector lines in a vector space of dimension $n + 1$. When a coordinate system has been chosen in the space of dimension $n + 1$, all the points of a line have the same set of coordinates, up to the multiplication by an element of k. This defines the homogeneous coordinates of a point of P^n as a sequence of $n + 1$ elements of the base field k, defined up to the multiplication by a nonzero element of k (the same for the whole sequence).

A polynomial in $n + 1$ variables vanishes at all points of a line passing through the origin if and only if it is homogeneous. In this case, one says that the polynomial *vanishes* at the corresponding point of P^n. This allows us to define a *projective algebraic set* in P^n as the set $V(f_1, ..., f_k)$, where a finite set of homogeneous polynomials $\{f_1, ..., f_k\}$ vanishes. Like for affine algebraic sets, there is a bijection between the projective algebraic sets and the reduced homogeneous ideals which define them. The *projective varieties* are the projective algebraic sets whose defining ideal is prime. In other words, a projective variety is a projective algebraic set, whose homogeneous coordinate ring is an integral domain, the *projective coordinates ring* being defined as the quotient of the graded ring or the polynomials in $n + 1$ variables by the homogeneous (reduced) ideal defining the variety. Every projective algebraic set may be uniquely decomposed into a finite union of projective varieties.

The only regular functions which may be defined properly on a projective variety are the constant functions. Thus this notion is not used in projective situations. On the other hand, the *field of the rational functions* or *function field* is a useful notion, which, similarly to the affine case, is defined as the set of the quotients of two homogeneous elements of the same degree in the homogeneous coordinate ring.

Real Algebraic Geometry

Real algebraic geometry is the study of the real points of algebraic geometry.

The fact that the field of the real numbers is an ordered field cannot be ignored in such a study. For example, the curve of equation $x^2 + y^2 - a = 0$ is a circle if $a > 0$, .but does not have any real point if $a < 0$. It follows that real algebraic geometry is not only the study of the real algebraic varieties, but has been generalized to the study of the *semi-algebraic sets*, which are the solutions of systems of polynomial equations and polynomial inequalities. For example, a branch of the hyperbola of equation $xy - 1 = 0$ is not an algebraic variety, but is a semi-algebraic set defined by $xy - 1 = 0$ and $x > 0$ or by $xy - 1 = 0$ and $x + y > 0$.

One of the challenging problems of real algebraic geometry is the unsolved Hilbert's sixteenth problem: Decide which respective positions are possible for the ovals of a nonsingular plane curve of degree 8.

Computational Algebraic Geometry

One may date the origin of computational algebraic geometry to meeting EUROSAM'79 (International Symposium on Symbolic and Algebraic Manipulation) held at Marseille, France in June 1979. At this meeting,

- Dennis S. Arnon showed that George E. Collins's Cylindrical algebraic decomposition (CAD) allows the computation of the topology of semi-algebraic sets,
- Bruno Buchberger presented the Gröbner bases and his algorithm to compute them,
- Daniel Lazard presented a new algorithm for solving systems of homogeneous polynomial equations with a computational complexity which is essentially polynomial in the expected number of solutions and thus simply exponential in the number of the unknowns. This algorithm is strongly related with Macaulay's multivariate resultant.

Since then, most results in this area are related to one or several of these items either by using or improving one of these algorithms, or by finding algorithms whose complexity is simply exponential in the number of the variables.

A body of mathematical theory complementary to symbolic methods called numerical algebraic geometry has been developed over the last several decades. The main computational method is homotopy continuation. This supports, for example, a model of floating point computation for solving problems of algebraic geometry.

Gröbner Basis

A Gröbner basis is a system of generators of a polynomial ideal whose computation allows the deduction of many properties of the affine algebraic variety defined by the ideal.

Given an ideal *I* defining an algebraic set *V*:

- *V* is empty (over an algebraically closed extension of the basis field), if and only if the Gröbner basis for any monomial ordering is reduced to {1}.
- By means of the Hilbert series one may compute the dimension and the degree of *V* from any Gröbner basis of *I* for a monomial ordering refining the total degree.
- If the dimension of *V* is 0, one may compute the points (finite in number) of *V* from any Gröbner basis of *I*.
- A Gröbner basis computation allows one to remove from *V* all irreducible components which are contained in a given hyper-surface.
- A Gröbner basis computation allows one to compute the Zariski closure of the image of *V* by the projection on the *k* first coordinates, and the subset of the image where the projection is not proper.
- More generally Gröbner basis computations allow one to compute the Zariski closure of the image and the critical points of a rational function of *V* into another affine variety.

Gröbner basis computations do not allow one to compute directly the primary decomposition of *I* nor the prime ideals defining the irreducible components of *V*, but most algorithms for this involve Gröbner basis computation. The algorithms which are not based on Gröbner bases use regular chains but may need Gröbner bases in some exceptional situations.

Gröbner bases are deemed to be difficult to compute. In fact they may contain, in the worst case, polynomials whose degree is doubly exponential in the number of variables and a number of polynomials which is also doubly exponential. However, this is only a worst case complexity, and the complexity bound of Lazard's algorithm of 1979 may frequently apply. Faugère F5 algorithm realizes this complexity, as it may be viewed as an improvement of Lazard's 1979 algorithm. It follows that the best implementations allow one to compute almost routinely with algebraic sets of degree more than 100. This means that, presently, the difficulty of computing a Gröbner basis is strongly related to the intrinsic difficulty of the problem.

Cylindrical Algebraic Decomposition (CAD)

CAD is an algorithm which was introduced in 1973 by G. Collins to implement with an acceptable complexity the Tarski–Seidenberg theorem on quantifier elimination over the real numbers.

This theorem concerns the formulas of the first-order logic whose atomic formulas are polynomial equalities or inequalities between polynomials with real coefficients. These formulas are thus the formulas which may be constructed from the atomic formulas by the logical operators *and* (∧), *or* (∨), *not* (¬), *for all* (∀) and *exists* (∃). Tarski's theorem asserts that, from such a formula, one may compute an equivalent formula without quantifier (∀, ∃).

The complexity of CAD is doubly exponential in the number of variables. This means that CAD allows, in theory, to solve every problem of real algebraic geometry which may be expressed by such a formula, that is almost every problem concerning explicitly given varieties and semi-algebraic sets.

While Gröbner basis computation has doubly exponential complexity only in rare cases, CAD has almost always this high complexity. This implies that, unless if most polynomials appearing in the input are linear, it may not solve problems with more than four variables.

Since 1973, most of the research on this subject is devoted either to improve CAD or to find alternate algorithms in special cases of general interest.

As an example of the state of art, there are efficient algorithms to find at least a point in every connected component of a semi-algebraic set, and thus to test if a semi-algebraic set is empty. On the other hand, CAD is yet, in practice, the best algorithm to count the number of connected components.

Asymptotic Complexity vs. Practical Efficiency

The basic general algorithms of computational geometry have a double exponential worst case complexity. More precisely, if d is the maximal degree of the input polynomials and n the number of variables, their complexity is at most $d^{2^{cn}}$ for some constant c, and, for some inputs, the complexity is at least $d^{2^{c'n}}$ for another constant c'.

During the last 20 years of 20th century, various algorithms have been introduced to solve specific subproblems with a better complexity. Most of these algorithms have a complexity $d^{O(n^2)}$.

Among these algorithms which solve a sub problem of the problems solved by Gröbner bases, one may cite *testing if an affine variety is empty* and *solving nonhomogeneous polynomial systems which have a finite number of solutions*. Such algorithms are rarely implemented because, on most entries Faugère's F4 and F5 algorithms have a better practical efficiency and probably a similar or better complexity (*probably* because the evaluation of the complexity of Gröbner basis algorithms on a particular class of entries is a difficult task which has been done only in a few special cases).

The main algorithms of real algebraic geometry which solve a problem solved by CAD are related to the topology of semi-algebraic sets. One may cite *counting the number of connected components*, *testing if two points are in the same components* or *computing a Whitney stratification of a real algebraic set*. They have a complexity of $d^{O(n^2)}$, but the constant involved by O notation is so high that using them to solve any nontrivial problem effectively solved by CAD, is impossible even if one could use all the existing computing power in the world. Therefore, these algorithms have never been implemented and this is an active research area to search for algorithms with have together a good asymptotic complexity and a good practical efficiency.

Abstract Modern Viewpoint

The modern approaches to algebraic geometry redefine and effectively extend the range of basic objects in various levels of generality to schemes, formal schemes, ind-schemes, algebraic spaces, algebraic stacks and so on. The need for this arises already from the useful ideas within theory of varieties, e.g. the formal functions of Zariski can be accommodated by introducing nilpotent elements in structure rings; considering spaces of loops and arcs, constructing quotients by group actions and developing formal grounds for natural intersection theory and deformation theory lead to some of the further extensions.

Most remarkably, in late 1950s, algebraic varieties were subsumed into Alexander Grothendieck's concept of a scheme. Their local objects are affine schemes or prime spectra which are locally ringed spaces which form a category which is antiequivalent to the category of commutative unital rings, extending the duality between the category of affine algebraic varieties over a field *k*, and the category of finitely generated reduced *k*-algebras. The gluing is along Zariski topology; one can glue within the category of locally ringed spaces, but also, using the Yoneda embedding, within the more abstract category of presheaves of sets over the category of affine schemes. The Zariski topology in the set theoretic sense is then replaced by a Grothendieck topology. Grothendieck introduced Grothendieck topologies having in mind more exotic but geometrically finer and more sensitive examples than the crude Zariski topology, namely the étale topology, and the two flat Grothendieck topologies: fppf and fpqc; nowadays some other examples became prominent including Nisnevich topology. Sheaves can be furthermore generalized to stacks in the sense of Grothendieck, usually with some additional representability conditions leading to Artin stacks and, even finer, Deligne-Mumford stacks, both often called algebraic stacks.

Sometimes other algebraic sites replace the category of affine schemes. For example, Nikolai Durov has introduced commutative algebraic monads as a generalization of local objects in a generalized algebraic geometry. Versions of a tropical geometry, of an absolute geometry over a field of one element and an algebraic analogue of Arakelov's geometry were realized in this setup.

Another formal generalization is possible to universal algebraic geometry in which every variety of algebras has its own algebraic geometry. The term *variety of algebras* should not be confused with *algebraic variety*.

The language of schemes, stacks and generalizations has proved to be a valuable way of dealing with geometric concepts and became cornerstones of modern algebraic geometry.

Algebraic stacks can be further generalized and for many practical questions like deformation theory and intersection theory, this is often the most natural approach. One can extend the Grothendieck site of affine schemes to a higher categorical site of derived affine schemes, by replacing the commutative rings with an infinity category of differential graded commutative algebras, or of simplicial commutative rings or a similar category with an appropriate variant of a Grothendieck topology. One can also replace presheaves of sets by presheaves of simplicial sets (or of infinity groupoids). Then, in presence of an appropriate homotopic machinery one can develop a notion of derived stack as such a presheaf on the infinity category of derived affine schemes, which is satisfying certain infinite categorical version of a sheaf axiom (and to be algebraic, inductively a sequence of representability conditions). Quillen model categories, Segal categories and quasicategories are some of the most often used tools to formalize this yielding the *derived algebraic geometry*, introduced by the school of Carlos Simpson, including Andre Hirschowitz, Bertrand Toën, Gabrielle Vezzosi, Michel Vaquié and others; and developed further by Jacob Lurie, Bertrand Toën, and Gabrielle Vezzosi. Another (noncommutative) version of derived algebraic geometry, using A-infinity categories has been developed from early 1990s by Maxim Kontsevich and followers.

Analytic Geometry

An analytic variety is defined locally as the set of common solutions of several equations involving analytic functions. It is analogous to the included concept of real or complex algebraic variety. Any complex manifold is an analytic variety. Since analytic varieties may have singular points, not all analytic varieties are manifolds.

Modern analytic geometry is essentially equivalent to real and complex algebraic geometry, as has been shown by Jean-Pierre Serre in his paper *GAGA*, the name of which is French for *Algebraic geometry and analytic geometry*. Nevertheless, the two fields remain distinct, as the methods of proof are quite different and algebraic geometry includes also geometry in finite characteristic.

Applications

Algebraic geometry now finds applications in statistics, control theory, robotics, error-correcting codes, phylogenetics and geometric modelling. There are also connections to string theory, game theory, graph matchings, solitons and integer programming.

Algebraic Curve

An algebraic variety of dimension one. An algebraic curve is the most frequently studied object in algebraic geometry. In the sequel, an algebraic curve means an irreducible algebraic curve over an algebraically closed field.

The simplest and clearest concept is that of a plane affine algebraic curve. This is a set of points in an affine plane A_k^2 satisfying the equation $f(x,y)=0$, where $f(x,y)$ is a polynomial with coefficients from an algebraically closed field k. The field of rational functions of an irreducible algebraic curve over k is the field of algebraic functions in one variable of the form $k(x,y)$, where x and y are connected by the equation $f(x,y)=0$, where $f(x,y)$ is a polynomial over k. This means that every algebraic curve is birationally isomorphic to a plane affine curve.

It has been known for a long time that even when studying affine curves fundamental relationships can only be revealed by considering points at an infinite distance, and by a detailed study of the singular points. In order to study all the points of an affine curve, the curve is imbedded into a projective space P^n, with subsequent closure in the Zariski topology. In this way a projective curve X is obtained, and the initial affine curve may be obtained from X by deleting a finite number of points. If Y is irreducible, then X and Y are birationally isomorphic. Every complete algebraic curve is projective. If X is a smooth projective curve, then all valuation rings of the field $k(x)$ are given by the local rings O_x, $x \in X$. If two smooth projective curves are birationally equivalent, then they are isomorphic. A normal algebraic curve is smooth. In particular, any irreducible algebraic curve is birationally equivalent to a smooth projective curve. The projective model of an algebraic curve, obtained by the process of normalization, lies in some space P^n. Any smooth projective curve is isomorphic to a curve situated in P^3. Any plane algebraic curve can be converted by a Cremona transformation into a curve with ordinary singular points.

Divisors on a smooth algebraic curve are represented by linear combinations of points with integer coefficients:

$$D=\sum_{x\in X} n_x x, \quad n_x \in Z$$

where $n_x = 0$ for almost all x. If all $n_x \geq 0$, the divisor D is called positive, or effective, which is written as $D \geq 0$. The degree of the divisor D is the number

$$\deg D = \sum_X n_X$$

The principal divisors form a subgroup $P(X)$ of the group Div X of all divisors on X. The quotient group Div (X) / P(X) is called the group of divisor classes and is denoted by $CI(X)$. The group $CI(X)$ is isomorphic to the group $Pic(X)$ of classes of one-dimensional vector bundles on X. The degree of the principal divisors on a smooth projective curve is zero, and thus all divisors in one class are of the same degree. In particular, one can speak of the degree of a divisor class, and of the subgroup $CI^{\circ}(X)$ of divisor classes of degree zero. The following equality is valid:

$$CI(X) / CI^{\circ}(X) = Z$$

For the line $P^1, CI(P^1) = Z, CI^{\circ}(P^1) = 0$, i.e. any divisor of degree zero is a principal divisor. This property is characteristic for rational smooth projective curves.

For any complete algebraic curve X, the number $\pi = \dim H^1(X, O_X)$ is known as the arithmetic genus of the algebraic curve X. If X is smooth, π is identical with the dimension of the space $H^{\circ}(X, \Omega_X^1)$ of all regular differential forms on X; this dimension is known as the genus of X. By definition, the genus of an algebraic curve is equal to the genus of its non-singular model. For any non-negative integer g there exists an algebraic curve of genus g. Rational curves are distinguished by the equality $g = 0$. If X is a projective plane curve of order m, then

$$\pi = \frac{(m-1)(m-2)}{2}$$

and its genus is given by the formula:

$$g = \frac{(m-1)(m-2)}{2} - d$$

where d is a non-negative integer which is a measure of the smoothness on X. If has only ordinary double points, d is simply the number of singular points.

In particular, the genus of a plane smooth projective curve is given by

$$g = \frac{(m-1)(m-2)}{2}$$

which means that not every smooth projective curve is plane. For a curve X in space the following estimate holds:

$$g \leq \begin{cases} \dfrac{(n-2)^2}{4} & \textit{for even} \quad n, \\ \dfrac{(n-1)(n-3)}{4} & \textit{for odd} \quad n, \end{cases}$$

where n is the degree of X. Curves of degree n of maximal genus exist for each value of n and lie on a quadric.

The degree of the canonical class K_X of a smooth projective curve X is connected with the genus of the curve by means of the formula $\deg K_X = 2g-2$. If a smooth projective curve X lies on a smooth algebraic surface F, the adjunction formula $K_X = X(X+K_F)$ holds. In particular $\deg K_X = (X)^2(X,K_F)$. For an arbitrary divisor D on X, one can consider the subset of the field $k(X)$ consisting of zero and of the functions f for which $(f)+D \geq 0$. This is a linear space over k of finite dimension $l(D)$. The dimension of the complete linear system defined by the divisor D is $l(D)-1$ and the calculation of $l(D)$ is an important task of the theory of algebraic curves. The strongest relevant result is the Riemann–Roch theorem. For smooth projective curves this theorem is the equality

$$l(D)-l(K-D) = \deg(D)-g+1$$

where g is the genus of the curve X. If $l(K-D)>0$ (or if $l(K-D)=0$), one says that D is special (or, respectively, non-special). For non-special divisors the Riemann–Roch theorem yields $l(D) = \deg(D)-g+1$. Each divisor of degree higher than $2g-2$ is non-special.

The class of divisors which are linearly equivalent to a divisor D on a smooth projective curve X defines a point on the Jacobi variety $J(X)$ of X. This variety is identical with the Albanese variety and with the Picard variety of X. Points which correspond to classes of special divisors are singular points of the Poincaré divisor on $J(X)$. If G_n^r denotes the subset of points of $J(X)$ which correspond to the classes of divisors D with $\deg D \quad n$ and $l(D)=r$, then G_n^r forms a subscheme in $J(X)$ and

$$\dim G_n^r \geq r(n-r+1)-(r-1)g$$

The Riemann–Brill–Noether theorem: This theorem has numerous applications, one of which will now be described. Any divisor D for which $l(D)\geq 1$ defines a rational mapping ϕ_D of the curve X into the projective space $P^l(D)-1$. The mapping ϕ_D depends on the class of D. If $\deg(D)\geq 2g+1$ then ϕ_D defines an isomorphic imbedding of X into P^m, while $\phi_D(X)$ is not contained in any proper subspace of the space $P_m (m \equiv l(D)-1)$. Mappings ϕ which correspond to a multiple nK of the canonical class of X are the most interesting from the point of view of the birational classification of curves. If $g>1$, the class $3K$ defines an isomorphic imbedding of the smooth projective curve into P^{5g-6}. Two X curves Y and are birationally equivalent if and only if their images and are obtained from each other by a projective transformation of P^{5g-6}. The study of the mapping yielded a more precise characteristic of curves of genus. For these curves is an isomorphic imbedding if and only if is not a hyper-elliptic curve. If ϕ_K is an isomorphism, the curve $\phi_K(X)$ is called canonical; it is defined uniquely up to projective transformations in P^{5g-1}. A very important task of the theory of algebraic curves is their classification up to a birational isomorphism. A number of important results have been obtained in this field, but an adequate solution of the problem is not available now (1977).

Smooth projective curves are subdivided into four classes:

1) Curves of genus 0, birationally equivalent to P^1;

2) Curves of genus 1 (elliptic curves), birationally equivalent to a smooth cubic curve in P^2;

3) Hyper-elliptic curves;

4) Non-hyper-elliptic curves of genus $g > 1$, birationally equivalent to a canonical curve in P^{5g-1} (algebraic curves of basic type).

The genus of a curve does not fully characterize the birational class of an algebraic curve. The only exception are curves of genus zero. If K is the field of complex numbers C, the set of classes of mutually-isomorphic elliptic curves is described by points in the quotient space H / G, where H is the upper half-plane and G is the modular group consisting of rational-linear transformations with integral coefficients and with determinant equal to $+1$. The space H / G has the structure of an analytic manifold isomorphic to C. Classes of birationally equivalent curves of genus $g > 1$ are described by points belonging to some algebraic variety Mg of dimension 3_{g-3}, which is called the moduli variety of curves of genus g. This variety is irreducible. According to one conjecture Mg is unirational; however, this has been proved (by F. Severi) for $g < 11$ only.

The following results are valid for the group $\text{Aut}(X)$ of automorphisms of a smooth projective curve X.) If X is P_k^1, then $\text{Aut}(X)$ is the group of rational-linear transformations $PGL(1,k)$.) If X is an elliptic curve, then $Aut(X)$ is an algebraic group, the connected component of the unit of which coincides with the group of points of $X(k)$.) If X is a curve of genus $g > 1$, then $Aut(X)$ is always a finite group. Its order is bounded by the number $84(g-1)$. Weierstrass points on X play an important role in the study of the group $Aut(X)$ in the latter case.

Another way of studying $Aut(X)$ is based on the fact that all smooth projective curves are finite (ramified) coverings of the projective line.

Let X be a smooth projective curve defined over the field C. The set of points of the curve $X(C)$ has the natural structure of a one-dimensional compact analytic manifold, which is also known as a compact Riemann surface. The converse is also true, i.e. any compact Riemann surface is obtained from some smooth projective curve. Usually one uses the same symbol X for the smooth projective curve and its corresponding one-dimensional complex manifold. Any connected complex manifold X can be represented as a quotient $\dot{X} / G$, where $\dot{X}$ is a simply-connected complex manifold, and G is a group of automorphisms $\dot{X}$ of which acts on $\dot{X}$ discretely and freely. It is noteworthy that there are only three one-dimensional simply-connected connected analytic manifolds, up to an isomorphism. These are the projective line CP^1 (the Riemann sphere), the affine straight line C (the finite plane) and the interior of the unit disc $D = \{z : |z| < 1\}$ (the Lobachevskii plane). All smooth projective curves are subdivided into three classes, depending on which one of the three types their universal covering belongs to.

The problem of classifying smooth projective curves of a given type can be reduced to the study of discrete groups of transformations of the universal coverings which act freely with a relatively compact fundamental domain. In the case of the projective line, G is the identity group; in the case of the affine straight line, G is isomorphic to a subgroup Ω of the additive group C which is a two-dimensional lattice in C; in the case of the interior of the unit disc, G is a subgroup of motions in the Lobachevskii plane which can be defined by some non-Euclidean bounded polygon. Thus, the first class above contains a unique curve P^1, the second class consists of complex tori C / Ω

that all have the structure of a one-dimensional Abelian variety (elliptic curve), the addition of points on the torus defining the group structure on the respective curve. All smooth elliptic curves are obtained in this way. The field of rational functions $C(X)$ on an elliptic curve $X; C/\Omega$ is isomorphic to the field of meromorphic doubly-periodic (elliptic) functions with period group Ω. If $f(x,y)=0$ is the equation of the affine model of the curve X, then there exists a parametrization $x=\phi(Z), Y=\psi(Z)$, by elliptic functions of it (a uniformization of X). The third class consists of all smooth projective curves X of genus $g>1$. In this case the field $C(X)$ is isomorphic to the field of meromorphic functions on D that are invariant with respect to the group G. Such functions are known as automorphic. Every algebraic curve of genus $g>1$ is uniformized by automorphic functions. The problem of the classification of elliptic curves also led to a study of the quotient D/G, but this situation is substantially different from the one just discussed. First, the group G has fixed points in D; secondly, the manifold D/G is non-compact, though containing a finite Lobachevskii plane. The study of the general case of such groups, and of the respective quotients, plays an important role in modern arithmetical research.

If an algebraic curve X is defined over a non-closed field k, one of the most important problems is that of the existence and location of the set of rational points $X(k)$ of X. In the case of a smooth projective curve over a finite field k a proof has been given of the inequality $|N-q-1|\leq 2g\sqrt{q}$, where N is the number of points on X that are rational over a finite extension L of k, is the number of elements of L and g is the genus of X. This inequality is equivalent to the Riemann hypothesis concerning the zeros of the ς-function of X, viz. that all zeros of the ς-function lie on the vertical line $\sigma=1/2$.

Now let X be an algebraic curve defined over the field of rational numbers X. For curves of genus zero the points of $X(Q)$ are relatively easily found, for elliptic curves the rational points constitute a finitely-generated group (if $X(Q)$ is non-empty), while for curves of genus $g\geq 2$ there is the Mordell conjecture to the effect that $X(Q)$ is finite.

If the ground field k is the field of rational functions $k_0(B)$ of a smooth projective curve B, each smooth projective curve X over k is isomorphic to the general fibre X_n of the morphism $f:V\to B$ of a smooth projective algebraic surface V over k_0. This morphism is uniquely defined if it is assumed that its fibres do not contain exclusively curves of genus 1. The set of rational points $X(k)$ is in one-to-one correspondence with the set of sections $V(B)$ of f, and $X(k)$ is finite for curves of genus $g>2$. Curves of genus 0 and 1 over the field $k_0(B)$ are studied in the theory of algebraic surfaces.

Remarkable Points of a Plane Curve

In this topic, we consider a plane algebraic curve defined by a bivariate polynomial $p(x, y)$ and its projective completion, defined by the homogenization $P(x,y,z)=^h p(x,y,z)$ of p.

Intersection with a Line

Knowing the points of intersection of a curve with a given line is frequently useful. The intersection with the axes of coordinates and the asymptotes are useful to draw the curve. Intersecting with a line parallel to the axes allows one to find at least a point in each branch of the curve. If an efficient root-finding algorithm is available, this allows to draw the curve by plotting the intersection point with all the lines parallel to the y-axis and passing through each pixel on the x-axis.

If the polynomial defining the curve has degree d, any line cuts the curve in at most d points. Bézout's theorem asserts that this number is exactly d, if the points are searched in the projective plane over an algebraically closed field (for example the complex numbers), and counted with their multiplicity. The method of computation that follows proves again this theorem, in this simple case.

To compute the intersection of the curve defined by the polynomial p with the line of equation $ax+by+c = 0$, one solves the equation of the line for x (or for y if $a = 0$). Substituting the result in p, one gets a univariate equation $q(y) = 0$ (or $q(x) = 0$, if the equation of the line has been solved in y), each of whose roots is one coordinate of an intersection point. The other coordinate is deduced from the equation of the line. The multiplicity of an intersection point is the multiplicity of the corresponding root. There is an intersection point at infinity if the degree of q is lower than the degree of p; the multiplicity of such an intersection point at infinity is the difference of the degrees of p and q.

Tangent at a Point

The tangent at a point (a, b) of the curve is the line of equation $(x-a)p'_x(a,b)+(y-b)p'_y(a,b)=0$, like for every differentiable curve defined by an implicit equation. In the case of polynomials, another formula for the tangent has a simpler constant term and is more symmetric:

$$xp'_x(a,b)+yp'_y(a,b)+p'_\infty(a,b)=0,$$

where $p'_\infty(x,y)=P'_z(x,y,1)$ is the derivative at infinity. The equivalence of the two equations results from Euler's homogeneous function theorem applied to P.

If $p'_x(a,b)=p'_y(a,b)=0$, the tangent is not defined and the point is a singular point.

This extends immediately to the projective case: The equation of the tangent of at the point of projective coordinates $(a{:}b{:}c)$ of the projective curve of equation $P(x, y, z) = 0$ is

$$xP'_x(a,b,c)+yP'_y(a,b,c)+zP'_z(a,b,c)=0,$$

and the points of the curves that are singular are the points such that

$$P'_x(a,b,c)=P'_y(a,b,c)=P'_z(a,b,c)=0.$$

(The condition $P(a, b, c) = 0$ is implied by these conditions, by Euler's homogeneous function theorem).

Asymptotes

Every infinite branch of an algebraic curve corresponds to a point at infinity on the curve, that is a point of the projective completion of the curve that does not belongs to its affine part. The corresponding asymptote is the tangent of the curve at that point. The general formula for a tangent to a projective curve may apply, but it is worth to make it explicit in this case.

Let $p=p_d+\cdots+p_0$ be the decomposition of the polynomial defining the curve into its homogeneous parts, where p_i is the sum of the monomials of p of degree i. It follows that:

$$P = {}^h p = p_d + zp_{d-1} + \cdots + z^d p_0$$

and

$$P'_z(a,b,0) = p_{d-1}(a,b).$$

A point at infinity of the curve is a zero of p of the form $(a, b, 0)$. Equivalently, (a, b) is a zero of p_d. The fundamental theorem of algebra implies that, over an algebraically closed field (typically, the field of complex numbers), p_d factors into a product of linear factors. Each factor defines a point at infinity on the curve: if $bx - ay$ is such a factor, then it defines the point at infinity $(a, b, 0)$. Over the reals, p_d factors into linear and quadratic factors. The irreducible quadratic factors define non-real points at infinity, and the real points are given by the linear factors. If $(a, b, 0)$ is a point at infinity of the curve, one says that (a, b) is an asymptotic direction. Setting $q = p_d$ the equation of the corresponding asymptote is

$$xq'_x(a,b) + yq'_y(a,b) + p_{d-1}(a,b) = 0.$$

If $q'_x(a,b) = q'_y(a,b) = 0$ and $p_{d-1}(a,b) \neq 0$, the asymptote is the line at infinity, and, in the real case, the curve has a branch that looks like a parabola. In this case one says that the curve has a *parabolic branch*. If

$$q'_x(a,b) = q'_y(a,b) = p_{d-1}(a,b) = 0$$

the curve has a singular point at infinity and may have several asymptotes. They may be computed by the method of computing the tangent cone of a singular point.

Singular Points

The singular points of a curve of degree d defined by a polynomial $p(x, y)$ of degree d are the solutions of the system of equations:

$$p'_x(x, y) = p'_y(x, y) = p(x, y) = 0.$$

In characteristic zero, this system is equivalent to

$$p'_x(x, y) = p'_y(x, y) = p'_\infty(x, y) = 0,$$

where, with the notation of the preceding topic, $p'_\infty(x, y) = P'_z(x, y, 1)$. The systems are equivalent because of Euler's homogeneous function theorem. The latter system has the advantage of having its third polynomial of degree d-1 instead of d.

Similarly, for a projective curve defined by a homogeneous polynomial $P(x,y,z)$ of degree d, the singular points have the solutions of the system.

$$P'_x(x, y, z) = P'_y(x, y, z) = P'_z(x, y, z) = 0$$

as homogeneous coordinates. (In positive characteristic, the equation $P(x,y,z)$ has to be added to the system).

This implies that the number of singular points is finite as long as $p(x, y)$ or $P(x, y, z)$ is square free. Bézout's theorem implies thus that the number of singular points is at most $(d-1)^2$, but this bound is not sharp because the system of equations is overdetermined. If reducible polynomials are allowed, the sharp bound is $d(d-1)/2$, this value being reached when the polynomial factors in linear factors, that is if the curve is the union of d lines. For irreducible curves and polynomials, the number of singular points is at most $(d-1)(d-2)/2$, because of the formula expressing the genus in term of the singularities. The maximum is reached by the curves of genus zero whose all singularities have multiplicity two and distinct tangents.

The equation of the tangents at a singular point is given by the nonzero homogeneous part of lowest degree in the Taylor series of the polynomial at the singular point. When one changes the coordinates to put the singular point at the origin, the equation of the tangents at the singular point is thus the nonzero homogeneous part of lowest degree of the polynomial, and the multiplicity of the singular point is the degree of this homogeneous part.

Analytic Structure

The study of the analytic structure of an algebraic curve in the neighborhood of a singular point provides an accurate information of the topology of singularities. In fact, near a singular point, a real algebraic curve is the union of a finite number of branches that intersect only at the singular point, and look either as a cusp or as a smooth curve.

Near a regular point, one of the coordinates of the curve may be expressed as an analytic function of the other coordinate. This is a corollary of the analytic implicit function theorem, and implies that the curve is smooth near the point. Near a singular point, the situation is more complicated, and involves Puiseux series, which provide analytic parametric equations of the branches.

For describing a singularity, it is worth to translate the curve for having the singularity at the origin. This consists of a change of variable of the form $X = x - a, Y = y - b,$ where a, b are the coordinates of the singular point. In the following, the singular point under consideration is always supposed to be at the origin.

The equation of an algebraic curve is $f(x, y) = 0$, where f is a polynomial in x and y. This polynomial may be considered as a polynomial in y, with coefficients in the algebraically closed field of the Puiseux series in x. Thus f may be factored in factors of the form $y - P(x)$, where P is a Puiseux series. These factors are all different if f is an irreducible polynomial, because this implies that f is square-free, a property which is independent of the field of coefficients.

The Puiseux series that occur here have the form

$$P(x) = \sum_{n=n_0}^{\infty} a_n x^{n/d},$$

where d is a positive integer, and n_0 is an integer that may also be supposed to be positive, because we consider only the branches of the curve that pass through the origin. Without loss of generality, we may suppose that d is coprime with the greatest common divisor of the n such that $a_n \neq 0$ (otherwise, one could choose a smaller common denominator for the exponents).

Let ω_d be a primitive dth root of unity. If the above Puiseux series occurs in the factorization of $f(x,y)=0$, then the d series

$$P_i(x)=\sum_{n=n_0}^{\infty} a_n \omega_d^i x^{n/d}$$

occur also in the factorization (a consequence of Galois theory). These d series are said conjugate, and are considered as a single branch of the curve, of *ramification* index d.

In the case of a real curve, that is a curve defined by a polynomial with real coefficients, three cases may occur. If none $P_i(x)$ has real coefficients, then one has a non-real branch. If some $P_i(x)$ has real coefficients, then one may choose it as $P_0(x)$. If d is odd, then every real value of x provides a real value of $P_0(x)$., and one has a real branch that looks regular, although it is singular if $d > 1$. If d is even, then $P_0(x)$ and $P_{d/2}(x)$ have real values, but only for $x \geq 0$. In this case, the real branch looks as a cusp (or is a cusp, depending of the definition of a cusp that is used).

For example, the ordinary cusp has only one branch. If it is defined by the equation $y^2 - x^3 = 0$, then the factorization is $(y - x^{3/2})(y + x^{3/2})$; the ramification index is 2, and the two factors are real and define each an half branch. If the cusp is rotated, it equation becomes $y^3 - x^2 = 0$, and the factorization is $(y - x^{2/3})(y - j^2 x^{2/3})(y - (j^2)^2 x^{2/3})$, with $j = (1 + \sqrt{-3})/2$ (the coefficient $(j^2)^2$ has not been simplified to j for showing how the above definition of $P_i(x)$ is specialized). Here the ramification index is 3, and only one factor is real; this shows that, in the first case, the two factors must be considered as defining the same branch.

Non-plane Algebraic Curves

An algebraic curve is an algebraic variety of dimension one. This implies that an affine curve in an affine space of dimension n is defined by, at least, $n-1$ polynomials in n variables. To define a curve, these polynomials must generate a prime ideal of Krull dimension 1. This condition is not easy to test in practice. Therefore, the following way to represent non-plane curves may be preferred.

Let $f, g_0, g_3, \ldots, g_n$ be n polynomials in two variables x_1 and x_2 such that f is irreducible. The points in the affine space of dimension n such whose coordinates satisfy the equations and inequations

$$\begin{aligned}
&f(x_1,x_2)=0\\
&g_0(x_1,x_2)\neq 0\\
&x_3=\frac{g_3(x_1,x_2)}{g_0(x_1,x_2)}\\
&\vdots\\
&x_n=\frac{g_n(x_1,x_2)}{g_0(x_1,x_2)}
\end{aligned}$$

are all the points of an algebraic curve in which a finite number of points have been removed. This curve is defined by a system of generators of the ideal of the polynomials h such that it exists an integer k such $g_0^k h$ belongs to the ideal generated by $f, x_3 g_0 - g_3, \ldots, x_n g_0 - g_n$. This representation is a rational equivalence between the curve and the plane curve defined by f. Every algebraic curve may be represented in this way. However, a linear change of variables may be needed in order to make almost always injective the projection on the two first variables. When a change of variables is needed, almost every change is convenient, as soon as it is defined over an infinite field.

This representation allows us to deduce easily any property of a non-plane algebraic curve, including its graphical representation, from the corresponding property of its plane projection.

For a curve defined by its implicit equations, above representation of the curve may easily deduced from a Gröbner basis for a block ordering such that the block of the smaller variables is (x_1, x_2). The polynomial f is the unique polynomial in the base that depends only of x_1 and x_2. The fractions g_i/g_0 are obtained by choosing, for $i = 3, ..., n$, a polynomial in the basis that is linear in x_i and depends only on x_1, x_2 and x_i. If these choices are not possible, this means either that the equations define an algebraic set that is not a variety, or that the variety is not of dimension one, or that one must change of coordinates. The latter case occurs when f exists and is unique, and, for $i = 3, ..., n$, there exist polynomials whose leading monomial depends only on x_1, x_2 and x_i.

Algebraic Function Fields

The study of algebraic curves can be reduced to the study of irreducible algebraic curves: those curves that cannot be written as the union of two smaller curves. Up to birational equivalence, the irreducible curves over a field F are categorically equivalent to algebraic function fields in one variable over F. Such an algebraic function field is a field extension K of F that contains an element x which is transcendental over F, and such that K is a finite algebraic extension of $F(x)$, which is the field of rational functions in the indeterminate x over F.

For example, consider the field C of complex numbers, over which we may define the field C(x) of rational functions in C. If $y^2 = x^3 - x - 1$, then the field C(x, y) is an elliptic function field. The element x is not uniquely determined; the field can also be regarded, for instance, as an extension of C(y). The algebraic curve corresponding to the function field is simply the set of points (x, y) in C^2 satisfying $y^2 = x^3 - x - 1$.

If the field F is not algebraically closed, the point of view of function fields is a little more general than that of considering the locus of points, since we include, for instance, "curves" with no points on them. For example, if the base field F is the field R of real numbers, then $x^2 + y^2 = -1$ defines an algebraic extension field of R(x), but the corresponding curve considered as a subset of R^2 has no points. The equation $x^2 + y^2 = -1$ does define an irreducible algebraic curve over R in the scheme sense (an integral, separated one-dimensional schemes of finite type over R). In this sense, the one-to-one correspondence between irreducible algebraic curves over F (up to birational equivalence) and algebraic function fields in one variable over F holds in general.

Two curves can be birationally equivalent (i.e. have isomorphic function fields) without being isomorphic as curves. The situation becomes easier when dealing with *nonsingular* curves, i.e. those that lack any singularities. Two nonsingular projective curves over a field are isomorphic if and only if their function fields are isomorphic.

Tsen's theorem is about the function field of an algebraic curve over an algebraically closed field.

Complex Curves and Real Surfaces

A complex projective algebraic curve resides in n-dimensional complex projective space CP^n. This has complex dimension n, but topological dimension, as a real manifold, $2n$, and is compact, connected, and orientable. An algebraic curve over C likewise has topological dimension two; in other words, it is a surface.

The topological genus of this surface, that is the number of handles or donut holes, is equal to the geometric genus of the algebraic curve that may be computed by algebraic means. In short, if one consider a plane projection of a nonsingular curve that has degree d and only ordinary singularities (singularities of multiplicity two with distinct tangents), then the genus is $(d - 1)(d - 2)/2 - k$, where k is the number of these singularities.

Compact Riemann Surfaces

A Riemann surface is a connected complex analytic manifold of one complex dimension, which makes it a connected real manifold of two dimensions. It is compact if it is compact as a topological space.

There is a triple equivalence of categories between the category of smooth irreducible projective algebraic curves over C (with non-constant regular maps as morphisms), the category of compact Riemann surfaces (with non-constant holomorphic maps as morphisms), and the opposite of the category of algebraic function fields in one variable over C (with field homomorphisms that fix C as morphisms). This means that in studying these three subjects we are in a sense studying one and the same thing. It allows complex analytic methods to be used in algebraic geometry, and algebraic-geometric methods in complex analysis, and field-theoretic methods to be used in both. This is characteristic of a much wider class of problems in algebraic geometry.

Singularities

Using the intrinsic concept of tangent space, points P on an algebraic curve C are classified as *smooth* (synonymous: *non-singular*), or else *singular*. Given $n-1$ homogeneous polynomials in $n+1$ variables, we may find the Jacobian matrix as the $(n-1)\times(n+1)$ matrix of the partial derivatives. If the rank of this matrix is $n-1$, then the polynomials define an algebraic curve (otherwise they define an algebraic variety of higher dimension). If the rank remains $n-1$ when the Jacobian matrix is evaluated at a point P on the curve, then the point is a smooth or regular point; otherwise it is a *singular point*. In particular, if the curve is a plane projective algebraic curve, defined by a single homogeneous polynomial equation $f(x, y, z) = 0$, then the singular points are precisely the points P where the rank of the $1\times(n+1)$ matrix is zero, that is, where

$$\frac{\partial f}{\partial x}(P)=\frac{\partial f}{\partial y}(P)=\frac{\partial f}{\partial z}(P)=0.$$

Since f is a polynomial, this definition is purely algebraic and makes no assumption about the nature of the field F, which in particular need not be the real or complex numbers. It should, of course, be recalled that (0,0,0) is not a point of the curve and hence not a singular point.

Similarly, for an affine algebraic curve defined by a single polynomial equation $f(x, y) = 0$, then the singular points are precisely the points P *of the curve* where the rank of the $1 \times n$ Jacobian matrix is zero, that is, where

$$f(P)=\frac{\partial f}{\partial x}(P)=\frac{\partial f}{\partial y}(P)=0.$$

The singularities of a curve are not birational invariants. However, locating and classifying the singularities of a curve is one way of computing the genus, which is a birational invariant. For this to work, we should consider the curve projectively and require F to be algebraically closed, so that all the singularities which belong to the curve are considered.

Classification of Singularities

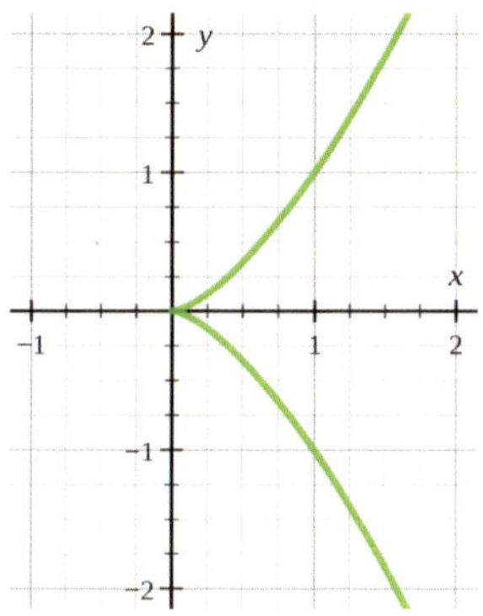

$x^3 = y^2$

Singular points include multiple points where the curve crosses over itself, and also various types of *cusp*, for example that shown by the curve with equation $x^3 = y^2$ at (0,0).

A curve C has at most a finite number of singular points. If it has none, it can be called *smooth* or *non-singular*. Commonly, this definition is understood over an algebraically closed field and for a curve C in projective space (i.e., *complete* in the sense of algebraic geometry). For example, the curve of equation $y-x^3=0$ is considered as singular, as having a singular point (a cusp) at infinity.

The singular points are classified by means of several invariants. The multiplicity m is defined as the maximum integer such that the derivatives of f to all orders up to $m - 1$ vanish (also the minimal intersection number between the curve and a straight line at P). Intuitively, a singular point has delta invariant δ if it concentrates δ ordinary double points at P. To make this precise, the blow up process produces so-called infinitely near points, and summing $m(m-1)/2$ over the infinitely near points, where m is their multiplicity, produces δ. For an irreducible and reduced curve and a point P we can define δ algebraically as the length of $\tilde{\mathcal{O}}_P / \mathcal{O}_P$ where $\mathcal{O}_P$ is the local ring at P and $\tilde{\mathcal{O}}_P$ is its integral closure.

The Milnor number μ of a singularity is the degree of the mapping $\frac{grad f(x,y)}{|grad\, f(x,y)|}$ on the small sphere of radius ε, in the sense of the topological degree of a continuous mapping, where grad f is the (complex) gradient vector field of f. It is related to δ and r by the Milnor-Jung formula,

$$\mu = 2\delta - r + 1.$$

Here, the branching number r of P is the number of locally irreducible branches at P. For example, $r = 1$ at an ordinary cusp, and $r = 2$ at an ordinary double point. Note that m is at least r, and that P is singular if and only if m is at least 2. Moreover, δ is at least $m(m\text{-}1)/2$.

Computing the delta invariants of all of the singularities allows the genus g of the curve to be determined; if d is the degree, then

$$g = \frac{1}{2}(d-1)(d-2) - \sum_P \delta_P,$$

where the sum is taken over all singular points P of the complex projective plane curve. It is called the genus formula.

Assign the invariants $[m, \delta, r]$ to a singularity, where m is the multiplicity, δ is the delta-invariant, and r is the branching number. Then an *ordinary cusp* is a point with invariants [2, 1, 1] and an *ordinary double point* is a point with invariants [2, 1, 2], and an ordinary m-multiple point is a point with invariants $[m, m(m-1)/2, m]$.

Examples of Curves

Rational Curves

A rational curve, also called a unicursal curve, is any curve which is birationally equivalent to a line, which we may take to be a projective line; accordingly, we may identify the function field of the curve with the field of rational functions in one indeterminate $F(x)$. If F is algebraically closed, this is equivalent to a curve of genus zero; however, the field of all real algebraic functions defined on the real algebraic variety $x^2+y^2 = -1$ is a field of genus zero which is not a rational function field.

Concretely, a rational curve embedded in an affine space of dimension n over F can be parameterized (except for isolated exceptional points) by means of n rational functions of a single parameter t; by reducing these rational functions to the same denominator, the n+1 resulting polynomials define a *polynomial parametrization* of the projective completion of the curve in the projective space. An example is the rational normal curve, where all these polynomials are monomials.

Any conic section defined over F with a rational point in F is a rational curve. It can be parameterized by drawing a line with slope t through the rational point, and intersection with the plane quadratic curve; this gives a polynomial with F-rational coefficients and one F-rational root, hence the other root is F-rational (i.e., belongs to F) also.

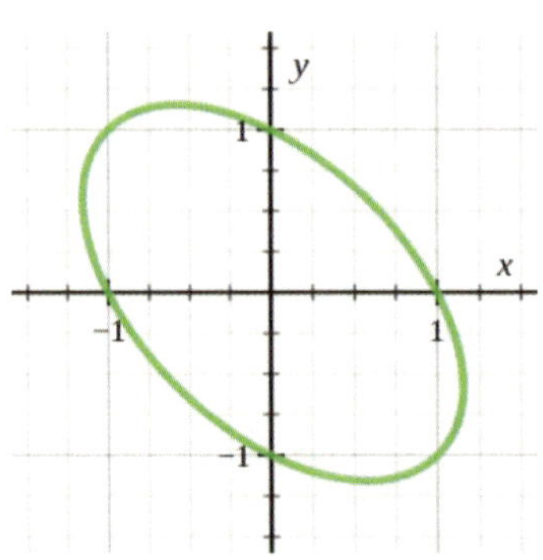

$x^2 + xy + y^2 = 1$

For example, consider the ellipse $x^2 + xy + y^2 = 1$, where $(-1, 0)$ is a rational point. Drawing a line with slope t from $(-1, 0)$, $y = t(x+1)$, substituting it in the equation of the ellipse, factoring, and solving for x, we obtain

$$x = \frac{1-t^2}{1+t+t^2}.$$

Then the equation for y is

$$y = t(x+1) = \frac{t(t+2)}{1+t+t^2},$$

which defines a rational parameterization of the ellipse and hence shows the ellipse is a rational curve. All points of the ellipse are given, except for $(-1, 1)$, which corresponds to $t = \infty$; the entire curve is parameterized therefore by the real projective line.

Such a rational parameterization may be considered in the projective space by equating the first projective coordinates to the numerators of the parameterization and the last one to the common denominator. As the parameter is defined in a projective line, the polynomials in the parameter should be homogenized. For example, the projective parameterization of above ellipse is

$$X = U^2 - T^2, \quad Y = T(T+2U), \quad Z = T^2 + TU + U^2.$$

Eliminating T and U between these equations we get again the projective equation of the ellipse

$$X^2 + XY + Y^2 = Z^2,$$

which may be easily obtained directly by homogenizing above equation.

Elliptic Curves

An elliptic curve may be defined as any curve of genus one with a rational point: a common model is a nonsingular cubic curve, which suffices to model any genus one curve. In this model the distinguished point is commonly taken to be an inflection point at infinity; this amounts to requiring that the curve can be written in Tate-Weierstrass form, which in its projective version is

$$y^2z + a_1xyz + a_3yz^2 = x^3 + a_2x^2z + a_4xz^2 + a_6z^3.$$

If the characteristic of the field is different from 2 and 3, then a linear change of coordinates allows putting $a_1 = a_2 = a_3 = 0$, which gives the classical Weierstrass form

$$y^2 = x^3 + px + q.$$

Elliptic curves carry the structure of an abelian group with the distinguished point as the identity of the group law. In a plane cubic model three points sum to zero in the group if and only if they are collinear. For an elliptic curve defined over the complex numbers the group is isomorphic to the additive group of the complex plane modulo the period lattice of the corresponding elliptic functions.

The intersection of two quadric surfaces is in general a nonsingular curve of genus one and degree four, and thus an elliptic curve, if it has a rational point. In special cases, the intersection either may be a rational singular quartic, or is decomposed in curves of smaller degrees which are not always distinct (either a cubic curve and a line, or two conics, or a conic and two lines, or four lines).

Curves of Genus Greater than One

Curves of genus greater than one differ markedly from both rational and elliptic curves. Such curves defined over the rational numbers, by Faltings's theorem, can have only a finite number of rational points, and they may be viewed as having a hyperbolic geometry structure. Examples are the hyperelliptic curves, the Klein quartic curve, and the Fermat curve $x^n + y^n = z^n$ when n is greater than three.

Algebraic Variety

In algebraic geometry, algebraic variety is a scheme which is integral, separated? and of finite type over an algebraically closed field k.

Classically, the term algebraic variety referred to a scheme as above which is further quasi-projective, i.e. admits a locally closed embedding into projective space. Thus, these were objects which locally are cut out inside projective space as the geometric locus of zeros of a set of polynomial equations in finitely many variables.

Given an algebraically closed field k, an algebraic k-variety usually means either a quasiprojective variety or an abstract variety (in the sense of Serre). 'Quasiprojective' unifies affine, quasiaffine, projective and embedded quasiprojective k-varieties. Many modern sources by a variety mean a reduced separated scheme of finite type over a field, often requiring also irreducibility (that is integral = reduced and irreducible).

- An embedded affine k-variety (or an affine algebraic set) is a set of zeros of a locus of common zeros of a set of polynomial equations in the affine space A_k^n . By the Hilbert Nullstellensatz there is a more invariant definition. Affine k-varieties are maximal spectra (= sets of maximal ideals) of finitely generated noetherian (commutative unital) k-algebras without nilpotents with the Zariski topology; the algebra can be recovered as the coordinate ring of the variety; this correspondence is an equivalence of categories, if the morphisms are properly defined.
- Affine varietes can be embedded as closed subvarieties into an affine space (in the sense of algebraic geometry). As topological spaces affine varieties are noetherian.
- Projective k-varieties are obtained in a similar way from graded k-algebras, or in embedded incarnation, as loci of zeros of a set of homogeneous polynomials in projective space P_k^n.

- Embedded quasiaffine k-varieties are Zariski-open subspaces of affine k-varieties.
- Embedded quasiprojective k-varieties are Zariski-open subspaces of projective k-varieties. We can remove the embedding by equipping them with the sheaf of regular functions and therefore considering them as locally ringed spaces. In the category of locally ringed spaces, projective, affine, and quasiaffine varieties are (isomorphic to) special cases of quasiprojective. Alternatively, we can put all 4 classes without sheaves into a category, by defining regular maps directly, and we get an isomorphic category of varieties.

In fact, by noticing that the affine k-space is Zariski open in a projective space of the same dimension, we see that the quasiprojective case includes all others.

Morphisms between varieties are so-called regular maps.

Sometimes a smooth algebraic variety may also be called algebraic manifold.

An abstract k-prevariety in the sense of Serre is a locally ringed space which is locally isomorphic to affine k-variety. The category of k-prevarieties has a product which is obtained by locally gluing products in the category of affine k-varieties. This enables defining a diagonal X→X→X; a prevariety is separated, or an abstract k-variety if the diagonal is closed in Zariski topology (which is, of course, not a product of Zariski topologies of factors).

Relation to Schemes

There is an equivalence of categories between the category of integral schemes of finite type over Speck, where k is an algebraically closed field, and the category of (irreducible) algebraic k-varieties.

Of course, given a variety the corresponding scheme and variety have different sets of points; the points in common are the closed points of the scheme. The remaining points are the generic points of subvarieties. Generic points were often used, without proper foundations, in other language, already in the works of the Italian school.

Some modern algebraic geometers mean, by varieties, objects of certain slightly bigger categories of relative S-schemes of finite type (where S is not necessarily Speck for k a field); typically they are required to be separated reduced S-schemes of finite type.

Examples

Subvariety

A subvariety is a subset of a variety that is itself a variety (with respect to the structure induced from the ambient variety). For example, every open subset of a variety is a variety.

Hilbert's Nullstellensatz says that closed subvarieties of an affine or projective variety are in one-to-one correspondence with the prime ideals or homogeneous prime ideals of the coordinate ring of the variety.

Affine Variety

Examples:

1. Let $k = C$, and A^2 be the two-dimensional affine space over C. Polynomials in the ring $C[x, y]$ can be viewed as complex valued functions on A^2 by evaluating at the points in A^2. Let subset S of $C[x, y]$ contain a single element $f(x, y)$:

   $$f(x,y) = x + y - 1.$$

The zero-locus of $f(x, y)$ is the set of points in A^2 on which this function vanishes: it is the set of all pairs of complex numbers (x, y) such that $y = 1 - x$. This is called a line in the affine plane. (In the classical topology coming from the topology on the complex numbers, a complex line is a real manifold of dimension two.) This is the set $Z(f)$:

$$Z(f) = \{(x, 1-x) \in \mathbf{C}^2\}.$$

Thus the subset $V = Z(f)$ of A^2 is an algebraic set. The set V is not empty. It is irreducible, as it cannot be written as the union of two proper algebraic subsets. Thus it is an affine algebraic variety.

2. Let $k = C$, and A^2 be the two-dimensional affine space over C. Polynomials in the ring C[x, y] can be viewed as complex valued functions on A^2 by evaluating at the points in A^2. Let subset S of $C[x, y]$ contain a single element $g(x, y)$:

   $$g(x,y) = x^2 + y^2 - 1.$$

The zero-locus of $g(x, y)$ is the set of points in A^2 on which this function vanishes, that is the set of points (x,y) such that $x^2 + y^2 = 1$. As $g(x, y)$ is an absolutely irreducible polynomial, this is an algebraic variety. The set of its real points (that is the points for which x and y are real numbers), is known as the unit circle; this name is also often given to the whole variety.

3. The following example is neither a hypersurface, nor a linear space, nor a single point. Let A^3 be the three-dimensional affine space over C. The set of points (x, x^2, x^3) for x in C is an algebraic variety, and more precisely an algebraic curve that is not contained in any plane. It is the twisted cubic. It may be defined by the equations

   $$y - x^2 = 0$$
   $$z - x^3 = 0$$

The irreducibility of this algebraic set needs a proof. One approach in this case is to check that the projection $(x, y, z) \to (x, y)$ is injective on the set of the solutions and that its image is an irreducible plane curve.

For more difficult examples, a similar proof may always be given, but may imply a difficult computation: first a Gröbner basis computation to compute the dimension, followed by a random linear change of variables; then a Gröbner basis computation for another monomial ordering to compute the projection and to prove that it is generically injective and that its image is a hypersurface, and finally a polynomial factorization to prove the irreducibility of the image.

Projective Variety

A projective variety is a closed subvariety of a projective space. That is, it is the zero locus of a set of homogeneous polynomials that generate a prime ideal.

Examples:

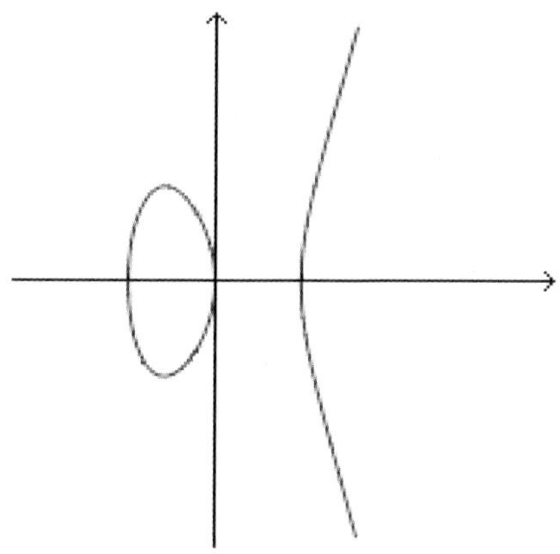

The affine plane curve $y^2 = x^3 - x$. The corresponding projective curve is called an elliptic curve.

1. A plane projective curve is the zero locus of an irreducible homogeneous polynomial in three indeterminates. The projective line P^1 is an example of a projective curve; it can be viewed as the curve in the projective plane $P^2 = \{[x, y, z]\}$ defined by $x = 0$. For another example, first consider the affine cubic curve

$$y^2 = x^3 - x.$$

in the 2-dimensional affine space (over a field of characteristic not two). It has the associated cubic homogeneous polynomial equation:

$$y^2 z = x^3 - xz^2,$$

which defines a curve in P^2 called an elliptic curve. The curve has genus one (genus formula); in particular, it is not isomorphic to the projective line P^1, which has genus zero. Using genus to distinguish curves is very basic: in fact, the genus is the first invariant one uses to classify curves.

2. Let V be a finite-dimensional vector space. The Grassmannian variety $G_n(V)$ is the set of all n-dimensional subspaces of V. It is a projective variety: it is embedded into a projective space via the Plücker embedding:

$$G_n(V) \mapsto \mathrm{P}(\wedge^n V), \langle b_1, \ldots, b_n \rangle \mapsto [b_1 \wedge \cdots \wedge b_n]$$

where b_i are any set of linearly independent vectors in V, $\wedge^n V$ is the n-th exterior power of V and the bracket $[w]$ means the line spanned by the nonzero vector w.

The Grassmannian variety comes with a natural vector bundle (or locally free sheaf in other terminology) called the tautological bundle, which is important in the study of characteristic classes such as Chern classes.

Non-affine and Non-projective Example

An algebraic variety can be neither affine nor projective. To give an example, let $X = P^1 \times A^1$ and p: $X \to A^1$ the projection. It is an algebraic variety since it is a product of varieties. It is not affine since P^1 is a closed subvariety of X (as the zero locus of p), but an affine variety cannot contain a projective variety of positive dimension as a closed subvariety. It is not projective either, since there is a nonconstant regular function on X; namely, p.

Another example of a non-affine non-projective variety is $X = A^2 - (0, 0)$.

Basic Results

- An affine algebraic set V is a variety if and only if $I(V)$ is a prime ideal; equivalently, V is a variety if and only if its coordinate ring is an integral domain.
- Every nonempty affine algebraic set may be written uniquely as a finite union of algebraic varieties (where none of the varieties in the decomposition is a subvariety of any other).
- The dimension of a variety may be defined in various equivalent ways.

Isomorphism of Algebraic Varieties

Let V_1, V_2 be algebraic varieties. We say V_1 and V_2 are isomorphic, and write $V_1 \cong V_2$, if there are regular maps $\varphi : V_1 \to V_2$ and $\psi : V_2 \to V_1$ such that the compositions $\psi \circ \varphi$ and $\varphi \circ \psi$ are the identity maps on V_1 and V_2 respectively.

1. A topological manifold of dimension n is a ringed space .V;OV / such that V is Hausdorff and every point of V has an open neighbourhood U for which $(U,OV|U)$is isomorphic to the ringed space of continuous functions on an open subset of $\mathbb{R}^n$.
2. A differentiable manifold of dimension n is a ringed space such that V is Hausdorff and every point of V has an open neighbourhood U for which $(U,OV|U)$ is isomorphic to the ringed space of smooth functions on an open subset of $\mathbb{R}^n$.
3. A complex manifold of dimension n is a ringed space such that V is Hausdorff and every point of V has an open neighbourhood U for which $(U,OV|U)$ is isomorphic to the ringed space holomorphic functions on an open subset of $\mathbb{C}^n$.

Regular Map

The geometry of three basic types of regular maps of varieties.

Definition: A regular map $\Phi : X \to Y$ of quasi-projective varieties is a closed embedding if Φ is an isomorphism from X to a closed subset $Z \subset Y$.

Examples: (a) A quasi-projective variety Y is affine (resp. projective), by definition, if there is a closed embedding $\Phi : Y \to C^n$ (resp. $\Phi : Y \to CP^n$).

(b) Let $\Phi : X \to Y$ be any regular map. The graph of Φ is the regular map $\gamma\Phi := (id_X, \Phi) : X \to X \times Y$, whose image is $\Gamma_\Phi := \{(x, \Phi(x)) \mid x \in X\}$. This is a closed subset of X × Y, by Proposition, since it is equal to the inverse image of the diagonal under the regular map:

$$(\Phi \circ \pi_X, \pi_Y): X \times Y \to Y \times Y$$

Moreover, the compositions: $X \overset{\Upsilon\Phi}{\to} \Gamma_\Phi \overset{\pi X}{\to} X$ and $\Gamma_\Phi \overset{\pi X}{\to} X \overset{\Upsilon\Phi}{\to} \Gamma_\Phi$ are clearly the identity maps, so Υ_Φ is an isomorphism from X to Γ_Φ. That is, the graph Υ_Φ is a closed embedding, and so every regular map Φ factors:

$$\Phi = \pi_Y \circ \Upsilon\Phi : X \to X \times Y \to Y$$

as a closed embedding (the graph) followed by a projection.

Proposition: A regular map of affine varieties $\Phi : X \to Y$ is a closed embedding if and only if $\Phi : C[Y] \to C[X]$ is surjective.

Proof: If Φ is an isomorphism from X to a closed subvariety $Z \subset Y$, then $C[Z] = C[Y]/I(Z)$, where *I(Z)* is the (prime) ideal of regular functions on Y that vanish on Z, so

$$\Phi^* : C[Y] \to C[Y]/I[Z] \tilde{\to} C[X]$$

is surjective. On the other hand, if Φ: X → Y is any regular map of affine varieties, and if Φ^*: C[Y] → C[X] is surjective, let I = ker(Φ^*) and consider $Z = V(I) \subset Y$. If $y \notin Z$, let $\bar{g} \in I$ be a regular function such that $\bar{g}(y) \neq 0$. Then y cannot be in the image of Φ, because if it were, say $\Phi(x) = y$, then $\Phi^*(\bar{g})(x) = \bar{g}(y) \neq 0$, and so $\bar{g} \in \ker(\Phi^*)$. Thus Φ factors through a map Φ : X → Z ⊂ Y, and the induced Φ^*: C[Z] → C[X] is an isomorphism, so Φ is an isomorphism from X to Z. That is, Φ : X → Y is a closed embedding.

So surjective homomorphisms of coordinate rings Φ^* : C[Y] → C[X] correspond to closed embeddings of affine varieties. It makes sense next to ask about the injective homomorphisms of coordinate rings.

Definition: A regular map Φ : X → Y of affine varieties is dominant if the map on coordinate rings Φ^*: C[Y] → C[X] is injective.

Examples: (a) (Projections) Recall that $C[X \times Y] \cong C[X] \otimes_C C[Y]$ for affine varieties. It follows that the two projections:

$$\pi_X : X \times Y \to X \text{ and } \pi_Y : X \times Y \to Y$$

are dominant, with corresponding injective maps:

$$\pi_X^* : C[X] \to C[X \times Y]; g \mapsto g \otimes 1 \text{ and } \pi_Y^* : C[Y] \to C[X \times Y]; h \mapsto 1 \otimes h$$

(b) (Open embeddings) Recall that a basic open set $U = X - V(\bar{g}) \subset X$ in a affine variety X is again an affine variety and that its affine coordinate ring is:

$$C[X] \to C[U] = C[X]_{\bar{g};} \ \bar{f} \mapsto \frac{\bar{f}}{1}$$

which is evidently injective. So the open embedding of U is a dominant map.

(c) (Blow Up) The regular map $\Phi : C^2 \to C^2$ given by:

$$\Phi(x,y) = (x,xy)$$

is dominant, since $\Phi^* : C[s, t] \to C[x, y]$ is given by $\Phi^*(s) = x$, $\Phi^*(t) = xy$, and there are no polynomial relations between x and xy. Notice that Φ induces an isomorphism of open sets:

$$\Phi : C^2 - V(x) \xrightarrow{\sim} C^2 - V(s)$$

but that Φ "blows down" the y-axis V (x) to the origin $(0, 0) \in C^2$. We will see in that the image of Φ is not isomorphic to a quasi-projective variety.

Proposition: $\Phi : X \to Y$ is dominant if and only if $\Phi(X) \subset Y$ is dense.

Proof: Suppose Φ is dominant, and $U = Y - V(\bar{g})$ is a basic open set. Since Φ^* is injective and $g \neq 0$ (always for basic open sets) then $\Phi^* g \neq 0$, which is to say that for some $x \in X$, $\Phi^*(\bar{g})(x) = g(\Phi(x)) \neq 0$. But this means that $\Phi(x) \in U$, and so $U \cap \Phi(X) \neq \emptyset$. Since this is true of all basic open sets, it follows that Φ (X) is dense.

On the other hand, if $\Phi(X)$ is dense then $\Phi(X)$ intersects each basic open set $U = Y - V(\bar{g})$, and so for each $\bar{g} \neq 0$, we can find an $x \in X$ so that $\Phi(x) = y$ and $\bar{g}(y) \neq 0$, which implies that $\Phi^*(\bar{g})(x) \equiv \bar{g}(y) \neq 0$, so $\Phi^*(\bar{g}) \neq 0$. Thus Φ is dominant.

Following the previous proposition, it makes sense to define in general:

Definition: A regular map $\Phi : X \to Y$ of quasi-projective varieties X, Y is dominant if Φ (X) is dense in Y.

Observation: If X is a projective variety, then the factorization of Φ:

$$\Phi = \pi_Y \circ \Upsilon_\Phi : X \times Y \to Y$$

as a composition of a closed embedding followed by a projection shows that Φ is a closed map, i.e. it takes closed sets to closed sets, since both the closed embedding Υ_Φ and the projection map π_Y (by Proposition above) are closed. Since the only subset of Y that is both dense and closed is Y.

Corollary: A dominant map of projective varieties is always surjective. It is time to begin to pay more attention to the fields of rational functions of a quasi-projective variety. Recall that when $Y \subset C^n$ is affine, then:

C(Y) = (the field of fractions of C[Y])

and when X is projective, then C(X) ⊂ (the field of fractions of C[X]) is the set of fractions that are a ratio of regular functions of the same degree.

If $Y \subset X$ is a quasi-projective variety (open in the projective variety X), we defined $O_Y(U)$ to be the elements of C(X) that are regular on U. Since any rational function on X is regular on some open set $U \subset X$, and then also on the (non-empty) intersection $V = U \cap Y \subset Y$, it follows that:

$$C(X) = \underset{UCX}{U} O_X(U) = \underset{V \subset Y}{U} O_Y(V)$$

and this will be our definition of C(Y), so that in particular C(X) = C(Y).

Proposition: A dominant map $\Phi : X \to Y$ of quasi-projective varieties induces a well-defined (injective!) pull-back of fields of rational functions:

$$\Phi^* : C(Y) \, a \;\; C(X)$$

Proof: Recall that by definition, Φ satisfies: Φ^*: O_Y (U) → O_X(Φ^{-1} (U)) for all open subsets U ⊂ Y. If φ ∈ C(Y) is nonzero, then the domain of φ is some such open set U, and because Φ is dominant, the inverse image Φ^{-1} (U) is non-empty. Thus $\Phi^*(\phi)$ is a nonzero(!) rational function on X, and the pull-back is defined on all rational functions (and injective).

Remarks: (a) When Φ is not dominant, the pull-back is similarly defined, but only for rational functions $\phi \in$ C(Y) whose domains intersect Φ (X).

(b) When X and Y are affine, then Φ *: $C(Y) \to C(X)$ is the extension of the injection Φ *: C[Y], → C[X] to an injection of the fields of fractions.

We can reverse Proposition above, but only if we allow for dominant rational maps.

Proposition: For quasi-projective varieties X and Y, each inclusion of fields $l : C(Y) \to C(X)$ gives rise to a rational map:

$$\Phi : X --\to Y$$

such that the regular map $\Phi|_U : U \to Y$ (from the domain of Φ) is dominant and the two inclusions of fields agree, i.e. $\Phi^* = l$.

Proof: $X \subset \overline{X} \subset CP^m$ and $Y \subset \overline{Y} \subset CP^n$, construct the rational map Φ out of the rational functions $l\left(\frac{\overline{y}_i}{\overline{y}_j}\right) \in C(X)$ as in Proposition.

Definition: A rational map $\Phi : X --> Y$ is birational if it is dominant (thought of as a regular map from its domain) and Φ^* is a field isomorphism. Examples: Examples (b) and (c) (the open embedding and blow-up) of dominant maps are also birational. Note that neither is an isomorphism! We will have more to say about birational maps. For now, I want to concentrate on another very important class of regular maps:

Definition: A regular map $\Phi : X \to Y$ of affine varieties is finite if the pull-back $\Phi \quad C[Y] \to C[X]$ makes C[X] a finitely generated C[Y]-module.

Examples: (a) A closed embedding $\Phi : X \to Y$ is finite, since in that case Φ^* is surjective (Proposition), so C[X] is generated by 1 as a C[Y]-module.

b) The open embedding Y – V ($\overline{g}$) =: U → Y of a basic open set in an affine variety is not finite because, as a module, C[U] = C[X]$_{\overline{g}}$ is freely generated by 1, $\overline{g}^{-1}, \overline{g}^{-2}, \overline{g}^{-3}, \ldots$

(c) A projection $\pi_X: X \times Y \to X$ is only finite if Y is a point.

(d) Let $X = V(y^2 - x) \subset C^2$ be the parabola (on its side) and let $\Phi : X \to C^1$ be the projection to the x-axis, which is two-to-one except at the origin. Then the homomorphism of coordinate rings is the inclusion:

$$\Phi^* : C[x] \to C[X] = C[x, y]/\langle y^2 - x\rangle$$

so Φ is dominant. It is finite since $C[X]$ is generated by 1, $\bar{y}$ as a $C[x]$-module.

(e) (Affine No ether Normalization) If X is any affine variety, then the Neither Normalization Lemma showed that there is a polynomial ring $C[y_1, ..., y_d] \subset C[X]$ so that C[X] is a finitely generated $C[y_1, ..., y_d]$-module. So the associated regular map $\Phi : X \to C^d$ is both dominant and finite.

Observation: If $\Phi : X \to Y$ is finite and dominant, then the inclusion of fields $\Phi^* : C(Y) \to C(X)$ makes C(X) into a finite field extension of C(Y). The generators of C[Y] as a C[X]-module span C(Y) as a C(X)-vector space.

Proposition: A finite map $\Phi : X \to Y$ of affine varieties always has the following geometric properties:

- The "fibers" Φ^{-1} (y) are finite sets for all $y \in Y$.
- Φ is a closed map (i.e. takes closed sets to closed sets)

Remark: Each of these requires a new lemma from commutative algebra, so we will precede each part of the proof with the relevant algebra result.

Chinese Remainder Lemma: If A is a commutative ring with 1 and if $m_1, ..., m_r \subset A$ are distinct maximal ideals, then the natural map:

$$A/m_1 \cap ... \cap m_r \to \times_{i=1}^{r} A/m_i$$

is an isomorphism. In particular, if A is a C-algebra of dimension r as a vector space over C, then A has at most r distinct maximal ideals.

Proof: We proceed by induction on r. If r = 1, the result is a tautology. Suppose the Lemma holds for r distinct maximal ideals. Since there are exactly r maximal ideals in the product of fields: $\times_{i=1}^{r} A/m_i$ it follows that $m_1, ..., m_r$ must be the only maximal ideals containing $m_1 \cap ... \cap m_r$. Thus if m_{r+1} is a new maximal ideal, then $m_{r+1} + (m_1 \cap ... \cap m_r) = A$, and by the usual Chinese Remainder Theorem, we get the desired isomorphism:

$$A/m_1 \cap ... \cap m_r \cap m_{r+1} \xrightarrow{\sim} \times_{i=1}^{r} A/m_i \times A/m_{r+1}$$

Remark: There may be fewer maximal ideals than the rank of A in the lemma. For example, $A = C[x]/\langle x^r\rangle$ has rank r as a vector space over C, but only one maximal ideal, namely $\langle \bar{x}\rangle$. This is no contradiction to the lemma, of course, which is a tautology in this case.

Proof of (a): If $\Phi : X \to Y$ is any regular map and $Z \subset Y$ is a closed set, then we always have $\Phi^{-1}(Z) = V(\langle \Phi^* I(Z)\rangle)$. That is because a point a $\in X$ satisfies $\Phi(a) \in Z$ if and only if for each regular function $\bar{g} \in I(Z)$ the pull-back $\Phi^* g$ vanishes at a. Thus we need to show that if Φ is a finite map, then $V\left(\langle \Phi^* m_b \rangle\right)$ is a finite set whenever m_b is the maximal ideal of functions vanishing at a point $b \in Y$. To see this, consider the following diagram:

$$\begin{array}{ccc} C[Y] & \overset{\Phi}{\rightarrow} & C[X] \\ \downarrow & & \\ C = C[Y]/m_b & \overset{\bar{\Phi}^*}{\rightarrow} & A = C[X]/\langle \Phi^* m_b \rangle \end{array}$$

Now either A = 0, in which case $\langle \Phi * mb \rangle = C[X]$, and there are no points in Φ^{-1} (b), or else $A \neq 0$, in which case A is a C-algebra and the generators $\bar{f}_1, \ldots, \bar{f}_r$ of C[X] as a C[Y]-module also span A as a C-vector space. Since points of Φ^{-1} (b) correspond to the maximal ideals of A by the Nullstellensatz, it follows from the Chinese Remainder Lemma that Φ^{-1} (b) in fact consists of at most r points.

Example: Consider again the example of the sideways parabola. Then

$$\Phi^{-1}(b) = V\left(\langle \bar{x} - b \rangle\right) = V\left(\langle \bar{y}^2 - b \rangle\right) \subset X, \text{ and } C[X] = C[x, y]/\langle y^2 - x \rangle \cong C[y].$$

Thus there are two possibilities:

$$A \cong C[y]/\langle y^2 - b \rangle has \begin{cases} 2 \max imal\, ideal\, if\; b \neq 0 \; and \\ 1 \max imal\, ideal\, if\; b = 0 \end{cases}$$

corresponding to the 2 square roots of b when $b \neq 0$ and the 1 when b = 0.

Localization: A subset $S \subset A$ of a commutative ring with 1 is multiplicative if $1 \in S$, $0 \notin S$ and S is closed under multiplication. If A is a domain, the localization of A at a multiplicative subset $S \subset A$ is the domain:

$$A \subset AS := \left\{ \frac{f}{s} \mid f \in A,\ s \in S \right\} \subset K$$

sitting between A and its field of fractions K (which is A_S for S = A – {0}).

Examples: (a) The set S = A–P is multiplicative when P is a prime ideal.

In this case, we always write AP instead of AS (by convention).

(b) The set $S = \{1, f, f^2, f^3, \ldots\} \subset A$ is multiplicative if $f^n \neq 0$ for all n, and in this case, we always write A_f instead of A_S (also by convention).

Note: We've already used this notation to write C[U] = C[X]$\bar{g}$ when U = X – V ($\bar{g}$) is a basic open subset of an affine variety X.

Rational Mapping

A generalization of the concept of a rational function on an algebraic variety. Namely, a rational mapping from an irreducible algebraic variety X to an algebraic variety Y (both defined over a field k) is an equivalent class of pairs (U,ϕ_U), where U is a non-empty open subset of X and ϕ_U is a morphism from U to Y . Two pairs (U,ϕ_U) and (V,ψ_V) are said to be equivalent if ϕ_U and ϕ_V coincide on $U \cap V$. In particular, a rational mapping from a variety X to an affine line is a rational function on X. For every rational mapping $\phi: X \to Y$ there is a pair $(\tilde{U},\phi_U)$ such that for all equivalent pairs and is the restriction of ϕ_U to U. The open subset $\tilde{U}$ is called the domain of regularity of the rational mapping ϕ, and $\phi\ (\tilde{U})$ is the image of the variety X (written) under ϕ.

If $\phi: X \to Y$ is a rational mapping of algebraic varieties and $\phi(X)$ is dense in Y , then ϕ determines an imbedding of fields, $\phi^*: k(Y) \to k(Y)$. Conversely, an imbedding of the fields of rational functions $\phi^*: k(Y) \to k(Y)$ determines a rational mapping from X to Y. If ϕ induces an isomorphism of the fields $k(X)$ and $k(Y)$ of rational functions, then ϕ is called a birational mapping.

The set of points of X at which the rational mapping $\phi: X \to Y$ is not regular has codimension 1, in general. But if Y is complete and X is smooth and irreducible, then this set has codimension at least 2. If X and Y are complete irreducible varieties over an algebraically closed field of characteristic 0, then the rational mapping $\phi: X \to Y$ can be included in a commutative diagram.

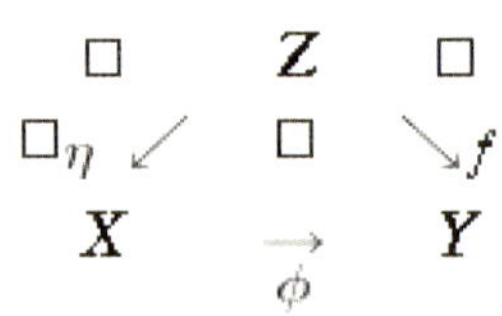

where η, f are morphisms of an algebraic variety Z and η is a composite of monoidal transformations (cf. Monoidal transformation). If is a birational mapping of complete non-singular surfaces, then there exists a diagram (*) in which both and are composites of monoidal transformations with non-singular centres (Zariski's theorem), that is, every birational mapping of complete non-singular surface decomposes into monoidal transformations with non-singular centres and their inverses. In the case when $\dim X \geq 3$, the question of whether every birational mapping can be decomposed in this way is open (1990).

1. Let t be the co-ordinate on A^1 and x,y the co-ordinates on A^2. Let $V = A^1$ and let $W \subset A^2$ be the circle $x^2+y^2-1=0$. Consider the intersections of the line $y=t(x-1)$ with the circle.

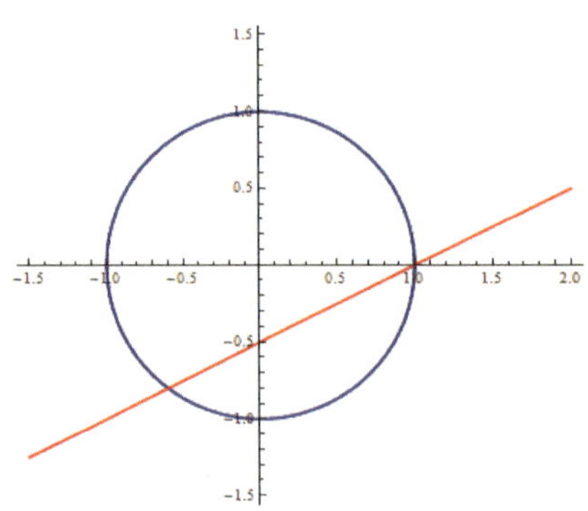

After substituting $y=t(x-1)$ into $x^2+y^2-1=0$ we get a quadratic for x, one solution is always $x=1$, the other is $x=\dfrac{t^2-1}{t^2-1}$. The first one gives the point (1, 0), the other $\left(\dfrac{t^2-1}{t^2-1}, -\dfrac{2t}{t^2-1}\right)$.

$\frac{t^2-1}{t^2+1}$ and $-\frac{2t}{t^2+1}$ are rational functions in t, so they are elements of C(V), and $\left(\frac{t^2-1}{t^2+1},-\frac{2t}{t^2+1}\right)$ $\in$ W by its construction whenever it is defined. Therefore $\phi(t)=\left(\frac{t^2-1}{t^2+1},-\frac{2t}{t^2+1}\right)$ defines a rational map $V \to W$. It can be verified purely algebraically, too, that the image lies in W, since $\left(\frac{t^2-1}{t^2+1}\right)^2+\left(-\frac{2t}{t^2+1}\right)^2-1=0$. To find an inverse for φ, we need solve $x=\frac{t^2-1}{t^2+1}, y=-\frac{2t}{t^2+1}$ for t as a rational function in x and y. From the geometric construction it is clear that t can be recovered as the slope of the line through (x, y) and (1, 0) therefore $t=y/(x-1), \psi(x,y)=y/(x-1)$ is a rational map $W \to V$, since $y/(x-1) \in K(W)$ and because the codomain is A^1, there are no equations to check. (As it is clear in all the examples that the maps are written in terms of polynomial or rational functions, this will not be mentioned specifically again).

It can be verified purely algebraically that ψ is the inverse of ϕ as a rational map.

$$\phi(\psi\circ\phi)(t)=\psi\left(\frac{t^2-1}{t^2+1},-\frac{2t}{t^2+1}\right)=\frac{-\frac{2t}{t^2+1}}{\frac{t^2-1}{t^2+1}-1}=\frac{-\frac{2t}{1+t^2}}{\frac{-2}{t^2+1}}=t \quad (t\neq\pm i),$$

$$(\psi\circ\phi)(x,y)=\phi\left(\frac{y}{x-1}\right)=\left(\frac{-1+\left(\frac{y}{x-1}\right)^2}{1+\left(\frac{y}{x-1}\right)^2},-\frac{2\left(\frac{y}{x-1}\right)}{1+\left(\frac{y}{x-1}\right)^2}\right)$$

$$=\left(\frac{-(x-1)^2+y^2}{(x-1)^2+y^2},-\frac{2y(x-1)}{\left(x-1)^2+y^2\right)}\right)$$

In *K[W]*, we have

$$(x-1)^2+y^2=x^2-2x+1+y^2=(x^2-2x+1+y^2)-(x^2+y^2-1)=2(1-x)$$
$$and$$
$$-(x-1)^2+y^2=-x^2+2x+1+y^2=(x^2-2x+1+y^2)-(x^2+y^2-1)=2x(1-x)$$

therefore

$$(\phi\circ\psi)(x,y)=\left(\frac{2x(1-x)}{2(1-x)},-\frac{2y(x-1)}{2(1-x)}\right)=(x,y)\ (x,y)\neq(1,0)$$

Hence ϕ and ψ mutually inverse birational maps between Vand W, therefore V and W are birationally equivalent and W is a rational variety.

Neither φ, nor ψ is a morphism. ϕ is already written in terms of rational functions with coprime numerator and denominator and $K[V]\cong K[t]$ is a unique factorisation domain, so ϕ is not defined at $t=\pm i$, where $1+t^2=0$. ψ is not defined at (1, 0). $y/(x-1)$ is a 0/0 type fraction, but $\frac{y}{(x-1)}=-\frac{x+1}{y}$

in *K(W)*, since $y^2=1-x^2$ in K[W], the numerator $x+1$ is not 0 at (1,0), but denominator y is 0.

A further point worth noting is that $(x-1)/y$ is not defined at (1, 0) as a rational map on A^2, but it is defined as a rational map on W, since $\frac{x-1}{y}=\frac{y}{(x+1)}$ on W and this latter fraction is defined and has value 0 at (1, 0).

Let V = A^1 and let W ⊂ A^2 be the cuspidal cubic $y^2-x^3 = 0$. We already saw that $\phi: V \to W$, $\phi(t) = (t^2, t^3)$ is a morphism $V \to W$. $\psi(x,y) = y/x$ is a rational map which is the inverse of ϕ considered as a rational map.

$$(\psi\circ\phi)(t)=\psi(t^2,t^3)=t^3/t^2=t(t\neq 0) and (\phi\circ\psi)(x,y)=\psi(y/x)$$
$$=\left(\left(\frac{y}{x}\right)^2,\left(\frac{y}{x}\right)^3\right)=\left(\frac{x^3}{x^2},\frac{y^3}{y^2}\right)=(x,y)((x,y)\neq(0,0)).$$

Geometrically, for $t\neq 0$, $\varphi(t)$ is the unique intersection point of W with the line $y=tx$ other than the origin for $t\neq 0$, therefore the inverse is clearly $t=y/x$.

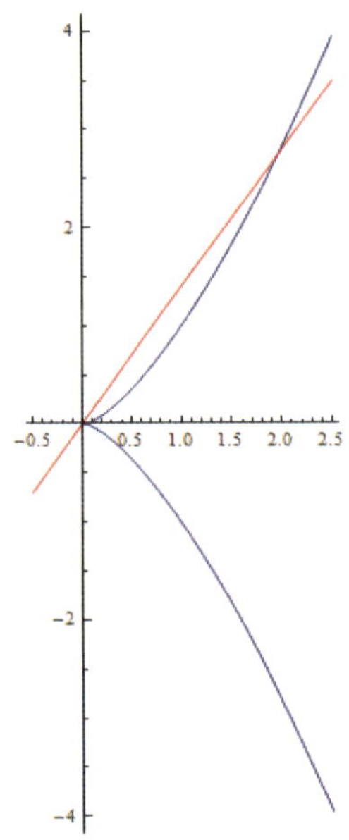

ψ is defined at every point of W except (0, 0), but it is not defined at (0, 0), since if it were, it would be a morphism but we proved earlier that ϕ has no inverse morphism.

The existence of the mutually inverse rational maps between V and W shows that they birationally equivalent, so W is rational. We can also prove this algebraically. We saw that K[W] is isomorphic to the subring S of K[t] consisting of polynomials with no degree 1 term. This implies that K(W) is isomorphic to a subfield of K(t). However, $t^2, t^3 \in S$, therefore $t = t^3/t^2$ is in the field of fractions of S, so the field of fractions is the whole of K (t) and therefore $K(W)\cong K(t)\cong K(V)$.

Let V=A^1 and let W ⊂ A^2 be the nodal cubic $y^2 - x^3 - x^2 = 0$. $\phi: V \to W$, $\phi(t) = (t^2-1, t(t^2-1))$ is a morphism $V \to W$, since $(t(t^2-1))^2-(t^2 - 1)^3 - (t^2 - 1)^2 = (t^2 - 1)2(t^2 - (t^2 - 1) - 1) = 0$.

The geometric description of $\phi(t)$ is the same as in the previous example, $\phi(t)$ is the intersection point of W with the line $y = tx$ other than the origin for for $t \neq \pm 1$, therefore the inverse is clearly $t = yx$.

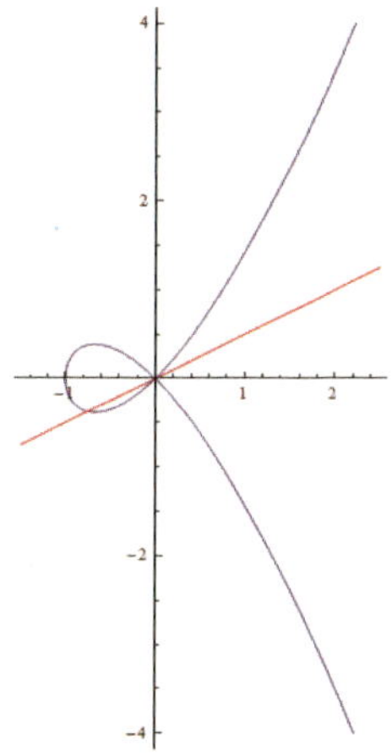

$\Psi(x, y) = y/x$ is a rational map which is the inverse of ϕ considered as a rational map since

$$(\psi \circ \varphi)(t) = \psi\left(t^2 - 1,\ t\left(t^2 - 1\right)\right) = t\left(t^2 - 1\right)/\left(t^2 - 1\right) = t \ \ (t \neq \pm 1)$$

$$and (\phi \circ \varphi)(x;y) = \left(\left(\frac{y}{x}\right)^2 - 1, \frac{y}{x}\left(\left(\frac{y}{x}\right)^2 - 1\right)\right) = \left(\frac{y^2 - x^2}{x^2}, \frac{y(y^2 - x^2)}{x^3}\right)$$

$$= \left(\frac{x^3}{x^2}, \frac{yx^3}{x^3}\right) = (x,y)\left((x,y) \neq (0,0)\right)$$

This shows that V and W are birationally equivalent, therefore W is rational. ψ is defined at every point of W except (0, 0), but it is not defined at (0, 0), since if it were, it would be a morphism, but ϕ is not injective, $\phi(-1) = \phi(1) = (0, 0)$, so it cannot have an inverse morphism.

Affine Variety

An affine variety V is an algebraic variety contained in affine space. For example,

$$\left\{(x,y,z) : x^2 + y^2 - z^2 = 0\right\}$$

is the cone, and

$$\left\{(x,y,z) : x^2 + y^2 - z^2 = 0, ax + by + cz = 0\right\}$$

is a conic section, which is a subvariety of the cone. The cone can be written $V\left(x^2 + y^2 - z^2\right)$ to indicate that it is the variety corresponding to $x^2 + y^2 - z^2 = 0$. Naturally, many other polynomials vanish on $V\left(x^2 + y^2 - z^2\right)$ in fact all polynomials in $I(C) = \left\{x^2 + y^2 - z^2\right\}$. The set *I(C)* is an ideal in the polynomial ring $C[x,y,z]$. Note also, that the ideal of polynomials vanishing on the conic section is the ideal generated by $x^2 + y^2 - z^2$ and $ax + by + cz$.

A morphism between two affine varieties is given by polynomial coordinate functions. For example, the map $\phi(x,y,z)=(x^2,y^2,z^2)$ is a morphism from $X=V(x^2+y^2+z^2)$ to $Y=V(x+y+z)$. Two affine varieties are isomorphic if there is a morphism which has an inverse morphism. For example, the affine variety $V(x^2+y^2+z^2)$ is isomorphic to the cone $V(x^2+y^2-z^2)$ via the coordinate change $\phi(x,y,z)=(x,y,iz)$.

Many polynomials f may be factored, for instance $f=x^2+y^2=(x+iy)(x-iy)$, and then $V(f)=V(x+iy)\cup V(x-iy)$. Consequently, only irreducible polynomials, and more generally only prime ideals p are used in the definition of a variety. An affine variety V is the set of common zeros of a collection of polynomials p_1, ..., p_k, i.e.,

$$V=\{x=(x_1,\ldots,x_n):p_1(x)=\ldots=p_k(x)=0\}$$

as long as the ideal $I=<p_1,\ldots,p_k>$ is a prime ideal. More classically, an affine variety is defined by any set of polynomials, i.e., what is now called an algebraic set. Most points in V will have dimension n-k, but V may have singular points like the origin in the cone.

When V is one-dimensional generically (at almost all points), which typically occurs when k=n-1, then V is called a curve. When V is two-dimensional, it is called a surface. In the case of CW-complex affine space, a curve is a Riemann surface, possibly with some singularities.

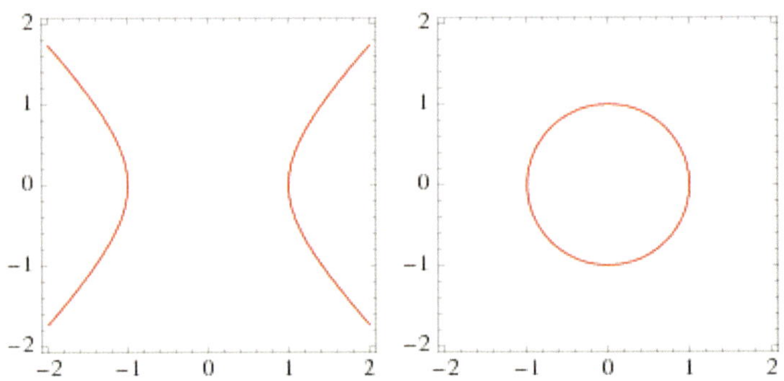

Definition: A non empty subset Y of a topological space X is irreducible if it can not be expressed as the union $Y = Y_1\cup Y_2$ of two proper subsets, each one of which is closed in Y. The empty set is not considered to be irreducible.

Example: A^1 is irreducible, because its only proper closed subsets are finite, yet it is infinite because K is algebraically closed, hence infinite.

Proposition: Any nonempty open subsets of an irreducible space is irreducible and dense.

Proof: Let X be irreducible topological space. Let U be a nonempty open subset of the irreducible space X.

To prove: U is dense in X, that is to prove that for any nonempty open subset V of X, $U\cap V\neq\varphi$. Suppose not, that is suppose $U\cap V=\varphi\Rightarrow(U\cap V)^c=(\varphi)^c\Rightarrow U^c\cup V^c = X$. U and V both are open sets ⇒ U^c and V^c both are closed sets.

Note that $U^c\neq X$ because $U\neq\varphi$, similarly $V^c\neq X$ because $V\neq\varphi\Rightarrow U$ and V^c are proper closed subsets of X therefore X = $U^c \cup V^c$, whichis a contradiction because X is given to be irreducible. Hence U is dense in X is given to be irreducible. Hence U is dense in X.

To prove: U is irreducible. Suppose not, Then $U = U_1 \cup U_2$ for some U_1, U_2 closed in U in the induced topology, which are proper subsets of U. U_1 is closed in $U \Rightarrow U_1 = V_1 \cap U$, where V_1 is closed in X. U_2 is closed in $U \Rightarrow U_2 = V_2 \cap U$ where V_2 is closed in X.

Consider the equation,

$$\begin{aligned} X &= U \cup U^C \\ &= (U_1 \cup U_2) \cup U^c \\ &= (V_1 \cap U) \cup (V_2 \cap U) \cup U^c \\ &= [(V_1 \cup V_2) \cap U] \cup U^c \\ &= [(V_1 \cup V_2) \cup U^C] \cap [U \cup U^c] \\ &= (V_1 \cup V_2) \cup U^c \end{aligned}$$

U is open in $X \Rightarrow U^c$ is closed in X. $U \neq \phi \Rightarrow U^c \neq X \Rightarrow U^c$ is a proper subset of X. V_1 and V_2 both are closed in $X \Rightarrow V_1 \cup V_2$ is closed in X. $V_1 \cup V_2 \neq X$ because if $V_1 \cup V_2 = X$, then $(V_1 \cup V_2)^c = \phi \Rightarrow V_1^c \cap V_2^c = \varphi \Rightarrow V_1^c$ and V_2^c are both open sets in X whose intersection is empty. But since X is irreducible, every open subset in X is dense $\Rightarrow V_1^c \cap V_2^c$ can not be empty.

So that $V_1 \cup V_2$ is a proper closed subset of X. $X = [V_1 \cup V_2] \cup U^c$ which is a contradiction, since given X is irreducible.

Proposition: If Y is an irreducible subset of X, then its closure $\bar{Y}$ in X is also irreducible. Proof: Suppose $\bar{Y}$ is not irreducible. Then $\bar{Y} = Y_1 \cup Y_2$ for two proper subsets of $\bar{Y}$, each one of which is closed in $\bar{Y}$. Then $Y_1 = V_1 \cap \bar{Y}$ and $Y_2 = V_2 \cap \bar{Y}$

for some V_1,V_2 proper and closed in X, Then

$$\begin{aligned} Y &= Y_1 \cup Y_2 \\ &= (V_1 \cap \bar{Y}) \cup (V_2 \cap \bar{Y}) \\ &= (V_1 \cup V_2) \cap \bar{Y} \end{aligned}$$

$$\begin{aligned} \textit{Also we have, } Y &= \bar{Y} \cup Y \\ &= (V_1 \cup V_2) \cap \bar{Y} \cap Y \\ &= ((V_1 \cup V_2) \cap Y \\ &= ((V_1 \cap Y) \cup (V_2 \cap Y), \textit{ where } (V_1 \cap Y) \textit{ and } (V_2 \cap Y) \end{aligned}$$

Are Proper closed subsets in Y. This is contradiction the fact that Y is irreducible. So $\bar{Y}$ is irreducible.

Definition: An affine algebraic variety is an irreducible closed subset of An.In other words, irreducible affine algebraic sets are called affine algebraic variety.

So for any subset $Y \subseteq A^n$, let us define the ideal of Y in A by

$$I(Y) = \{f \in A \mid f(p) = 0 \forall p \in Y\}$$

Now we have a function Z which maps subsets of A to algebraic sets, and a function I which maps subsets of A^n to ideals.

Quasi-projective Variety

A quasi-projective variety is a locally closed subset of $\mathbb{P}^n$, considered with the Zariski topology induced from $\mathbb{P}^n$. Recall that a locally closed set of any topological space is a closed subset of an open subspace, in other words, an intersection of an open set and a closed set.

The class of quasi-projective varieties includes all projective varieties, all affine varieties, and all Zariski open subsets of these. The set of quasi projective varieties is closed under taking open or closed subsets. For brevity, we often say "variety," instead of quasi-projective variety.

We can define a morphism between quasi-projective varieties in exactly the same way we defined a morphism between projective varieties. To be precise, if $V \subseteq \mathbb{P}^n$ and $W \subseteq \mathbb{P}^m$ are quasi-projective varieties, then a morphism $V \xrightarrow{F} W$ is a map such that for each $p \in V$, there exist homogeneous polynomials $F_o,\ldots,F_m$ in n+1 variables such that the "map"

$$\begin{aligned} V &\rightarrow \mathbb{P}^m \\ q &\mapsto [F_o(q):\ldots:f_m(q)] \end{aligned}$$

is well-defined at p and agrees with F on some nonempty open set containing p.

Example: Let $U = \mathbb{A}^1 - \{0\}$ and let $V = \mathbb{V}(xy-1) \subseteq \mathbb{A}^2$. Both U and V are quasi-projective varieties, and we have a well-defined map

$$\begin{aligned} U &\xrightarrow{F} V, \\ t &\rightarrow t\left(t, \frac{1}{t}\right). \end{aligned}$$

We claim that this is a morphism of quasi-projective varieties. To see this, we identify U with a locally closed (in fact, open) subset of $\mathbb{P}^1$ by the map $T \mapsto [t:1]$. Likewise, we identify V with a locally closed subset of $\mathbb{P}^2$ defined by $xy - Z_2 = 0$ and $z \neq 0$ via the map $(x, y) \mapsto [x:y:1]$. It is easy to check that the map F agrees everywhere on U with the morphism.

$$\begin{aligned} &\mathbb{P}^1 \xrightarrow{F} \mathbb{P}^2, \\ &[a:b] \mapsto [a^2:b^2:ab] \end{aligned}$$

Indeed, on U neither a nor b is zero, so setting $t=\frac{a}{b}$, we see that $\overline{F}$ sends $[t:1]$ to

$$[t^2:1:t] = \left[t:\frac{1}{t}:1\right]$$

The image of this map is precisely $\subseteq$, and the map obviously agrees with the original map F.

By enlarging our world to include all quasi-projective varieties, we gain flexibility. However, we must redefine the concept of an affine variety.

Definition: A quasi-projective variety is said to be affine if it is isomorphic as a quasi-projective variety to some affine algebraic variety. From now on, when we want to refer to the original definition of an "affine variety in $\mathbb{A}^n$," we will call it a Zariski-closed subset of $\mathbb{A}^n$.

Thus an affine algebraic variety is a variety that can be embedded in affine space as a Zariski-closed subset. With this refined definition of an affine variety, we emphasize the intrinsic nature of the variety, rather than the extrinsic feature of a particular embedding in affine space.

Hilbert's Nullstellensatz

Let k be an uncountable algebraically closed field.

Proposition (Hilbert's Nullstellensatz I): Let k be as above. Then an ideal $I \subset k[X_1,...X_n]$ is maximal iff $I = < X_1 - a_1,..,X_n - a_n >$ for some $a_i \in k$. In other words, maximal ideals are in 1-1 correspondence with points of k^n.

Proof: For n = 1, the result is true

First, let us convince ourselves that all the ideals $I = < X_1 - a_1,...X_n - a_n >$ are indeed maximal. For this we need to show that A/I is a field, where $A = k[X_1,...,X_n]$ is any polynomial in A, one can rewrite it as $f(X_1 - a_1) + a_1,...,(X_n - a_n) + a_n)$. Now any power $((X_i - a_i) + a_i)n_i$ can be written in the form $(X_i - a_i)h_i + a_i^{ni}$ (by binomial expansion), where h_i is a polynomial in X_i. Substituting these in f, we can 'Taylor expand' f:

$$f(X_1,...,X_n) = f(a_1,...a_n) + \sum_{i=1}^{n} g_i(X_1,...X_n)(X_i - a_i)$$

Where $g_i \in A$ are some polynomials. The second term on the right hand side is clearly in the ideal $I = < X_1 - a_1,...,X_n - a_n >$. This shows that every element in $A = k[X_1,...X_n]$ is congruent to the element $f(a_1,...a_n) \in k\,(\mathrm{mod}\, I)$, Also the element a $a \in k \subset A$ is clearly congruent only to itself $(\mathrm{mod}\, I)$, and no other element of k (since $I \cap k = \{0\}$), Thus A/I is just the field k, and I is maximal.

It is in the converse that the algebraic assumptions on k are required. Let $I \subset A = k[X_1,...X_n]$ be a maximal ideal. Then by definition A/I is a k-algebra of finite type, contains k, and is a field Denote it by K. Thus $k \subset K$ is a field extension $k = K$, Denoting the equivalence classes of X_i by $\bar{X}_i \in K = A/I$, it follows that $\bar{X}_i = a_i \in k$ for all $i = 1,...,n$. Saying that $\bar{X}_i = a_i$ in K, by definition, means $X_i - a_i \in I$ for all i. Thus the ideal $< X_1 - a_1,...,X_n - a_n > \subset I$. However, in the last para we saw that the ideal $< X_1 - a_1,...,X_n - a_n >$ is maximal. Thus it must be equal to I, which is a proper ideal (being a maximal ideal). This proves the proposition.

Corollary: Let k be as above, and $\{f_i\}_{i=1}^{m}$ be some set of polynomials in $A = k[X_1,...X_n]$. Then the algebraic set $V(f_1,...f_m)$ is empty if and only if there exist polynomials $h_i \in A$ for $i=1,...,m$ such that $\sum_{i=1}^{m} h_i f_i = 1$.

Proof: Consider the ideal $J=<f_1,...,fm>$. Then either J is a proper ideal or $J=A$. If J were a proper ideal, it would follow that $J \subset I$ for some maximal ideal *I*. By the small nullstellensatz I above, $I=<X_1-a_1,...,X_n-a_n>$. Thus every element of J would be of the form $\sum_{i=1}^{n} g_i(X_i-a_i)$, and would thus vanish at the point $(a_1,...,a_n)$, In particular $V(J)=V(f_1,..,f_m)$ would contain $(a_1,...,a_n)$, and be non-empty. Thus $J=A$, and hence $1 \in J=<f_1,..,f_m>$. In other words, $V(f_1,..,f_m)=\phi$ implies $1=\sum_{i=1}^{m} h_i f_i$ for some $h_i \in A$. The converse is obvious (since $1 \neq 0$).

Note how crucial it is for k to be algebraically closed for the corollary above because, for example, $V(X^2+1) \subset R$ is empty, but you cannot multiply X^2+1 by any polynomial $h \in R[X]$ to get 1.

The Big Nullstellensatz

Theorem (Hilbert's Nullstellensatz II): Let k be as assumed at the outset of this section (i.e. uncountable and algebraically closed). If a polynomial $f \in A=k[X_1,...X_n]$ vanishes identically at all points of $V(I)$ for some ideal $I \subset A$, then $f^r \in I$ for some r.

Proof: Write $I=<f_1,..,f_m>$. If $f=0$, there is nothing to prove, so assume $f \neq 0$. The trick is to add an extra variable, and "invert f" (called the Rabinowitch trick). Indeed, all the polynomials in A can be regarded as elements of the bigger ring $B=k[X_1,..,X_{n+1}]$. Consider the ideal $J \subset B$ generated by the elements $f_i(X_1,..,X_n)$ for $1 \leq i \leq m$ and the extra element $X_{n+1}f(X_1,..,X_n)-1$. Claim that $V(J) \subset k^{n+1}$ is empty. If not, there would be a point $(a_1,..,a_n+1) \in V(J)$. Since all the $f_i=f_i(X_1,..,X_n)$ would have to vanish at this point, it would follow that. Also since $X_{n+1}f-1$ would have to vanish at this point, we would have $a_{n+1}f(a_1,..,a_n)=1$ But then since f vanishes identically on $V(I)$, and $(a_1,..,a_n) \in V(I)$, we have $f(a_1,..,a_n)=0$. Thus $a_{n+1}0=1$, i.e. 0 = 1, a contradiction. This proves the claim.

So, by the corollary above (applied to $J \subset B$), there must be polynomials $h_i(X_1,..,X_{n+1}) \in B$ such that:

$$1=h_{m+1}(X_1,..,X_{n+1})(X_{n+1}f(X_1,..,X_n)-1+$$
$$\sum_{i=1}^{m} h_i(X_1,..,X_{n+1})f_i(X_1,...X_n)$$

The above identity holds in the polynomial ring

$$B=k[X_1,..,X_{n+1}]$$

Substituting $X_i=X_i$ for $1 \leq i \leq n$ and $X_{n+1}=1/f$ in this identity gives us an identity in the field $k(X_1,..,X_n)$, since $f \neq 0$. This substitution kills the first term, and we get:

$$1=\sum_{i=1}^{m} h_i(X_1,..,X_n,\frac{1}{f})f_i(X_1,..,X_n)$$

as an identity in $k(X_1,..,X_n)$. Clearly, by using a high enough power f^r as the common denominator on the right hand side, (e.g. r = maximum of the X^{n+1} degrees of all the h_i), and cross-multiplying, we have:

$$f^r = \sum_{i=1}^{m} P_i(X_1,..,X_n) f_i(X_1,..,X_n)$$

where P_i are some polynomials. This last identity holds in the field $k(X_1,..,X_n)$, and both sides of the identity are in $A = k[X_1,..,X_n]$. Since A sits as a subring in $k(X_1,..,X_n)$, the identity holds in A. Thus $f^r = \sum_i P_i f_i \in I$ as asserted.

Zariski's Main Theorem

Zariski's main theorem states is succint: a quasi-finite morphism of finite presentation between separated noetherian schemes factors as a composite of an open immersion and a finite morphism. This is a fairly big deal, as the condition of quasi-finiteness is seemingly rather weak—it is a condition on the fibers—while open immersions and finite morphisms are very nice. For instance, Zariski's Main Theorem is used to prove the local structure theory for étale and unramified morphisms of rings. It also immediately implies the theorem of Chevalley that a proper quasi-finite morphism is finite.

Main theorem: If W is an irreducible fundamental variety on V of a birational correspondence T between V and V' and if T has no fundamental elements on V', then - under the assumption that V is locally normal at W - each irreducible component of the transform *T[W]* is of higher dimension than *W*.

Zariski's objective was to systematically introduce concepts and results of the theory of birational correspondences. An important role there belongs to normal varieties, which were introduced as a class of varieties more general than nonsingular ones, such that many known properties of birational correspondences still hold true.

Rational Maps and Birational Transformations

In this topic we will define rational maps for integral schemes X, Y as equivalence classes of morphisms defined on some open subset $U \subset X$. Let us start with the following topological result.

Lemma: Let X be an irreducible scheme. Then every open subset of X is either empty or contains the generic point of X. In particular, every nonempty open of X is dense and the intersection of any two nonempty opens is nonempty.

Proof: This follows from the fact that every open subset U of X is stable under generization: if $z \in U$ is a point and $z' \in X$ such that $z \in \overline{\{z'\}}$, then we have $z' \in U$.

Let X, Y be integral schemes over some base scheme S. Consider the set of all pairs (U, f) where $U \subset X$ is a nonempty open and $f : U \to Y$ an S-morphism. Define an equivalence relation on this set by considering two pairs (U, f), (V, g) equivalent if there exists a nonempty open $W \subset U \cap V$ such that $f|W = g|W$. An S-rational map $f : X - - \to Y$ is given by an equivalence class $(U, \bar{f})$ and denote the set of all such maps by Rat S(X, Y). The morphism $\tilde{f}: U \to X$ is called a representative for f. Note that there exists a natural map $Hom_s(X,Y) \to Rat_s(X,Y)$. This map is not injective in general, since two different morphisms might agree on an nonempty open subset. However, when Y is separated, we have the following result:

Lemma: Let X, Y be two schemes over a scheme S, with Y separated over S. Let $U \subset X$ be a dense open subset and $f, g: X \to Y$ two S-morphisms that agree on U. Then $f\big|_{X_{red}=g}\big|_{X_{red}}$. In particular, if X is reduced we have $f = g$.

Let $f: X - - \to Y$ be a rational S-map. Denote by $\mathrm{dom}_S(f)$ the set of points $x \in X$ such that there exists a open $U \subset X$ with $x \in U$ and an S-morphism $\tilde{f}: U \to Y$ which is a representative of f. Clearly $\mathrm{dom}_S(f)$ itself is open. We call it the domain of definition of f and say that $x \in X$ is regular for f if $x \in dom_S(f)$; otherwise we call x a fundamental point. From the previous lemma we obtain immediately:

Proposition: Let $f: X - - \to Y$ be a rational S-map between integral schemes and assume that Y is separated over S. Then there exists a unique representative $\tilde{f}: dom_S(f) \to Y$ for f. In particular, the natural map $Hom_s(X,Y) \to Rat_s(X,Y)$ is injective.

Proof: By definition, the open set $dom_S(f)$ is covered by opens U_i such that there exists a representative $\tilde{f}_i: U_i \to Y$ for f. By Lemma the $\tilde{f}_i$ agree on the intersections of the U_i, hence we can glue them to obtain a unique representative $\tilde{f}: dom_S(f) \to Y$.

In general it does not make sense to compose rational maps: let $f: X - - \to Y$ and $g: Y - - \to Z$ be rational maps, then $f(dom_S(f))$ and $dom_S(g)$ might not intersect (and hence the composition $g \circ f$ is defined nowhere). This happens if $f(dom_S(f))$ is not dense anymore. In order to be able to build a category we restrict to integral schemes and rational maps between them which allow a dominant representative. We call a rational S-map $f: X - - \to Y$ dominant if there exists a dominant representative $\tilde{f}: U \to Y$ of f. Since every nonempty open subset of X contains the generic point ξ of X, f is dominant if and only if $f(\xi)$ is the generic point of Y. In particular, every representative of a dominant S-rational map is dominant.

Now let $f: X - - \to Y$ and $g: Y - - \to Z$ be rational S-maps between integral schemes and assume that f is dominant. Choose representatives $\tilde{f}: U \to Y$ and $\tilde{g}: V \to Z$ for f and g. Since f is dominant, we have that $\tilde{f}^{-1}(V)$ contains the generic point of X and hence $U \cap \tilde{f}^{-1}(V)$ is nonempty. We define the composition $g \circ f$ to be the rational map given by the representative $\tilde{g} \circ \tilde{f}: U \cap \tilde{f}^{-1}(V) \to Z$. It is easy to check that this definition is independent of the choice of representatives for f and g.

Let $f: X - - \to Y$ be a dominant rational S-map. Then f is called birational if f is an isomorphism in the category of integral S-schemes together with dominant rational S-maps as morphisms. The following lemma gives some characterizations for a rational map to be birational.

Lemma: Let $f: X - - \to Y$ be a dominant S-rational map between integral S-schemes. The following are equivalent:

1. f is birational.

2. For every representative $\tilde{f}:U \rightarrow Y$ there exists an S-morphism $\tilde{g}:U \rightarrow Y$ with $V \subset Y$ nonempty open such that $\tilde{f} \circ \tilde{g} = id$ and $\tilde{g} \circ \tilde{f} = id$, where they are defined.
3. There exists a representative $\tilde{f}:U \rightarrow Y$ which induces an isomorphism of U with some open $V \subset Y$.

Proof: If f is birational, then this means that there exists a rational S-map $g:Y --\rightarrow X$ such that $g \circ f = id$ (where id denotes the rational map represented by $id_X : X \rightarrow X$). In order to prove that (1) implies (2), take any representative $\tilde{g}:V \rightarrow X$.

Since f is dominant we have that $U \cap \tilde{f}^{-1}(V)$ is nonempty. By assumption, there exists a nonempty open $U' \subset U \cap \tilde{f}^{-1}(V)$ such that $\tilde{g} \circ \tilde{f} \equiv id$ on U'. After shrinking V we get that $\tilde{f} \circ \tilde{g} = id$ where the composition is defined. Repeat the same argument to obtain $\tilde{f} \circ \tilde{g} = id$.

Now assume (2). Choose a representative $\tilde{f}:U' \rightarrow Y$, then by assumptions there exists a morphism $\tilde{g}:V' \rightarrow X$ such that $\tilde{g} \circ \tilde{f} = id$ where they are defined. Set $U = U' \cap \tilde{f}^{-1}(V')$ and $V = V' \cap \tilde{g}^{-1}(U')$, then it is easy to check that $\tilde{f}|_U : U \rightarrow V$ is an isomorphism and again a representative for f.

Finally, let us prove that (3) implies (1). By assumption there exists a nonempty open $V \subset Y$ and a morphism $\tilde{g}:V \rightarrow X$ such that $\tilde{g} \circ \tilde{f} = id$ on U, and $\tilde{f} \circ \tilde{g} = id$ on V. Hence the S-rational map g given by $\tilde{g}$ is a two-sided inverse for f and we are done.

Given any S-rational $f : X --\rightarrow Y$ map as above, denote by f' the rational map represented by the morphism $\tilde{g}$ in (2) and call it the inverse of f.

Remark: Let X,Y be integral separated S-schemes and $f : X --\rightarrow Y$ a birational S-map. Denote by $\tilde{f}$ and $\tilde{f}$ the unique representatives for f and $\tilde{f}$ from Lemma.

Set

$$U := \{x \in X : x \text{ is regular for } f, f(x) \text{ is regular for } f'\}.$$

Then $U = dom_S(f) \cap \tilde{f} - 1(dom_S(f'))$ and U is the maximal open set such that there exists a representative for f with property (3) in Lemma. Any birational S-map $f : X --\rightarrow Y$ between integral S-schemes induces an isomorphism of function fields $K(X) \simeq K(Y)$. The converse does not hold true in general. A necessary condition is that the generic points of X and Y lie over the same point in S and $Spec(K(X)) \rightarrow Spec(K(Y))$ factors over S. However, there exist examples where such a morphism does not induce an isomorphism of open subset. It turns out that a sufficient additional condition for X and Y is to be locally of finite presentation over S.

Proposition: Let X, Y be S-schemes and $x \in X$, $y \in Y$ points lying over the same point $s \in S$.

1. Suppose that Y is locally of finite type over S. Let $f, f' : X \rightarrow Y$ be Smorphisms with $f(x) = f'(x) = y$ and such that local ring maps $O_{Y',y} \rightarrow O_{X,x}$ coincide. Then f and f' coincide on an open neighborhood of x.
2. Suppose that Y is locally of finite presentation over S. Let $\varphi_x : \mathcal{O}_{Y',y} \rightarrow \mathcal{O}_{X',x}$ be a local $\mathcal{O}_{S,s} - map$. Then there exists an open neighborhood U of x and an S-morphism $f : U \rightarrow Y$ with $f(x) = y$ and such that the local ring map $\mathcal{O}_{Y,y} \rightarrow \mathcal{O}_{X',x}$ induced by f is given by ϕx.

3. If in (2) in addition X is locally of finite type over S, then one can find U and f such that f is of finite type.

4. If in (2) in addition X is locally of finite presentation over S and φ_x is an isomorphism, then one can find U and f such that f is an isomorphism of U onto an open neighborhood of Y.

We will now apply the proposition to the special case of integral schemes which are locally of finite presentation over an affine scheme S and obtain the following result:

Proposition: Let X and Y be integral schemes locally of finite presentation over an affine base scheme $S = Spec(A)$. Assume the generic points of X and Y lie over the same base point in S. Then:

1. Any A-algebra map $\varphi : K(X) \to K(Y)$ induces a rational map $f : X --\to Y$. If ϕ is an isomorphism, then f is birational.

2. Let $f : X --\to Y$ be a birational map and $x \in X$ a point. Then x is regular for f if and only if there exists $y \in Y$ such that the local rings $O_{X,x}$ dominates $O_{Y,y}$, where both are seen as local subrings of $K(X) \simeq K(Y)$.

Proof: The first part follows directly from Proposition. For the second part, we only have to show that x and y lie over the same point $s \in S$. But this follows from the fact that, by our assumptions, the structure ring map $A \to O_{X,x}$ factors as $A \to O_{Y,y} \to O_{X,x}$.

We now want to extend the definition of birational maps: let X, X' be arbitrary integral schemes over some base scheme S. A birational transformation $T : X --\to X'$ is defined as an S-isomorphism $Spec(K(X)) \to Spec(K(Y))$. We say that a point x of X is regular for T if there exists a point x′ of X′ and an S-morphism $Spec(O_{X,x}) \to Spec(O_{X',x})$. The point x′ is called the image of x. If there does not exist such an S-morphism we say that x is fundamental for T. If X, X' and S fulfill the conditions of Proposition, then any birational transformation T between X and X' induces a birational map $f : X --\to X'$ and $x \in X$ is regular/fundamental for T if and only if x is regular/fundamental for f.

Using the above notions, it is possible to give a slightly more general definition of birational morphisms: if f $X \to Y$ is an S-morphism, then, from now on, we call f birational if there exists a diagram of S-morphisms

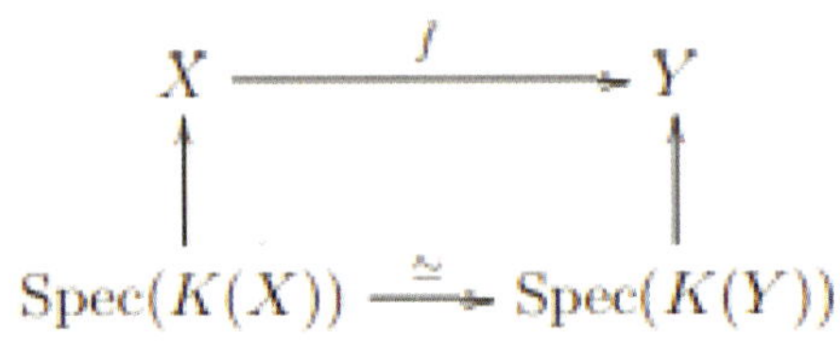

with the vertical arrows given by the canonical maps and the lower horizontal arrow an isomorphism. In particular, a birational morphism induces a birational transformation $T : X --\to Y$ which is regular at all $x \in X$ and such that the image of x under T is $f(x)$.

Finally, note that, if X is an integral scheme over S, then the normalization $\tilde{X} \to X$ is always birational under the above definition.

Zariski's Main Theorem for Quasifinite Morphisms

In EGA III, Grothendieck calls the following statement which does not involve connectedness a "Main theorem" of Zariski Grothendieck Théorème:

> If f: $X \rightarrow Y$ is a quasi-projective morphism of Noetherian schemes then the set of points that are isolated in their fiber is open in X. Moreover the induced scheme of this set is isomorphic to an open subset of a scheme that is finite over Y.

In EGA IV, Grothendieck observed that the last statement could be deduced from a more general theorem about the structure of quasi-finite morphisms, and the latter is often referred to as the "Zariski's main theorem in the form of Grothendieck". It is well known that open immersions and finite morphisms are quasi-finite. Grothendieck proved that under the hypothesis of separatedness all quasi-finite morphisms are compositions of such Grothendieck Théorème.

> If Y is a quasi-compact separated scheme and $f : X \rightarrow Y$ is a separated, quasi-finite, finitely presented morphism then there is a factorization into $X \rightarrow Z \rightarrow Y$, where the first map is an open immersion and the second one is finite.

The relation between this theorem about quasi-finite morphisms and Théorème of EGA III quoted above is that if f: $X \rightarrow Y$ is a projective morphism of varieties, then the set of points that are isolated in their fiber is quasifinite over Y. Then structure theorem for quasi-finite morphisms applies and yields the desired result.

Zariski's Main Theorem for Commutative Rings

Zariski (1949) reformulated his main theorem in terms of commutative algebra as a statement about local rings. Grothendieck Théorème generalized Zariski's formulation as follows:

> If B is an algebra of finite type over a local Noetherian ring A, and n is a maximal ideal of B which is minimal among ideals of B whose inverse image in A is the maximal ideal m of A, then there is a finite A-algebra A' with a maximal ideal m' (whose inverse image in A is m) such that the localization B_n is isomorphic to the A-algebra $A'_{m'}$.

If in addition A and B are integral and have the same field of fractions, and A is integrally closed, then this theorem implies that A and B are equal. This is essentially Zariski's formulation of his main theorem in terms of commutative rings.

Zariski's Main Theorem: Topological Form

A topological version of Zariski's main theorem says that if x is a (closed) point of a normal complex variety it is unibranch; in other words there are arbitrarily small neighborhoods U of x such that the set of non-singular points of U is connected.

The property of being normal is stronger than the property of being unibranch: for example, a cusp of a plane curve is unibranch but not normal.

Zariski's Main Theorem: Power Series Form

A formal power series version of Zariski's main theorem says that if x is a normal point of a variety then it is analytically normal; in other words the completion of the local ring at x is a normal integral domain.

Bézout's Theorem

Bézout's theorem states that the number of common points of two algebraic plane curves is either infinite or equal to the product of their degrees. The theorem holds if we count points at infinity in the projective plane and intersection multiplicities.

Theorem (Bézout's theorem): Let K be a field, and let $P,\ Q \in k[x,y]$ be non-zero polynomials in two variables x,y with no common factor. Then the two curves $\{(x,y) \in k^2 : P(x,y) = 0\}$ and $\{(x,y) \in k^2 : Q(x,y) = 0\}$ have no common components, and intersect in at most $\deg(P)\deg(Q)$ points.

This theorem can be proven by a number of means, for instance by using the classical tool of resultants. It has many strengthenings, generalisations, and variants; Bézout's theorem asserts a fundamental algebraic dichotomy, of importance in combinatorial incidence geometry: any two algebraic curves either share a common component, or else have a bounded finite intersection; there is no intermediate case in which the intersection is unbounded in cardinality, but falls short of a common component. This dichotomy is closely related to the integrality gap in algebraic dimension: an algebraic set can have an integer dimension such as 0 or 1, but cannot attain any intermediate dimensions such as 1/2. This stands in marked contrast to sets of analytic, combinatorial, or probabilistic origin, whose "dimension" is typically not necessarily constrained to be an integer.

Bézout's inequality tells us, roughly speaking, that the intersection of a curve of degree D_1 and a curve of degree D_2 forms a set of at most points. One can consider the converse question: given a set {S} of {N} points in the plane {k^2}, can one find two curves of degrees D_1, D_2 with $D_1, D_2 = O(N)$ and no common components, whose intersection contains these points?

A model example that supports the possibility of such a converse is a grid $S = A \times B$ that is a Cartesian product of two finite subsets A, B of k with $|A||B| = N$. In this case, one can take one curve to be the union of $|A|$ vertical lines, and the other curve to be the union of $|B|$ horizontal lines, to obtain the required decomposition. Thus, if the proposed converse to Bézout's inequality held, it would assert that any set of $|N|$ points was essentially behaving like a "nonlinear grid" of size N.

Unfortunately, the naive converse to Bézout's theorem is false. A counterexample can be given by considering a set $S = S_1 \cup S_2$ of $2N$ points for some large perfect square N, where P_1 is a $\sqrt{N}$ by $\sqrt{N}$ grid of the form described above, and S_2 consists of N points on an line l (e.g. a $1 \times N\}$ or $N \times 1$ grid). Each of the two component sets S_1, S_2 can be written as the intersection between two curves whose degrees multiply up to N; in the case of S_1, we can take the two families of parallel lines (viewed as reducible curves of degree $\sqrt{N}$ as the curves, and in the case of S_2, one can take as one curve, and the graph of a degree N polynomial on l vanishing on S_2 for the second l curve.

But, if N is large enough, one cannot cover S by the intersection of a single pair $\Upsilon_1\Upsilon_2$ of curves with no common components whose degrees D_1D_2 multiply up to $D_1D_2 = O(N)$. Indeed, if this were the case, then without loss of generality we may assume that $D_1 \leq D_2$, so that $D_1D_2 \leq O\left(\sqrt{N}\right)$. By Bézout's theorem, Υ_1 either contains l, or intersects l in at most $O(D_1) = O\left(\sqrt{N}\right)$ points. Thus, in order for Υ_1 to capture all of S, it must contain l, which forces Υ^2 to not contain l. But Υ^2 has to intersect l in N points, so by Bézout's theorem again we have $D_2 \geq N$, thus $D_1 = O(1)$. But then (by more applications of Bézout's theorem) Υ_1 can only capture $O\left(\sqrt{N}\right)$ of the N points of S_1, a contradiction.

But the above counterexample suggests that even if an arbitrary set of N (or $2N$) points cannot be covered by the single intersection of a pair of curves with degree multiplying up to $O(N)$, one may be able to cover such a set by a small number of such intersections. The purpose of this post is to record the simple observation that this is, indeed, the case:

Theorem (Partial converse to Bézout's theorem): Let k be a field, and let S be a set of N points in k for some $N > 1$. Then one can find { $m = O(log\ N)$ and pairs $P_i, Q_i \in k[x,y]$ of coprime non-zero polynomials for $i = 1,\ldots,m$ such that

$$S \subset \bigcup_{i=1}^{m}\left\{(x,y) \in k^2 : P_i(x,y) = Q_i(x,y) = 0\right\}$$

and

$$\sum_{i=1}^{m} \deg(P_i)\deg(Q_i) = O(N)$$

Informally, every finite set in the plane is (a dense subset of) the union of logarithmically many nonlinear grids. The presence of the logarithm is necessary, as can be seen by modifying the $P_1 \cup P_2$ example to be the union of logarithmically many Cartesian products of distinct dimensions, rather than just a pair of such products.

References

- Danilov, V.I. (2001) [1994], "Zariski theorem", in Hazewinkel, Michiel, Encyclopedia of Mathematics, Springer Science+Business Media B.V. / Kluwer Academic Publishers, ISBN 978-1-55608-010-4
- Algebraic-geometry, science: britannica.com, Retrieved 31 March 2018
- Zariski, Oscar (1943), "Foundations of a general theory of birational correspondences.", Trans. Amer. Math. Soc., 53 (3): 490542, doi:10.2307/1990215, MR 0008468
- Algebraic-curve: encyclopediaofmath.org, Retrieved 11 July 2018
- Raynaud, Michel (1970), Anneaux locaux henséliens, Lecture Notes in Mathematics, 169, Berlin, New York: Springer-Verlag, doi:10.1007/BFb0069571, ISBN 978-3-540-05283-8, MR 0277519
- Bezouts-theorem: wordpress.com, Retrieved 22 June 2018
- Zariski, Oscar (1949), "A simple analytical proof of a fundamental property of birational transformations.", Proc. Natl. Acad. Sci. U.S.A., 35 (1): 62–66, doi:10.1073/pnas.35.1.62, JSTOR 88284, MR 0028056, PMC 1062959

5

Convex Geometry: An Overview

Convex geometry is a branch of geometry that studies convex sets in a Euclidean space. The concepts of convex functions, convex sets, convex bodies, etc as well as the principles of convex analysis, Busemann-Petty problem and Carathéodory's theorem have been carefully analyzed in this chapter.

Convexity

A term used in various branches of mathematics and indicating properties which generalize some properties of convex sets in Euclidean spaces. The applicability of a number of methods of study is connected with the term "convexity".

The following two basic definitions are almost equivalent in E^n. A set is convex:

1) If it is the intersection of open half-spaces; or

2) If it contains with two arbitrary points the segment connecting these points. Both definitions of convexity are applied to the case of vector spaces L.

Definition 2 is generalized to include sets in spaces with geodesics (spaces with a connection; locally compact metric spaces, in particular Riemannian and Finsler spaces). The geodesics play the role of segments, but if the two points can be connected by non-unique geodesics or by shortest curves, the concept of convexity branches into several possible meanings. In Riemannian geometry, in particular, the following modifications of convexity are employed:

- A set M is convex if any two points of M can be connected by a unique shortest curve and if this shortest curve is contained in M;
- A set M is locally convex if each point of M has a neighborhood in M which is convex in the sense of;
- A set M is weakly convex if any two points can be connected by at least one shortest curve lying in M; and
- A set M is absolutely convex if all geodesics connecting any two points are located in.

The boundary (or part of the boundary) of an n-dimensional convex body in E^n is called a convex hyper surface; if $n = 3$, it is called a convex surface, if $n = 2$, it is said to be a convex curve.

For a function of a real variable convexity signifies the convexity of its super graph. The convexity of a functional f on L is defined in a similar manner.

For convex sets in one can speak of the convexity of a family $\mathfrak{U}$ of sets: $\mathfrak{U}$ is convex if $(1-\alpha) M_1 + \alpha M_2 \in \mathfrak{U}$ provided that $M_1,\ M_2 \in \mathfrak{U}$, $0 \le \alpha \le 1$. Convex (and concave) functional $\Phi(M)$ can be defined on convex families $\mathfrak{U}$. Convexity of a functional is defined by the requirement

$$\Phi((1-\alpha) M_1 + \alpha M_2) \le (1-\alpha)\Phi(M_1) + \alpha\,\Phi(M_2).$$

The term "convexity" as applied to univalent functions of a complex variable has a special meaning the property of mapping the unit disc onto a convex domain.

The R-convexity of a compactum M, meaning that each point with distance to M less than R has a unique nearest point in M, is one generalization of convexity in E^n which has been studied .

In the theory of linear differential operators the term "convexity" is connected with certain properties of homology groups. This is connected with the possibility of touching the boundary from inside a domain by a hyper surface for which a certain number of principal curvatures is positive. In the theory of functions of several complex variables an important role is played by holomorphic convexity, which is connected with the impossibility of touching the boundary of a domain from inside with an analytic surface. The last-named concept is a special case of K-convexity. To the concept of K-convexity one can transfer many given convexity properties.

The concept of H-convexity is employed in convex analysis; it is a generalization of the representability of a convex function as the supremum of a family of linear functions.

In the theory of metric spaces, (Menger) convexity of a metric is defined as follows: For any points $x \ne y$ there exists a third point z for which $\rho(x,y) = \rho(x,z) + \rho(z,x)$. d-convexity of a set M means that moreover z belongs to M if $x, y \in M$. Definitions of convexity in ordered spaces are very similar.

To almost each definition of convexity there corresponds a concept of local convexity. However, for the class of locally convex topological vector spaces, the term "local convexity" has a special meaning, namely, that for each point there exists a basis system of convex neighborhoods.

Convex Set

Convex sets play a very important role in geometry.

Let us consider a set $S \subset \mathbb{R}$ and suppose $x, x' \in S$ with $x \le x'$: The set $[x, x']$ is then a set of real numbers between (and including) x and x', or visually, the set of all points on the line joining x to x'. Now it may or may not be that $[x, x'] \subset S$. If it is the case that for all $x, x' \in S$ with $x \le x'$ we have that $[x, x'] \subset S$, then we say that S is a convex set.

Definition 1: A set $S \subset \mathbb{R}$ is convex if, $\forall \lambda \in [0,1]$ and $\forall x, x' \in S$,

$$\lambda x + (1-\lambda)x' \in S.$$

If $x, x' \in \mathbb{R}$ and $\lambda \in [0,1]$, then $\lambda x + (1-\lambda)x'$ is said to be a convex combination of x and x'

Definition 1 simply says that is a convex set if any convex combination of every two elements of S is also in S.

Example: Consider the interval $[a,b] \subset \mathbb{R}$. We next show that this interval is a convex set. Let $x, x' \in [a,b]$ be two arbitrary elements. We need to prove that $\lambda x + (1-\lambda) x' \in [a,b]$ for all $\lambda \in [0,1]$.

Select an arbitrary λ in $[0,1]$. Since $x, x' \in [a,b]$, then $x, x' \leq b$. As $\lambda \in [0,1]$, it follows that $\lambda x + (1-\lambda)x' \leq b$. Using a similar argument $\lambda x + (1-\lambda)x' \geq a$. As λ, x and x' were arbitrarily chosen, then $\lambda x + (1-\lambda)x' \geq \in [a,b]$ $\forall x, x' \in [a,b]$ and $\forall \lambda \in [0,1]$.

Let us move to higher dimensions. The Cartesian product of n sets $A_1, A_2, ..., A_n$, written $A_1 \times A_2 \times .... \times A_n$, is the set of all ordered tuples $(x_1, x_2, ..., x_n)$ such that $x_1 \in A_1, x_2 \in A_2, ...,$ and $x_n \in A_n$. A : A convex set in $\mathbb{R}^n$ can be visualized in the same way as in $\mathbb{R}$; $S \subset \mathbb{R}^n$ is convex if and only if the line joining any two points in the set lies entirely within the Set.

Definition 2: A set $S \subset \mathbb{R}^n$ is convex if, $\forall \lambda \in [0,1]$ and $\forall \mathrm{x}, \mathrm{x}' \in S$,

$$\lambda \mathrm{x} + (1-\lambda)x' \in S.$$

Non-convex Set

A set that is not convex is called a *non-convex set*. A polygon that is not a convex polygon is sometimes called a concave polygon, and some sources more generally use the term *concave set* to mean a non-convex set, but most authorities prohibit this usage.

The complement of a convex set, such as the epigraph of a concave function, is sometimes called a *reverse convex set*, especially in the context of mathematical optimization.

Properties

If S is a convex set in n-dimensional space, then for any collection of $r, r > 1$, n-dimensional vectors $u_1, ..., u_r$ in S, and for any nonnegative numbers $\lambda_1, ..., \lambda_r$ such that $\lambda_1 + ... + \lambda_r = 1$, then one has:

$$\sum_{k=1}^{r} \lambda_k u_k \in S.$$

A vector of this type is known as a convex combination of $u_1, ..., u_r$.

Intersections and Unions

The collection of convex subsets of a vector space has the following properties:

1. The empty set and the whole vector-space are convex.

2. The intersection of any collection of convex sets is convex.
3. The *union* of a non-decreasing sequence of convex subsets is a convex set. For the preceding property of unions of non-decreasing sequences of convex sets, the restriction to nested sets is important: The union of two convex sets need *not* be convex.

Closed Convex Sets

Closed convex sets are convex sets that contain all their limit points. They can be characterised as the intersections of *closed half-spaces* (sets of point in space that lie on and to one side of a hyperplane).

From what has just been said, it is clear that such intersections are convex, and they will also be closed sets. To prove the converse, i.e., every convex set may be represented as such intersection, one needs the supporting hyperplane theorem in the form that for a given closed convex set *C* and point *P* outside it, there is a closed half-space *H* that contains *C* and not *P*. The supporting hyperplane theorem is a special case of the Hahn–Banach theorem of functional analysis.

Convex Sets and Rectangles

Let *C* be a convex body in the plane. We can inscribe a rectangle *r* in *C* such that a homothetic copy *R* of *r* is circumscribed about *C*. The positive homothety ratio is at most 2 and:

$$\tfrac{1}{2}\cdot \text{Area}(R) \leq \text{Area}(C) \leq 2\cdot \text{Area}(r)$$

Convex Hulls and Minkowski Sums

Convex Hulls

Every subset *A* of the vector space is contained within a smallest convex set (called the convex hull of *A*), namely the intersection of all convex sets containing *A*. The convex-hull operator Conv() has the characteristic properties of a hull operator:

$$\textit{extensive} \quad S \subseteq Conv(S),$$

$$\textit{non-decreasing} \quad S \subseteq T \textit{ implies that } Conv(S) \subseteq Conv(T), \textit{ and}$$

$$\textit{idempotent} \quad Conv(Conv(S)) = Conv(S).$$

The convex-hull operation is needed for the set of convex sets to form a lattice, in which the "*join*" operation is the convex hull of the union of two convex sets

$$Conv(S) \vee Conv(T) = Conv(S \cup T) = Conv(Conv(S) \cup Conv(T)).$$

The intersection of any collection of convex sets is itself convex, so the convex subsets of a (real or complex) vector space form a complete lattice.

Minkowski Addition

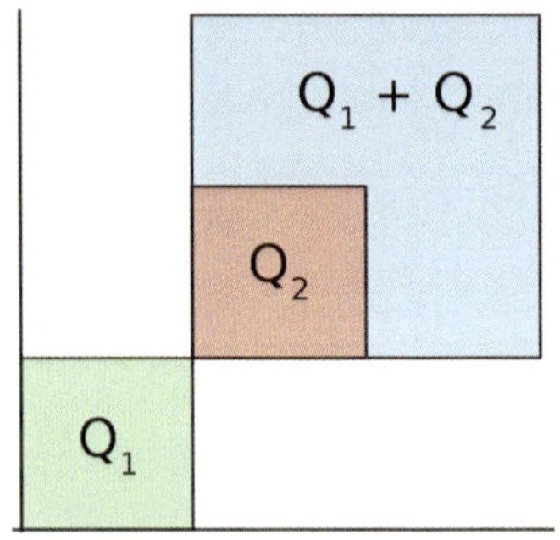

Minkowski addition of sets. The sum of the squares $Q_1=[0,1]^2$ and $Q_2=[1,2]^2$ is the square $Q_1+Q_2=[1,3]^2$.

In a real vector-space, the *Minkowski sum* of two (non-empty) sets, S_1 and S_2, is defined to be the set S_1+S_2 formed by the addition of vectors element-wise from the summand-sets

$$S_1+S_2=\{x_1+x_2 : x_1\in S_1, x_2\in S_2\}.$$

More generally, the *Minkowski sum* of a finite family of (non-empty) sets S_n is the set formed by element-wise addition of vectors

$$\sum_n S_n=\left\{\sum_n x_n : x_n\in S_n\right\}.$$

For Minkowski addition, the *zero set* $\{0\}$ containing only the zero vector 0 has special importance: For every non-empty subset S of a vector space

$$S+\{0\}=S;$$

in algebraic terminology, $\{0\}$ is the identity element of Minkowski addition (on the collection of non-empty sets).

Convex Hulls of Minkowski Sums

Minkowski addition behaves well with respect to the operation of taking convex hulls, as shown by the following proposition:

Let S_1, S_2 be subsets of a real vector-space, the convex hull of their Minkowski sum is the Minkowski sum of their convex hulls

$$\text{Conv}(S_1+S_2)=\text{Conv}(S_1)+\text{Conv}(S_2).$$

This result holds more generally for each finite collection of non-empty sets:

$$\text{Conv}\left(\sum_n S_n\right)=\sum_n \text{Conv}(S_n).$$

In mathematical terminology, the operations of Minkowski summation and of forming convex hulls are commuting operations.

Minkowski Sums of Convex Sets

The Minkowski sum of two compact convex sets is compact. The sum of a compact convex set and a closed convex set is closed.

Generalizations and Extensions for Convexity

The notion of convexity in the Euclidean space may be generalized by modifying the definition in some or other aspects. The common name "generalized convexity" is used, because the resulting objects retain certain properties of convex sets.

Star-convex Sets

Let C be a set in a real or complex vector space. C is star convex if there exists an x_0 in C such that the line segment from x_0 to any point y in C is contained in C. Hence a non-empty convex set is always star-convex but a star-convex set is not always convex.

Orthogonal Convexity

An example of generalized convexity is orthogonal convexity.

A set S in the Euclidean space is called orthogonally convex or ortho-convex, if any segment parallel to any of the coordinate axes connecting two points of S lies totally within S. It is easy to prove that an intersection of any collection of orthoconvex sets is orthoconvex. Some other properties of convex sets are valid as well.

Non-Euclidean Geometry

The definition of a convex set and a convex hull extends naturally to geometries which are not Euclidean by defining a geodesically convex set to be one that contains the geodesics joining any two points in the set.

Order Topology

Convexity can be extended for a space X endowed with the order topology, using the total order $<$ of the space.

Let $Y \subseteq X$. The subspace Y is a convex set if for each pair of points a, b in Y such that $a<b$, the interval $(a, b) = \{x \in X : a<x<b\}$ is contained in Y. That is, Y is convex if and only if for all a,b in Y, $a<b$ implies $(a, b) \subseteq Y$.

Convexity Spaces

The notion of convexity may be generalised to other objects, if certain properties of convexity are selected as axioms.

Given a set X, a convexity over X is a collection $\mathcal{C}$ of subsets of X satisfying the following axioms:

1. The empty set and X are in $\mathcal{C}$.

2. The intersection of any collection from $\mathcal{C}$ is in $\mathcal{C}$.
3. The union of a chain (with respect to the inclusion relation) of elements of $\mathcal{C}$ is in $\mathcal{C}$.

The elements of $\mathcal{C}$ are called convex sets and the pair $(X, \mathcal{C})$ is called a convexity space. For the ordinary convexity, the first two axioms hold, and the third one is trivial.

Convex Function

Consider a function $y = f(x)$, which is assumed to be continuous on the interval $[a,b]$. The function $y = f(x)$ is called *convex downward* (or just *convex*) if for any two points x_1 and x_2 in $[a,b]$ the following inequality holds:

$$f\left(\frac{x_1 + x_2}{2}\right) \leq \frac{f(x_1) + f(x_2)}{2}.$$

If this inequality is strict for any $x_1, x_2 \in [a,b]$, such that $x_1 \neq x_2$, then the function $f(x)$ is called *strictly convex downward* on the interval $[a,b]$.

Similarly, we define a concave function. A function $f(x)$ is called *convex upward* (or *concave*) if for any two points x1 and x2 in the interval $[a,b]$ the following inequality is valid:

$$f\left(\frac{x_1 + x_2}{2}\right) \geq \frac{f(x_1) + f(x_2)}{2}.$$

If this inequality is strict for any $x_1, x_2 \in [a,b]$, such that $x_1 \neq x_2$, then the function $f(x)$ is called *strictly convex upward* on the interval $[a,b]$.

Geometric Interpretation of Convexity

The introduced concept of convexity has a simple geometric interpretation.

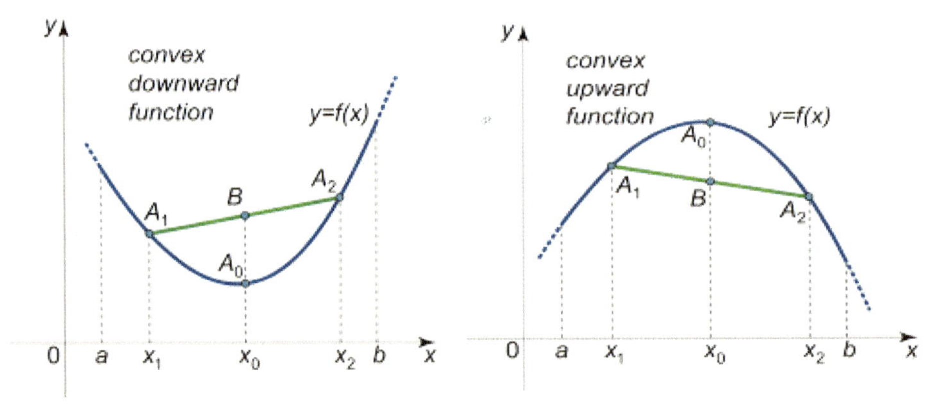

If a function is *convex downward*, the midpoint B of each chord A_1A_2 lies *above* the corresponding point A_0 of the graph of the function or coincides with this point.

Similarly, if a function is *convex upward*, the midpoint B of each chord A_1A_2 is located *below* the corresponding point A_0 of the graph of the function or coincides with this point.

Convex functions have another obvious property, which is related to the location of the *tangent* to the graph of the function. The function $f(x)$ is *convex downward* on the interval if and only if its graph does not lie below the tangent drawn to it at any point x_0 of the segment $[a,b]$.

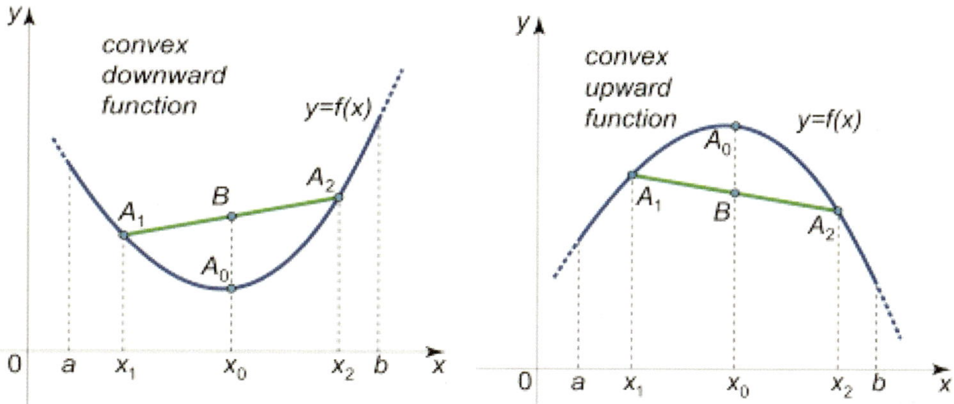

Accordingly, the function $f(x)$ is *concave* or *convex upward* on the interval $[a,b]$ if and only if its graph does not lie above the tangent drawn to it at any point x_0 of the segment $[a,b]$. These properties represent a theorem and can be proved using the definition of convexity.

Sufficient Conditions for Convexity

Suppose that the first derivative $f'(x)$ of a function $f(x)$ exists in a closed interval $[a,b]$ and the second derivative $f''(x)$ exists in an open interval $(a\ b)$. Then the following sufficient conditions for convexity are valid:

- If $f''(x) \geq 0$ for all $x \in (a,b)$, then the function $f(x)$ is *convex downward* on the interval $[a,b]$;
- If $f''(x) \leq 0$ for all $x \in (a,b)$, then the function $f(x)$ is *convex upward* on the interval $[a,b]$.

In the cases where the second derivative is strictly greater (or less) than zero, we say, respectively, about the *strict convexity downward* (or *strict convexity upward*).

We prove the theorem for the case of convexity downward. Let the function $f(x)$ have a non-negative second derivative on the interval $(a,b)\, f''(x) \geq 0$. Let x_0 be the midpoint of the interval $[x_1, x_2]$. Suppose that the length of this interval is equal to 2h. Then the coordinates x_1 and x_2 can be written as

$$x_1 = x_0 - h, \quad x_2 = x_0 + h$$

Expand the function $f(x)$ at x_0 in the Taylor series with the remainder in the Lagrange form. We obtain the following expressions:

$$f(x_1) = f(x_0 - h) = f(x_0) - f'(x_0)h + \frac{f''(\xi_1)h^2}{2!},$$

$$f(x_2) = f(x_0 + h) = f(x_0) - f'(x_0)h + \frac{f''(\xi_1)h^2}{2!},$$

Where $x_0 - h < \xi_1 < x_0, x_0 < \xi_2 < x_0 + h$.

Add the two equations:

$$f(x_1) + f(x_2) = 2f(x_0) + \frac{h^2}{2}\left[f''(\xi_1) + f''(\xi_2)\right].$$

Since $\xi_1, \xi_2 \in (a, b)$, then the second derivatives in the right-hand side are non-negative. Consequently,

$$f(x_1) + f(x_2) \geq 2f(x_0)$$

or

$$f\left(\frac{x_1 + x_2}{2}\right) \leq \frac{f(x_1) + f(x_2)}{2},$$

i.e. according to the definition, the function $f(x)$ is *convex downward.*

Note that the necessary condition for convexity (for example, the implication when from the convexity downwards it follows that $f''(x) \geq 0$) holds only for the non-strict inequality. In the case of strict convexity, the necessary condition is generally not valid. For example, the function $f(x) = x^4$ is strictly convex downward. However, its second derivative is zero at $x = 0$, i.e. the strict inequality $f''(x) > 0$ does not hold in this case.

Properties

Functions of one Variable

1. Suppose *f* is a function of one real variable defined on an interval, and let

$$R(x_1, x_2) = \frac{f(x_1) - f(x_2)}{x_1 - x_2}$$

(note that $R(x_1, x_2)$ is the slope of the purple line in the above drawing; note also that the function R is symmetric in (x_1, x_2)). *f* is convex if and only if $R(x_1, x_2)$ is monotonically non-decreasing in x_1, for every fixed x_2 (or vice versa). This characterization of convexity is quite useful to prove the following results.

2. A convex function f of one real variable defined on some open interval C is continuous on C and Lipschitz continuous on any closed subinterval. f admits left and right derivatives,

and these are monotonically non-decreasing. As a consequence, f is differentiable at all but at most countably many points. If C is closed, then f may fail to be continuous at the endpoints of C.

3. A differentiable function of one variable is convex on an interval if and only if its derivative is monotonically non-decreasing on that interval. If a function is differentiable and convex then it is also continuously differentiable.

4. A differentiable function of one variable is convex on an interval if and only if the function lies above all of its tangents:

   $$f(x) \geq f(y) + f'(y)(x-y)$$

   for all x and y in the interval. In particular, if $f'(c) = 0$, then c is a global minimum of $f(x)$.

5. A twice differentiable function of one variable is convex on an interval if and only if its second derivative is non-negative there; this gives a practical test for convexity. Visually, a twice differentiable convex function "curves up", without any bends the other way (inflection points). If its second derivative is positive at all points then the function is strictly convex, but the converse does not hold. For example, the second derivative of $f(x) = x^4$ is $f(x) = 12x^2$, which is zero for $x = 0$, but x^4 is strictly convex.

6. If f is a convex function of one real variable, and $f(0) \leq 0$, then f is superadditive on the positive reals.

Proof: Since f is convex, letting $y = 0$ we have

$$f(tx) = f(tx + (1-t)\cdot 0) \leq tf(x) + (1-t)f(0) \leq tf(x), \quad \forall t \in [0,1].$$

From this we have:

$$f(a) + f(b) = f\left((a+b)\frac{a}{a+b}\right) + f\left((a+b)\frac{b}{a+b}\right) \leq \frac{a}{a+b}f(a+b) + \frac{b}{a+b}f(a+b) = f(a+b)$$

Functions of n Variables

1. A function is midpoint convex on an interval C if

   $$\forall x_1, x_2 \in C: \qquad f\left(\frac{x_1 + x_2}{2}\right) \leq \frac{f(x_1) + f(x_2)}{2}.$$

   This condition is only slightly weaker than convexity. For example, a real-valued Lebesgue measurable function that is midpoint-convex will be convex by the Sierpinski theorem. In particular, a continuous function that is midpoint convex will be convex.

2. A twice continuously differentiable function of several variables is convex on a convex set if and only if its Hessian matrix of second partial derivatives is positive semidefinite on the interior of the convex set.

3. Any local minimum of a convex function is also a global minimum. A *strictly* convex function will have at most one global minimum.
4. For a convex function *f*, the sublevel sets $\{x \mid f(x) < a\}$ and $\{x \mid f(x) \leq a\}$ with $a \in \mathrm{R}$ are convex sets. However, a function whose sublevel sets are convex sets may fail to be a convex function. A function whose sublevel sets are convex is called a *quasiconvex function.*
5. Jensen's inequality applies to every convex function *f*. If *X* is a random variable taking values in the domain of *f*, then $E(f(X)) \geq f(E(X))$. (Here E denotes the mathematical expectation.)
6. A first-order homogeneous function of two positive variables *x* and *y* (i.e. $f(ax, ay) = a f(x,y)$ for each $a,x,y > 0$) that is convex in one variable must be convex in the other variable.

Operations that Preserve Convexity

- $-f$ is concave if and only if f is convex.
- Nonnegative weighted sums:
  - If $w_1,\ldots,w_n \geq$ and $f_1,\ldots,f_n$ are all convex, then so is $w_1 f_1 + \cdots + w_n f_n$. In particular, the sum of two convex functions is convex.
  - This extends to infinite sums, integrals and expected values as well (provided that they exist).
- Element wise maximum: let $\{f_i\}_{i \in I}$ be a collection of convex functions. Then $g(x) = \sup_{i \in I} f_i(x)$ is convex. The domain of $g(x)$ is the collection of points where the expression is finite. Important special cases:
  - If $f_1,\ldots,f_n$ are convex functions then so is $g(x) = \max\{f_1(x),\ldots,f_n(x)\}$.
  - If $f(x,y)$ is convex in *x* then $g(x) = \sup_{y \in C} f(x,y)$ is convex in *x* even if *C* is not a convex set.
- Composition:
  - If *f* and *g* are convex functions and *g* is non-decreasing over a univariate domain, then $h(x) = g(f(x))$ is convex. As an example, if f is convex, then so is $e^{f(x)}$, because e^x is convex and monotonically increasing.
  - If *f* is concave and *g* is convex and non-increasing over a univariate domain, then $h(x) = g(f(x))$ is convex.
  - Convexity is invariant under affine maps: that is, if *f* is convex with domain $D_f \subseteq \mathrm{R}^m$, then so is $g(x) = f(Ax+b)$, where $A \in \mathrm{R}^{m \times n}, b \in \mathrm{R}^m$ with domain $\mathrm{D_g} \subseteq \mathrm{R}^n$.
- Minimization: If $f(x,y)$ is convex in (x,y) then $g(x) = \inf_{y \in C} f(x,y)$ is convex in *x*, provided that *C* is a convex set and that $g(x) \neq -\infty$.
- If f is convex, then its perspective $g(x,t) = tf\left(\frac{x}{t}\right)$ with domain $\left\{(x,t) \middle| \frac{x}{t} \in \mathrm{Dom}(f), t > 0\right\}$ is convex.

Strongly Convex Functions

The concept of strong convexity extends and parameterizes the notion of strict convexity. A strongly convex function is also strictly convex, but not vice versa.

A differentiable function *f* is called strongly convex with parameter $m > 0$ if the following inequality holds for all points *x*, *y* in its domain:

$$(\nabla f(x) - \nabla f(y))^T (x - y) \geq m \| x - y \|_2^2$$

or, more generally,

$$\langle \nabla f(x) - \nabla f(y), x - y \rangle \geq m \| x - y \|^2$$

where $\|\cdot\|$ is any norm. Some authors, refer to functions satisfying this inequality as elliptic functions.

An equivalent condition is the following:

$$f(y) \geq f(x) + \nabla f(x)^T (y - x) + \frac{m}{2} \| y - x \|_2^2$$

It is not necessary for a function to be differentiable in order to be strongly convex. A third definition for a strongly convex function, with parameter *m*, is that, for all *x*, *y* in the domain and $t \in [0, 1]$,

$$f(tx + (1-t)y) \leq tf(x) + (1-t)f(y) - \frac{1}{2} mt(1-t) \| x - y \|_2^2$$

Notice that this definition approaches the definition for strict convexity as $m \rightarrow 0$, and is identical to the definition of a convex function when $m = 0$. Despite this, functions exist that are strictly convex but are not strongly convex for any $m > 0$.

If the function *f* is twice continuously differentiable, then *f* is strongly convex with parameter *m* if and only if $\nabla^2 f(x) \succeq mI$ for all *x* in the domain, where *I* is the identity and $\nabla^2 f$ is the Hessian matrix, and the inequality $\succeq$ means that $\nabla^2 f(x) - mI$ is positive semi-definite. This is equivalent to requiring that the minimum eigenvalue of $\nabla^2 f(x)$ be at least *m* for all *x*. If the domain is just the real line, then $\nabla^2 f(x)$ is just the second derivative $f''(x)$, so the condition becomes $f''(x) \geq m$. If *m* = 0, then this means the Hessian is positive semi definite (or if the domain is the real line, it means that $f''(x) \geq 0$), which implies the function is convex, and perhaps strictly convex, but not strongly convex.

Assuming still that the function is twice continuously differentiable, one can show that the lower bound of $\nabla^2 f(x)$ implies that it is strongly convex. Start by using Taylor's Theorem:

$$f(y) = f(x) + \nabla f(x)^T (y - x) + \frac{1}{2} (y - x)^T \nabla^2 f(z)(y - x)$$

for some (unknown) $z \in \{tx + (1-t)y : t \in [0,1]\}$. Then

$$(y-x)^T \nabla^2 f(z)(y-x) \geq m(y-x)^T (y-x)$$

by the assumption about the eigenvalues, and hence we recover the second strong convexity equation above.

A function *f* is strongly convex with parameter m if and only if the function $x \mapsto f(x) - \frac{m}{2}\|x\|^2$ is convex.

The distinction between convex, strictly convex, and strongly convex can be subtle at first glance. If *f* is twice continuously differentiable and the domain is the real line, then we can characterize it as follows:

f convex if and only if $f''(x) \geq 0$ for all x.

f strictly convex if $f''(x) > 0$ for all x (this is sufficient, but not necessary).

f strongly convex if and only if $f''(x) \geq m > 0$ for all x.

For example, consider a function *f* that is strictly convex, and suppose there is a sequence of points (x_n) such that $f''(x_n) = \frac{1}{n}$. Even though $f''(x_n) > 0$, the function is not strongly convex because $f''(x)$ will become arbitrarily small.

A twice continuously differentiable function *f* on a compact domain X that satisfies $f''(x) > 0$ for all $x \in X$ is strongly convex. The proof of this statement follows from the extreme value theorem, which states that a continuous function on a compact set has a maximum and minimum.

Strongly convex functions are in general easier to work with than convex or strictly convex functions, since they are a smaller class. Like strictly convex functions, strongly convex functions have unique minima on compact sets.

Lemma: Strong convexity $\Rightarrow$ Strict convexity $\Rightarrow$ Convexity.

(But the converse of neither implication is true.)

Proof: The fact that strict convexity implies convexity is obvious.

To see that strong convexity implies strict convexity, note that strong convexity of f implies

$$f(\lambda x + (1-\lambda)y) - \alpha\|\lambda x + (1-\lambda)y\|^2 \leq \lambda f(x) + (1-\lambda) f(y) - \lambda\alpha\|x\|^2 (1-\lambda)\alpha\|y\|^2$$

But

$$\lambda\alpha\|x\|^2 (1-\lambda)\alpha\|y\|^2 - \alpha\|\lambda x + (1-\lambda)y\|^2 > 0, \forall x, y\ x \neq y, \forall\lambda \in (0,1),$$

Because $\|x\|^2$ is strictly convex (why?). The claim follows.

To see that the converse statements are not true, observe that $f(x) = x$ is convex but not strictly convex and $f(x) = x^4$ is strictly convex but not strongly convex.

Uniformly Convex Functions

A uniformly convex function, with modulus ϕ, is a function f that, for all x, y in the domain and $t \in [0, 1]$, satisfies

$$f(tx + (1-t)y) \le tf(x) + (1-t)f(y) - t(1-t)\phi(\| x - y \|)$$

where ϕ is a function that is non-negative and vanishes only at 0. This is a generalization of the concept of strongly convex function; by taking $\phi(\alpha) = \frac{m}{2}\alpha^2$ we recover the definition of strong convexity.

Examples

Functions of one Variable

- The function $f(x) = x^2$ has $f''(x) = 2 > 0$, so f is a convex function. It is also strongly convex (and hence strictly convex too), with strong convexity constant 2.
- The function $f(x) = x^4$ has $f''(x) = 12x^2 \ge 0$, so f is a convex function. It is strictly convex, even though the second derivative is not strictly positive at all points. It is not strongly convex.
- The absolute value function $f(x) = |x|$ is convex (as reflected in the triangle inequality), even though it does not have a derivative at the point $x = 0$. It is not strictly convex.
- The function $f(x) = |x|^p$ for $p \ge 1$ is convex.
- The exponential function $f(x) = e^x$ is convex. It is also strictly convex, since $f''(x) = e^x > 0$, but it is not strongly convex since the second derivative can be arbitrarily close to zero. More generally, the function $g(x) = e^{f(x)}$ is logarithmically convex if f is a convex function. The term "superconvex" is sometimes used instead.
- The function f with domain [0,1] defined by $f(0) = f(1) = 1$, $f(x) = 0$ for $0 < x < 1$ is convex; it is continuous on the open interval (0, 1), but not continuous at 0 and 1.
- The function x^3 has second derivative $6x$; thus it is convex on the set where $x \ge 0$ and concave on the set where $x \le 0$.
- Examples of functions that are monotonically increasing but not convex include $f(x) = \sqrt{x}$ and $g(x) = \log x$.
- Examples of functions that are convex but not monotonically increasing include $h(x) = x^2$ and $k(x) = -x$.
- The function $f(x) = \frac{1}{x}$ has $f''(x) = \frac{2}{x^3}$ which is greater than 0 if $x > 0$, so $f(x)$ is convex on the interval $(0, \infty)$. It is concave on the interval $(-\infty, 0)$.
- The function $f(x) = \frac{1}{x^2}$ with $f(0) = \infty$ is convex on the interval $(0, \infty)$ and convex on the interval $(-\infty, 0)$ but not convex on the interval $(-\infty, \infty)$ because of the singularity at $x = 0$.

Functions of n Variables

- The function $-\log \det(X)$ on the domain of positive-definite matrices is convex.

- Every real-valued linear transformation is convex but not strictly convex, since if f is linear, then $f(a+b)=f(a)+f(b)$. This statement also holds if we replace "convex" by "concave".
- Every real-valued affine function, i.e., each function of the form $f(x)=a^T x+b$, is simultaneously convex and concave.
- Every norm is a convex function, by the triangle inequality and positive homogeneity.
- The spectral radius of a nonnegative matrix is a convex function of its diagonal elements.

Theorem: *A function* $f:\mathbb{R}^n \to \mathbb{R}$ *is convex if and only if the function* $g:\mathbb{R} \to \mathbb{R}$ *given by* $g(t) = f(x + ty)$ *is convex (as a univariate function) for all x in domain of f and all* $y \in \mathbb{R}^n$*. (The domain of g here is all t for which* $x + ty$ *is in the domain of* f*.)*

Proof: This is straightforward from the definition.

- The theorem simplifies many basic proofs in convex analysis but it does not usually make verification of convexity that much easier as the condition needs to hold for all lines (and we have infinitely many).
- Many algorithms for convex optimization iteratively minimize the function over lines. The statement above ensures that each subproblem is also a convex optimization problem.

First and second order characterizations of convex functions

Theorem: Suppose $f:\mathbb{R}^n \to \mathbb{R}$ is twice differentiable over an open domain. Then, the following are equivalent:

i. f is convex.

ii. $f(y) \geq f(x) + \nabla f(x)^T (y-x)$, *for all* $x, y \in dom(f)$.

iii. $\nabla^2 f(x) \succeq 0$, *for all* $x \in dom(f)$.

Intepretation:

Condition (ii): The first order Taylor expansion at any point is a global under estimator of the function.

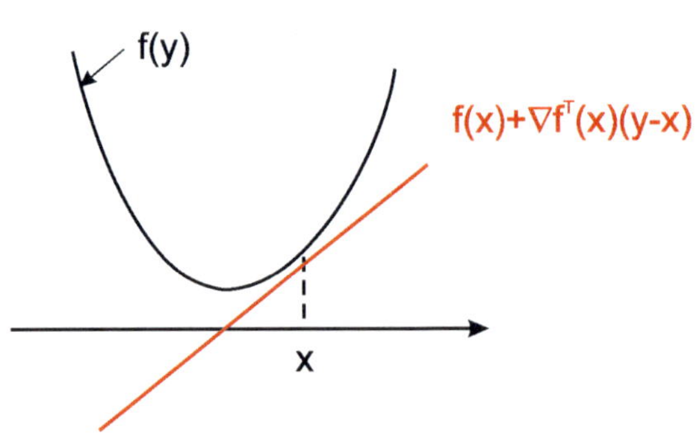

Condition (iii): The function f has nonnegative curvature everywhere. (In one dimension $f''(x) \geq 0, \forall x \in dom(f)$).

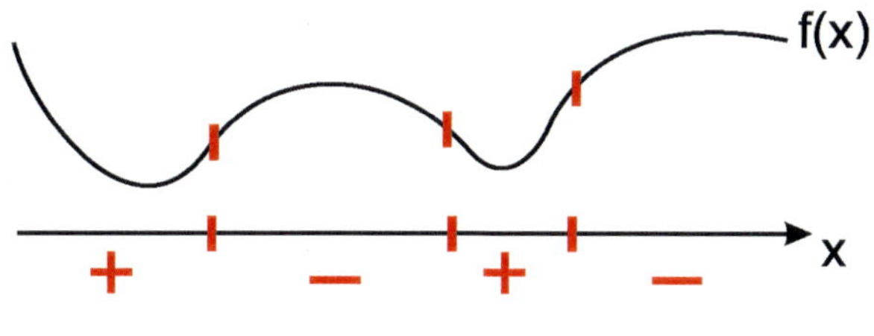

Sign of the curvature
(sign of f'')

Proof:

We prove $(i) \Leftrightarrow (ii)$ *then* $(ii) \Leftrightarrow (iii)$.

$(i) \Rightarrow (ii)$ If f is convex, by definition

$$f(\lambda y + (1-\lambda)x) \leq \lambda f(y) + (1-\lambda)f(x), \forall \lambda \in [0, 1], x, y \in dom(f).$$

After rewriting, we have

$$f(x + \lambda(y - x)) \leq f(x) + \lambda(f(y) - f(x))$$

$$\Rightarrow f(y) - f(x) \geq \frac{f(x + \lambda(y-x)) - f(x)}{\lambda}, \forall \lambda \in (0, 1]$$

As $\lambda \downarrow 0$, we get

$$f(y) - f(x) \geq \nabla f^T(x)(y - x).$$

$(ii) \Rightarrow (i)$ Suppose $f(y) - f(x) \geq \nabla f^T(x)(y-x)$. holds $\forall x, y \in dom(f)$. Takeany $x, y \in dom(f)$ and let

$$z = \lambda x + (1-\lambda)y.$$

We have

$$f(x) \geq f(z) + \nabla f^T(z)(x - z)$$

$$f(y) \geq f(z) + \nabla f^T(z)(x - z)$$

Multiplying $(f(x) \geq f(z) + \nabla f^T(z)(x-z))$ by λ, $(f(y) \geq f(z) + \nabla f\ T(z)(y-z))$ by $(1-\lambda)$ and adding, we get

$$\lambda f(x) + (1-\lambda)f(y) \geq f(z) + \nabla f^T(z)(\lambda x + (1-\lambda)y - z)$$

$$= f(z)$$

$$= f(\lambda x + (1-\lambda)y).$$

$(ii) \Leftrightarrow (iii)$ We prove both of these claims first in dimension 1 and then generalize.

$(ii) \Rightarrow (iii)$ $(n = 1)$ Let $x, y \in dom(f),\ y > x$. We have

$$f(y) \geq f(x) + f'(x)(y - x)$$

$$f(x) \geq f(y) + f'(y)(x - y)$$

$$\Rightarrow f'(x)(y - x) \leq f(y) - f(x) \leq f'(y)(y - x)$$

using $(f(y) \geq f(x) + f'(x)(y - x))$ then $(f(x) \geq f(y) + f'(y)(x - y))$. Dividing LHS and RHS by $(y - x)^2$ gives

$$\frac{f'(y) - f'(x)}{y - x} \geq 0, \forall x, y, x \neq y.$$

As we let $y \rightarrow x$, we get

$$f''(x) \geq 0, \forall x \in dom(f).$$

$(iii) \Rightarrow (ii)$ $(n = 1))$ Suppose $f''(x) \geq 0, \forall x \in dom(f)$. By the mean value version of Taylor's theorem we have

$$f(y) = f(x) + f'(x)(y - x) + \frac{1}{2} f''(z)(y - x)^2, \textit{for some } z \in [x, y].$$

$$\Rightarrow f(y) \geq f(x) + f'(x)(y - x).$$

Now to establish $(ii) \Leftrightarrow (iii)$ in general dimension, we recall that convexity is equivalent to convexity along all lines; i.e., $f : \mathbb{R}^n \rightarrow \mathbb{R}$ is convex if $g(\alpha) = f(x_0 + \alpha v)$) is convex $\forall x_0 \in dom(f)$ and $\forall v \in \mathbb{R}^n$. We just proved this happens if

$$g''(\alpha) = v^T \nabla^2 f(x_0 + \alpha v)v \geq 0,$$

$\forall x_0 \in dom(f), \forall v \in \mathbb{R}^n$ and $x_0 + \alpha v \in dom(f)$. Hence, f is convex if $\nabla^2 f(x) \succeq 0$ for all $x \in dom(f)$.

Corollary: Consider an unconstrained optimization problem

$$\min f(x)$$

$$s.t.\, x \in \mathbb{R}^n ,$$

where f is convex and differentiable. Then, any point $\bar{x}$ that satisfies $\nabla f(\bar{x}) = 0$ is a global minimum.

Proof: From the first order characterization of convexity, we have

$$f(y) \geq f(x) + \nabla f^T(x)(y - x), \forall x, y$$

In particular,

$$f(y) \geq f(\bar{x}) + \nabla f^T(\bar{x})(y - x), \forall y.$$

Since $\nabla f(\bar{x}) = 0$, we get

$$f(y) \geq f(\bar{x}), \forall y$$

Remarks:

- Recall that $\nabla f(x) = 0$ is always a necessary condition for local optimality in an unconstrained problem. The theorem states that for convex problems, $\nabla f(x) = 0$ is not only necessary, but also sufficient for local and global optimality.
- In absence of convexity, $\nabla f(x) = 0$ is not sufficient even for local optimality (e.g., think of $f(x) = x^3$ and $\bar{x} = 0$).
- Another necessary condition for (unconstrained) local optimality of a point x was $\nabla^2 f(x) \succeq 0$. Note that a convex function automatically passes this test.

Strict Convexity and Uniqueness of Optimal Solutions

Characterization of Strict Convexity

Recall that a fuction $f : \mathbb{R}^n \to \mathbb{R}$ is strictly convex if $\forall x, y, x \neq y, \forall \lambda \in (0, 1)$,

$$f(\lambda x + (1-\lambda)y) < \lambda f(x)(1-\lambda) f(y).$$

Like we mentioned before, if f is strictly convex, then f is convex (this is obvious from the definition) but the converse is not true $(e.g., f(x) = x, x \in \mathbb{R})$.

Second order sufficient condition:

$$\nabla^2 f(x) \succ 0, \forall x \in \Omega \Rightarrow f \text{ strictly convex on } \Omega.$$

The converse is not true though

First order characterization: A function f is strictly convex on $\Omega \subseteq \mathbb{R}^n$ if and only if

$$f(y) > f(x) + \nabla f^T(x)(y - x), \forall x, y \in \Omega, x \neq y.$$

- There are similar characterizations for strongly convex functions. For example, f is strongly convex if and only if there exists $m > 0$ such that

  $$\nabla^2 f(x) \succeq mI, \forall x \in dom(f)$$

- One of the main uses of strict convexity is to ensure uniqueness of the optimal solution.

Strict Convexity and Uniqueness of Optimal Solutions

Theorem: Consider an optimization problem

$$\min f(x)$$

$$s.t.\ x \in \Omega,$$

where $f:\mathbb{R}^n \to \mathbb{R}$ is strictly convex on Ω and Ω is a convex set. Then the optimal solution (assuming it exists) must be unique.

Proof: Suppose there were two optimal solutions $x, y \in \mathbb{R}^n$. This means that $x, y \in \Omega$ and

$$f(x) = f(y) \leq f(z), \forall z \in \Omega.$$

But consider $z = \frac{x+y}{2}$. By convexity of Ω, we have $z \in \Omega$. By strict convexity, we have

$$f(z) = f\left(\frac{x+y}{2}\right)$$

$$< \frac{1}{2} f(x) + \frac{1}{2} f(y)$$

$$= \frac{1}{2} f(x) + \frac{1}{2} f(x) = f(x)$$

But this contradicts ($f(x) = f(y) \leq f(z), \forall z \in \Omega$).

Convex Body

Convex Body is a geometric body having the property that any segment joining two of its points is entirely contained within it. Thebody shown in figure a is convex and that shown in figure b is not. A sphere, a cube, a spherical segment, or a half-space are examples of convex bodies. Anycon-nected part of a boundary of a convex body is called aconvex surface. Through every point of the boundary of aconvex body passes at least one plane of support having apoint (or segment or part of a plane) in common with theboundary of the body but not intersecting it. At points where the boundary of the convexbody is a smooth surface, the plane of support will be tangential. At points where the smoothness of the surfaceis violated (as at the corner of a cube), an infinite numberof planes of support can be drawn.

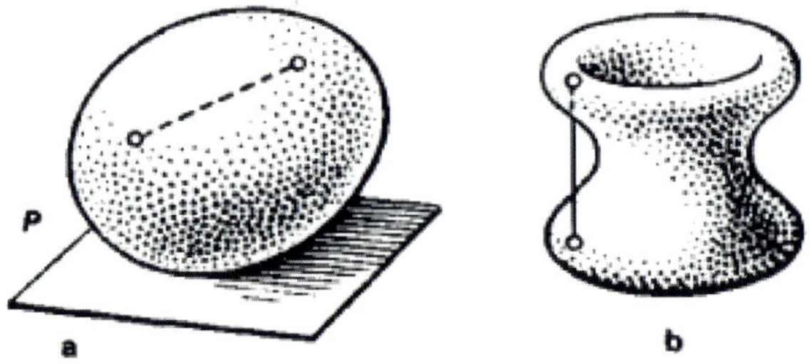

There are five types of convex bodies: finite (the boundary is a closed convex surface), infinite (the boundary is oneinfinite surface; for example, a convex body bounded by aparaboloid), cylinders infinite in both directions (theboundary is a closed convex cylindrical surface; for example, an infinite circular cylinder), layers between pairsof parallel surfaces; and all space. Convex bodies can be defined by a support function expressing the distance fromthe origin of coordinates to the support surface as afunction of the outer normal to the convex body (that is, of a unit vector that is perpendicular to the support surface and directed toward that one of two half-spaces that isdefined by this surface and which contains no points of the convex body).

Convex polyhedrons are the simplest convex bodies; they are convex bodies bounded by a finite number of polygons.For any finite convex body it is possible to construct polyhedrons approximating it to any degree. This makes it possible to solve many problems of convex bodies as follows: the problem is solved for convex polyhedrons, andthen by going to the limit, the corresponding result isestablished for any convex body. For example, the areas of convex surfaces and the volumes of any convex bodies are determined in this way. In particular, it can be proved that if one finite convex body encompasses another, thenthe surface area of the first is greater than the surface areaof the second. The method described was worked out indetail by A. D. Aleksandrov and is used to solve a variety of new problems in the theory of convex bodies.

The general theory of convex bodies and convex surfaces constitutes the so called geometry of convex bodies. Problems in the geometry of convex bodies cover a widerange of questions: the general properties of convex bodies (theorems on support surfaces, classification of convex bodies, and approximation by polyhedrons), extreme properties of convex bodies (for example, of all convex bodies with a given volume, the sphere has the minimum surface), existence and uniqueness theorems about convex bodies with given properties (for example, the theorem on the existence of a convex polyhedron with given directions and areas of faces), properties of variousclasses of convex bodies (for example, of bodies ofconstant width), general properties of convex surfaces, existence and uniqueness theorems for convex surfaces, the internal geometry of convex surfaces, and others. Theconcept of convex bodies naturally arises in the geometry of spaces of constant curvature. Many of the problemslisted above have been formulated and solved for convexbodies in such spaces.

Theorem: Let L be a convex body in $\mathbb{R}^n$ containing o in its interior. The intersection body $I(L)$ of L, defined by its radial function $\rho I(L)$ on the sphere S^{n-1}, which is

$$_{\rho I(L)}(u) = \mathrm{vol}(L \cap u^{\perp})$$

and the cross-section body $C(L)$ of L, defined by

$$_{\rho C(L)}(u) = \max_{t} vol((L \cap (tu + u^{\perp}))$$

are not, in general, convex bodies, although they are identical and, moreover, convex in the case when L is centrally symmetric. This follows from Brunn–Minkowski theorem and from Busemann's theorem. We introduce here a new convex body associated with L, generalizing both the intersection body and the cross-section body. More precisely, we define the convex intersection body $CI(L)$ of L by its radial function.

$$_{\rho CI(L)}(u) = \min_{z \in P_u\left(L^{*g(L)}\right)} vol\left(\left[Pu\left(L^{*g(L)}\right)^{*z}\right]\right),$$

where $g(L)$ denotes here the centroid of L, and if E is a linear subspace of n and M is a convex body in E, we define for $z \in E$,

$$M^{*z} = \{y \in E; \langle y - z, x - z \rangle \le 1 \, for \; every \; x \in M\}.$$

Thus (1) means: First apply duality with respect to the point $g(L)$, then project onto $u^{\perp}$, finally apply duality with respect to z and minimize the $(n-1)$-dimensional volume over z.

Convex Combination

Convex combinations are simply affine combinations where the constants in the combination are limited to be in the interval $[0,1]$.

Given a set of points $P_0, P_1, ..., P_n$, we can form an affine combination of these points by selecting constants $\alpha_0, \alpha_1, ..., \alpha_n$, (where, $\alpha_0 + \alpha_1 + \cdots + \alpha_n = 1$) and write

$$P = \alpha_0 P_0 + \alpha_1 P_1 + \cdots + \alpha_n P_n$$

If each α_i is such that $0 \le \alpha_i \le 1$, then the point P is called a *convex combination* of the points $P_0, P_1, ..., P_n$.

Let $\mathscr{X}$ be a real vector space. A set $\mathscr{A} \subseteq \mathscr{X}$ is affine if it contains the lines passing through all pairs of its points (all binomial affine combinations in $\mathscr{A}$ i.e., the combinations $p_1 P_1 + p_2 P_2$ of points $P_1, P_2 \in \mathscr{A}$ and coefficients $p_1, p_2 \in \mathbb{R}$ of the sum $p_1 + p_2 = 1$). A function $f : \mathscr{A} \to \mathbb{R}$ is affine if it satisfies the equality $f(p_1 P_1 + p_2 P_2) = p_1 f(P_1) + p_2 f(P_2)$ for all binomial affine combinations in $\mathscr{A}$.

A set $\mathscr{C} \subseteq \mathscr{X}$ is convex if it contains the line segments connecting all pairs of its points (all binomial convex combinations in $\mathscr{C}$, i.e., the combinations $p_1 P_1 + p_2 P$ of points $P_1, P_2 \in \mathscr{C}$ and nonnegative coefficients $p_1, p_2 \in \mathbb{R}$ of the sum $p_1 + p_2 = 1$). A function $f : \mathscr{C} \to \mathbb{R}$ is convex if it satisfies the inequality $f(p_1 P_1 + p_2 P_2) = p_1 f(P_1) + p_2 f(P_2)$ for all binomial convex combinations in $\mathscr{C}$.

Using the mathematical induction, it can be proved that every affine function $f : \mathscr{A} \to \mathbb{R}$ satisfies the equality

$$f\left(\sum_{i=1}^{n} p_i P_i\right) = \sum_{i=1}^{n} p_i f(P_i)$$

for all affine combinations in $\mathscr{S}$ and that every convex function $f : \mathscr{C} \to \mathbb{R}$ satisfies the Jensen inequality

$$f\left(\sum_{i=1}^{n} p_i P_i\right) \le \sum_{i=1}^{n} p_i f(P_i)$$

for all convex combinations in $\mathscr{C}$.

For an affine or a convex combination $P=\sum_{i=1}^{n} p_i P_i$ the point P itself is called the combination center, and it is important to mathematical inequalities.

Convex Combinations of the Line

If $a, b \in \mathbb{R}$ are different numbers, say $a<b$, then every number $x \in \mathbb{R}$ can be uniquely presented as the affine combination

$$x=\frac{b-x}{b-a} a+\frac{x-a}{b-a} b .$$

The above binomial combination is convex if and only if the number x belongs to the interval $[a \ b]$. Given the function $f: \mathbb{R} \rightarrow \mathbb{R}$, let $f_{\{a,b\}}^{line}: \mathbb{R} \rightarrow \mathbb{R}$ be the function of the line passing through the points $(a, f(a))$ and $(b, f(b))$ of the graph of f. Using the affinity of $f_{\{a,b\}}^{line}$ we get the equation

$$f_{\{a,b\}}^{\text{line}}(x)=\frac{b-x}{b-a} f(a)+\frac{x-a}{b-a} f(b).$$

If the function f is convex, then, using the definition of convexity, we obtain the inequality

$$f(x) \leq f_{\{a,b\}}^{line}(x), \qquad \text{if } x \in[a, b],$$

and the reverse inequality

$$f(x) \geq f_{\{a,b\}}^{line}(x), \qquad \text{if } x \notin(a, b)$$

By the end of this section we will use an interval $\mathscr{I} \subseteq \mathbb{R}$ with the nonempty interior $\mathscr{I}^0$.

The following lemma deals with two convex combinations having the same center. Assigning the convex function to such convex combinations, we obtain the following Jensen type inequality.

Lemma: Let $a, b \in \mathscr{I} \subseteq \mathbb{R}$ be points such that $a \leq b$. Let $\sum_{i=1}^{n} p_i x_i$ be a convex combination with points $x_i \in[a, b]$, and let $\sum_{j=1}^{m} q_j y_j$ be a convex combination with points $y_j \in \mathscr{I} \backslash(a, b)$.

If the above convex combinations have the same center

$$\sum_{i=1}^{n} p_i x_i=\sum_{j=1}^{m} q_j y_j,$$

then every convex function $a, b \in \mathscr{I} \subseteq \mathbb{R}$ satisfies the inequality

$$\sum_{i=1}^{n} p_i f(x_i)=\sum_{j=1}^{m} q_j f(y_j).$$

If f is concave, then the reverse inequality is valid in ($\sum_{i=1}^{n} p_i f(x_i)=\sum_{j=1}^{m} q_j f(y_j)$).

Proof: Assume f is convex. If $a < b$, the right-hand side follows from the series of inequalities

$$\sum_{i=1}^{n} p_i f(x_i) \le \sum_{i=1}^{n} p_i f_{\{a,b\}}^{\text{line}}(x_i) = f_{\{a,b\}}^{\text{line}}\left(\sum_{i=1}^{n} p_i x_i\right)$$
$$= f_{\{a,b\}}^{\text{line}}\left(\sum_{j=1}^{m} q_j\, y_j\right) = \sum_{j=1}^{m} q_j\, f_{\{a,b\}}^{\text{line}}(y_j)$$
$$\le \sum_{j=1}^{m} q_j\, f(y_j)$$

derived applying the inequality in ($f(x) \le f_{\{a,b\}}^{\text{line}}(x)$, if $x \in [a,b],$) to x_i and the inequality in ($f(x) \ge f_{\{a,b\}}^{\text{line}}(x)$, if $x \notin (a,b)$) to y_j. If $a = b$, we use any support line $f_{\{a\}}^{\text{line}}$ instead of the chord line $f_{\{a,b\}}^{\text{line}}$.

Lemma is the generalization of Jensen's inequality: applying this lemma to the convex combination center equality

$$1c = \sum_{i=1}^{n} p_i x_i,$$

with the assumption $a = b = c$, we come to the Jensen inequality

$$f\left(\sum_{i=1}^{n} p_i x_i\right) = 1f(c) \le \sum_{i=1}^{n} p_i f(x_i).$$

So, the discrete form of the famous Jensen inequality can be derived applying the convexity definition and the affinity of the chord or support line.

Theorem: Let $a, b, a_1, b_1 \in \mathcal{I} \subseteq \mathbb{R}$ be points such tha $a_1 < a < b < b_1$. Let $c = \sum_{i=1}^{n} p_i x_i$ be a convex combination of the points $x_i \in [a_1, b_1] \setminus (a,b)$ with the center $c \in (a,b)$.

Then there exist two binomial convex combinations $\alpha a + \beta b$ and $\alpha_1 a_1 + \beta_1 b_1$ so that

$$\alpha a + \beta b = \sum_{i=1}^{n} p_i x_i = \alpha_1 a_1 + \beta_1 b_1,$$

and consequently, every convex function $f : \mathcal{I} \to \mathbb{R}$ satisfies the inequality

$$\alpha f(a) + \beta f(b) \le \sum_{i=1}^{n} p_i f(x_i) \le \alpha_1\, f(a_1) + \beta\, f(b_1).$$

If f is concave, then the reverse inequality is valid in

$$(\alpha f(a) + \beta f(b) \le \sum_{i=1}^{n} p_i f(x_i) \le \alpha_1\, f(a_1) + \beta\, f(b_1)).$$

Proof: We use the formula in ($x=\frac{b-x}{b-a}a+\frac{x-a}{b-a}b$) to calculate the coefficients $\alpha=(b-c)/(b-a)$ and $\alpha=(b-c)/(b-a)$ that satisfy $c=\alpha a+\beta b$ and also for α_1 and β_1. Now we need to apply Lemma to both sides of the obtained equality in

$$(\alpha a+\beta b=\sum_{i=1}^{n}p_ix_i=\alpha_1a_1+\beta_1b_1).$$

The graphical representation of the equality in ($\alpha a+\beta b=\sum_{i=1}^{n}p_ix_i=\alpha_1a_1+\beta_1b_1$) and the inequality in ($\alpha f(a)+\beta f(b)\le\sum_{i=1}^{n}p_if(x_i)\le\alpha_1 f(a_1)+\beta f(b_1)$) is shown in figure below.

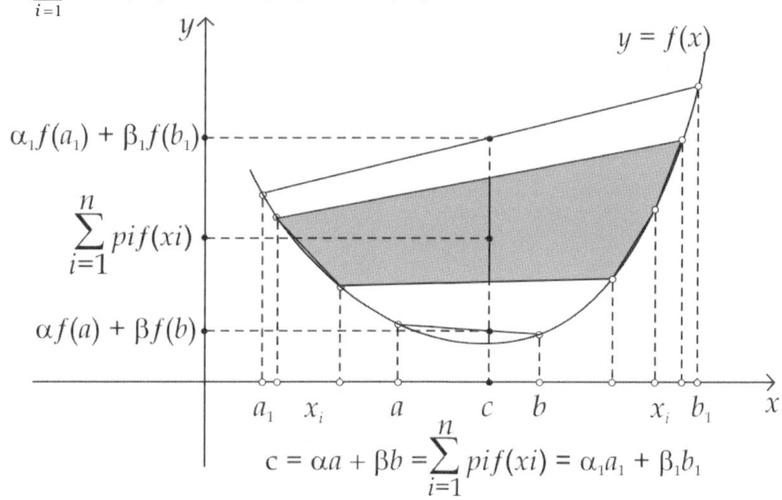

Figure: Line convex combinations with the same center c.

Binomial convex combinations are included into the definition of convexity. The following corollary demonstrates how the binomial combinations may be assigned to each convex combination.

Corollary: Let $\sum_{i=1}^{n}p_ix_i$ be a convex combination in $\mathcal{I}\subseteq\mathbb{R}$. Then there exist two binomial convex combinations $p_0x_0+q_0y_0$ and $px+qy$ with points x_0, y_0, x, y from the set $\{x_1,\dots,x_n\}$ so that

$$p_0x_0+q_0y_0=\sum_{i=1}^{n}p_ix_i=px+qy,$$

and consequently, every convex function $f:\mathcal{I}\to\mathbb{R}$ satisfies the inequality

$$p_0f(x_0)+q_0f(y_0)\le\sum_{i=1}^{n}p_if(x_i)\le pf(x)+qf(y).$$

Proof: Put $c=\sum_{i=1}^{n}p_ix_i$ and suppose $x_1\le\cdots\le x_n$. Take $x=x_1$ and $y=x_n$. If $x=y$, we take $p=q=1/2$. If $x<y$, we calculate the coefficients p and q by the formula in

($x=\frac{b-x}{b-a}a+\frac{x-a}{b-a}b$) to get $c=px+qy$.

If c is equal to some x_{i_0}, then we take $x_0 = y_0 = x_{i_0}$ and $p_0 = q_0 = 1/2$. Otherwise, it must be $c \in (x_i, x_{i+1})$ for some pair, in which case we take $x_0 = x_i$ and $y_0 = x_{i+1}$. Calculating p_0 and q_0 by

$$(x = \frac{b-x}{b-a} a + \frac{x-a}{b-a} b)\text{, we also get } c = p_0 x_0 + q_0 y_0.$$

It remains to apply Lemma to both sides of the obtained equality in

$$(p_0 x_0 + q_0 y_0 = \sum_{i=1}^{n} p_i x_i = px + qy).$$

Respecting the Jensen inequality, the formula in

$(p_0 f(x_0) + q_0 f(y_0) \le \sum_{i=1}^{n} p_i f(x_i) \le pf(x) + qf(y))$ can be expressed in the extended form:

$$f\left(\sum_{i=1}^{n} p_i x_i\right) \le p_0 f(x_0)\ q_0\ f(y_0)$$
$$\le \sum_{i=1}^{n} p_i f(x_i) \le pf(x) + q\, f(y).$$

Let us show how the intermediate member in $(\alpha f(a) + \beta f(b) \le \sum_{i=1}^{n} p_i f(x_i) \le \alpha_1 f(a_1) + \beta f(b_1))$ can be transformed into integral. Let $\mathscr{L} \subseteq \mathscr{I}$ be a bounded set with the length $|\mathscr{L}| > 0$, and let f be an integrable (in the sense of Riemann or Lebesgue) function on $\mathscr{L}$. Given the positive integer n , we employ a partition

$$\mathscr{L} = \bigcup_{i=1}^{n} \mathscr{L}_{ni},$$

where each of pairwise disjoint subsets $\mathscr{L}_{ni}$ contracts to the point as n approaches infinity. Take one point $x_{ni} \in \mathscr{L}_{ni}$ for every $i = 1, \ldots, n$ and then compose the convex combination c_n of the points $f(x_{ni})$ with the coefficients $p_{ni} = |\mathscr{L}_{ni}| / |\mathscr{L}|$; that is,

$$c_n = \frac{1}{|\mathscr{L}|} \sum_{i=1}^{n} |\mathscr{L}_{ni}|\, f(x_{ni})$$

Letting n to infinity, the sequence $(c_n)_n$ approaches the point

$$c = \frac{1}{|\mathscr{L}|} \int_{\mathscr{L}} f(x)\, dx.$$

Using the integral method with convex combinations, we obtain the mixed discrete-integral form of Theorem above as follows:

Corollary: Let $a, b, a_1, b_1 \in \mathcal{I} \subseteq \mathbb{R}$ be points such that $a_1 < a < b < b_1$. Let the barycenter c of the set $[a_1, b_1] \backslash (a,b)$ belongs to (a,b); that is,

$$c = \frac{\int_{[a_1,b_1]\backslash(a,b)} x\, dx}{b_1 - a_1 - b + a} \in (a,b).$$

Then there exist two binomial convex combinations $\alpha a + \beta b$ and $\alpha_1 a_1 + \beta_1 b_1$ so that

$$\alpha a + \beta b = c = \alpha_1 a_1 + \beta_1 b_1,$$

and consequently, every convex function $f : \mathcal{I} \to \mathbb{R}$ satisfies the inequality

$$\begin{aligned} \alpha f(a) \beta f(b) &\leq \frac{\int_{[a_1,b_1]\backslash(a,b)} f(x)\, dx}{b_1 - a_1 - b + a} \\ &\leq \alpha_1 f(a_1) + \beta_1 f(b_1). \end{aligned}$$

If f is concave, then the reverse inequality is valid in ($\alpha f(a) \beta f(b) \leq \frac{\int_{[a_1,b_1]\backslash(a,b)} f(x)\, dx}{b_1 - a_1 - b + a}$).

The summarizing Jensen's functional of a function $f : \mathcal{I} \to \mathbb{R}$ for the given convex combination $\sum_{i=1}^{n} p_i x_i$ in $\mathcal{I}$ is defined with

$$J_{p_1 x_1 + \cdots + p_n x_n}(f) = \sum_{i=1}^{n} p_i f(x_i) - f\left(\sum_{i=1}^{n} p_i x_i\right).$$

If the conditions of theorem hold, we get the functional inequality

$$J_{\alpha a + \beta b}(f) \leq J_{p_1 x_1 + \cdots + p_n x_n}(f) \leq J_{\alpha_1 a_1 + \beta_1 b_1}(f).$$

The integrating Jensen's functional of an integrable function $f : \mathcal{I} \to \mathbb{R}$ for the given subset $\mathcal{L} \subseteq \mathcal{I}$ with the length $|\mathcal{L}| > 0$, is defined with

$$J_{\mathcal{L}}(f) = \frac{1}{|\mathcal{L}|} \int_{\mathcal{L}} f(x)\, dx - f\left(\frac{1}{|\mathcal{L}|} \int_{\mathcal{L}} x\, dx\right).$$

If the conditions of Corollary above are satisfied, we get

$$J_{\alpha a + \beta b}(f) \leq J_{[a_1,b_1]\backslash(a,b)}(f) \leq J_{\alpha_1 a_1 + \beta_1 b_1}(f).$$

The inequalities in ($J_{\alpha a + \beta b}(f) \leq J_{p_1 x_1 + \cdots + p_n x_n}(f) \leq J_{\alpha_1 a_1 + \beta_1 b_1}(f)$) and ($J_{\alpha a + \beta b}(f) \leq J_{[a_1,b_1]\backslash(a,b)}(f) \leq J_{\alpha_1 a_1 + \beta_1 b_1}(f)$) offer the local bounds of Jensen's functionals. The global bounds of Jensen's summarizing functional were investigated in equation. Some new Jensen type inequalities were obtained in [1$c = \sum_{i=1}^{n} p_i x_i$].

Convex Combinations of the Plane

We assume that $\mathbb{R}^2$ is the real vector space treating its points as the vectors with the standard coordinate addition $(x_1, y_1)+(x_2, y_2)=(x_1+x_2+y_1+y_2)$ and the scalar multiplication $\alpha(x,y)=(\alpha x, \alpha y)$.

If $A(x_A, y_A), B(x_B, y_B)$, and $C(x_C, y_C)$ are the planar points that do not belong to one line, respectively, the convex hull $\{A, B, C\}$ is a real triangle, and then every point $P(x,y)\in\mathbb{R}^2$ can be presented by the unique affine combination

$$P = \alpha A + \beta B + \gamma C,$$

Where

$$\alpha = \frac{\begin{vmatrix} x & y & 1 \\ x_B & y_B & 1 \\ x_C & y_C & 1 \end{vmatrix}}{\begin{vmatrix} x_A & y_A & 1 \\ x_B & y_B & 1 \\ x_C & y_C & 1 \end{vmatrix}} \qquad \beta = -\frac{\begin{vmatrix} x & y & 1 \\ x_A & y_A & 1 \\ x_C & y_C & 1 \end{vmatrix}}{\begin{vmatrix} x_A & y_A & 1 \\ x_B & y_B & 1 \\ x_C & y_C & 1 \end{vmatrix}} \qquad \gamma = \frac{\begin{vmatrix} x & y & 1 \\ x_A & y_A & 1 \\ x_B & y_B & 1 \end{vmatrix}}{\begin{vmatrix} x_A & y_A & 1 \\ x_B & y_B & 1 \\ x_C & y_C & 1 \end{vmatrix}}$$

The above trinomial combination is convex if and only if the point P belongs to the triangle $\text{conv}\{A, B,C\}$.

Given the triangle with vertices A B, and C, the convex cone $\mathscr{C}_A$ with the vertex at A is the set spanned by the vectors $A-B$ and $A-C$ (similarly $\mathscr{C}_B$ and $\mathscr{C}_C$; all three cones can be viewed in figure); that is,

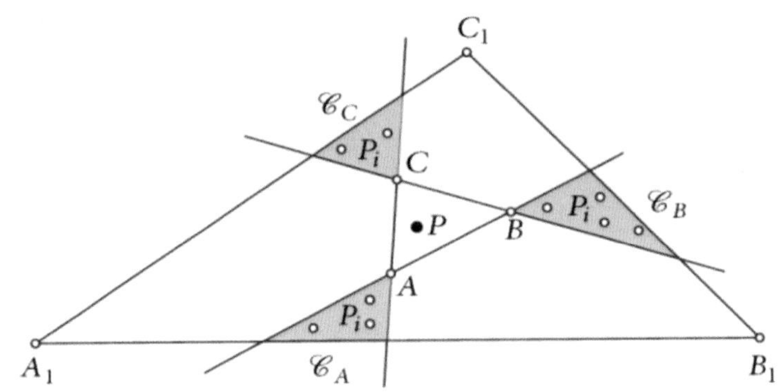

$$P = \alpha A + \beta B + \gamma C = \sum_{i=1}^{n} p_i P_i = \alpha_1 A_1 + \beta_1 B_1 + \gamma_1 C_1$$

Figure: Plane convex combinations with the same center P.

Given the function $f:\mathbb{R}^2 \to \mathbb{R}$ let $f^{plane}_{\{A,B,C\}}:\mathbb{R}^2 \to \mathbb{R}$ be the function of the plane passing through the points $(A, f(A)), (B, f(B))$, and $(C, f(C))$ of the graph of f. Due to the affinity of $f^{plane}_{\{A,B,C\}}$, it follows

$$f^{plane}_{\{A,B,C\}}(P) = \alpha f(A) + \beta f(B) + \gamma f(C).$$

For a convex function f, using the convexity definition, we get the inequality

$$f(P) \le f^{plane}_{\{A,B,C\}}(P), \quad \text{if } P \in \text{conv}\{A,B,C\},$$

and using the affinity of $f^{plane}_{\{A,B,C\}}$, the reverse inequality

$$f(P) \ge f^{plane}_{\{A,B,C\}}(P), \quad \text{if } P \in \mathscr{C}_A \cup \mathscr{C}_B \cup \mathscr{C}_C.$$

By the end of the section we will use a planar convex set $\mathscr{C} \subseteq \mathbb{R}^2$ with the nonempty interior $\mathscr{C}^0$. The area of a planar set $\mathscr{A}$ will be denoted with $ar(\mathscr{A})$.

Lemma: Let $A, B, C \in \mathscr{C} \subseteq \mathbb{R}^2$ be points, $\Delta = \text{conv}\{A, B, C\}$, and $\zeta_\Delta = C_A \cup C_B \cup C_C$ be the cone union. Let $\sum_{i=1}^{n} \alpha_i A_i$ be a convex combination of the points $A_i \in \Delta$ and let $\sum_{j=1}^{m} \beta_j B_j$ be a convex combination of the points $B_j \in \mathscr{C}_\Delta \cap \mathscr{C}$.

If the above convex combinations have the same center

$$\sum_{i=1}^{n} \alpha_i A_i = \sum_{j=1}^{m} \beta_j B_j$$

then every convex function $f : \mathscr{C} \to \mathbb{R}$ satisfies the inequality

$$\sum_{i=1}^{n} \alpha_i f(A_i) \le \sum_{j=1}^{m} \beta_j f(B_j)$$

Proof: If the set $\text{conv}\{A, B, C\}$ is a real triangle, we can apply the proof of Lemma using $f^{\text{plane}}_{\{A,B,C\}}$ instead of $f^{\text{line}}_{\{a,b\}}$ respecting the plane inequalities. If $\text{conv}\{A,B,C\} = \text{conv}\{A,B\}$, then we rely on the proof of Lemma with the chord line $f^{\text{line}}_{\{a,b\}}$. If $\text{conv}\{A,B,C\} = \{A\}$ we use any support line $f^{\text{line}}_{\{A\}}$ at the point A.

Theorem: Let $A, B, C; A_1, B_1, C_1 \in \mathscr{C} \subseteq \mathbb{R}^2$ be points such that

$$\Delta = \text{conv}\{A, B, C\} \subset \text{conv}\{A_1, B_1, C_1\} = \Delta_1$$

With $\Delta^0 \ne \emptyset$, and let $\mathscr{C}_\Delta = \mathscr{C}_A \cup \mathscr{C}_B \cup \mathscr{C}_C$ be the cone union. Let $P = \sum_{i=1}^{n} p_i P_i$ be a convex combination of the points $P_i \in \mathscr{C}_\Delta \cap \Delta_1$ with the center $P \in \Delta^0$.

Then there exist two trinomial convex combinations $\alpha A + \beta B + \gamma C$ and $\alpha_1 A_1 + \beta_1 B_1 + \gamma_1 C_1$ so that

$$\alpha A + \beta B + \gamma C = \sum_{i=1}^{n} p_i P_i = \alpha_1 A_1 + \beta_1 B_1 + \gamma_1 C_1,$$

and consequently, every convex function $f : \mathscr{C} \to \mathbb{R}$ satisfies the inequality

$$\alpha f(A) + \beta f(B) + \gamma f(C) \le \sum_{i=1}^{n} p_i f(P_i)$$
$$\le \alpha_1 f(A_1) + \beta_1 f(B_1) + \gamma_1 f(C_1).$$

Proof: First we calculate the coefficients by the formula in (Convex trinomial combination) to get the equality in ($\alpha A+\beta B+\gamma C=\sum_{i=1}^{n} p_i P_i=\alpha_1 A_1+\beta_1 B_1+\gamma_1 C_1$) and then apply Lemma to its both sides.

The graphical representation of the equality in ($\alpha A+\beta B+\gamma C=\sum_{i=1}^{n} p_i P_i=\alpha_1 A_1+\beta_1 B_1+\gamma_1 C_1$).

Applying the integral method with convex combinations, we obtain the form as follows:

Corollary: Let $A, B, C; A_1, B_1, C_1 \in \mathscr{C} \subseteq \mathbb{R}^2$ be points such that

$$\Delta=\operatorname{conv}\{A, B, C\} \subset \operatorname{conv}\{A_1, B_1, C_1\}=\Delta_1$$

With $\Delta^0 \neq \emptyset$, and let $\mathscr{C}_\Delta=\mathscr{C}_A \cup \mathscr{C}_B \cup \mathscr{C}_C$ be the cone union. Let the barycenter P of the set $\mathscr{C}_\Delta \cap \Delta_1$ belong to Δ^0; that is,

$$P\left(\frac{\int_{\mathscr{C}_\Delta \cap \Delta_1} x\,dx\,dy}{\operatorname{ar}(\mathscr{C}_\Delta \cap \Delta_1)}, \frac{\int_{\mathscr{C}_\Delta \cap \Delta_1} y\,dx\,dy}{\operatorname{ar}(\mathscr{C}_\Delta \cap \Delta_1)}\right) \in \Delta^0.$$

Then there exist two trinomial convex combinations $\alpha A+\beta B+\gamma C$ and $\alpha_1 A_1+\beta_1 B_1+\gamma_1 C_1$ so that

$$\alpha A+\beta B+\gamma C=P=\alpha_1 A_1+\beta_1 B_1+\gamma_1 C_1,$$

and consequently, every convex function $f: \mathscr{C} \rightarrow \mathbb{R}$ satisfies the inequality

$$\begin{aligned}
&\alpha f(A)+\beta f(B)+\gamma f(C) \\
&\leq \frac{\int_{\mathscr{C}_\Delta \cap \Delta_1} f(x,y)\,dx\,dy}{ar(\mathscr{C}_\Delta \cap \Delta_1)} \\
&\leq \alpha_1 f(A_1)+\beta_1 f(B_1)+\gamma_1 f(C_1).
\end{aligned}$$

If f is concave, then the reverse inequality is valid in $\left(\leq \frac{\int_{C_\Delta \cap \Delta_1} f(x,y)\,dx\,dy}{ar(C_\Delta \cap \Delta_1)}\right)$.

If the conditions of Theorem above are valid, then using the summarizing Jensen functional of f or the given convex combination $\sum_{i=1}^{n} p_i P_i$,

$$J_{p_1 P_1+\cdots+p_n P_n}(f)=\sum_{i=1}^{n} p_i f(P_i)-f\left(\sum_{i=1}^{n} p_i P_i\right).$$

we get the functional inequality

$$\begin{aligned}
J_{\alpha A+\beta B+\gamma C}(f) &\leq J_{p_1 P_1+\cdots+p_n P_n}(f) \\
&\leq J_{\alpha_1 A_1+\beta_1 B_1+\gamma_1 C_1}(f).
\end{aligned}$$

If the conditions of Corollary above are satisfied, then applying another integrating Jensen functional of f given subset $\mathscr{A} \subseteq \mathscr{C}$ with the area $\operatorname{ar}(\mathscr{A}) > 0$,

$$J_{\mathscr{A}}(f) = \frac{\int_{\mathscr{A}} f(x,y)\,dx\,dy}{\operatorname{ar}(\mathscr{A})} - f\left(\frac{\int_{\mathscr{A}} x\,dx\,dy}{\operatorname{ar}(\mathscr{A})}, \frac{\int_{\mathscr{A}} y\,dx\,dy}{\operatorname{ar}(\mathscr{A})}\right)$$

we have the inequality

$$J_{\alpha A+\beta B+\gamma C}(f) \leq J_{\mathscr{C}_\Delta \cap \Delta_1}(f) \leq J_{\alpha_1 A_1+\beta_1 B_1+\gamma_1 C_1}(f).$$

Generalization

If $A_1, \ldots, A_{m+1} \in \mathbb{R}^2$ are points such that the vectors $A_1 - A_2, \ldots, A_1 - A_{m+1}$ are linearly independent, then the convex hull

$$\Delta = \operatorname{conv}\{A_1, \ldots, A_{m+1}\}$$

is called the m-simplex with the vertices $A_1, \ldots, A_{m+1}$. All the simplex vertices can not belong to the same hyperplane in $\mathbb{R}^m$. Any point $P \in \mathbb{R}^m$ can be presented by the unique affine combination

$$P = \sum_{j=1}^{m+1} \alpha_j A_j .$$

If we use the point coordinates, then the coefficients α_j can be determined by the generalized coefficient formula. The combination in ($P = \sum_{j=1}^{m+1} \alpha_j A_j$) is convex if and only if the point P belongs to the m-simplex Δ.

Given the m-simplex with vertices $A_1, \ldots, A_{m+1}$, let $\mathscr{C}_{A_1}$ be the convex cone with the vertex at A_1 spanned by the vectors $A_1 - A_2, \ldots, A_1 - A_{m+1}$ (similarly $\mathscr{C}_{A_2}, \ldots, \mathscr{C}_{A_{m+1}}$); that is,

$$\mathscr{C}_{A_1} \left\{ A_1 \sum_{k=2}^{m+1} p_k (A_1 - A_k) : p_k \in \mathbb{R}, p_k \geq 0 \right\}.$$

Given the function $f : \mathbb{R}^m \to \mathbb{R}$ let $f_\Delta^{hp} : \mathbb{R}^m \to \mathbb{R}$ be the function of the hyperplane $\left(\text{in } \mathbb{R}^{m+1}\right)$ passing through the points $\left(A_j, f\left(A_j\right)\right)$ of the graph of f. Applying the affinity of f_Δ^{hp} to the combination in $P = \sum_{j=1}^{m+1} \alpha_j A_j$, it follows

$$f_\Delta^{hp}(P) \sum_{j=1}^{m+1} \alpha_j f\left(A_j\right).$$

If we use the convex function f then we get the inequality

$$f(P) \leq f_{\Delta}^{hp}(P), \qquad \text{if } P \in \Delta,$$

and the reverse inequality

$$f(P) \leq f_{\Delta}^{hp}(P), \qquad \text{if } P \in \bigcup_{j=1}^{m+1} \mathscr{C}_{Aj}$$

which can be proved in the same way as the inequality in $f(P) \geq f_{\{A,B,C\}}^{plane}(P), \qquad \text{if } P \in \mathscr{C}_A \cup \mathscr{C}_B \cup \mathscr{C}_C.$

By the end of the section we will use a convex set $\mathscr{C} \subseteq \mathbb{R}^m$ with the nonempty interior $\mathscr{C}^0$ The volume of a set $\mathscr{V} \subset \mathbb{R}^m$ will be denoted with $Vol(\mathscr{V})$.

The generalization of Theorem above applied to m-simplexes is as follows.

Theorem: Let $A_1, ..., A_{m+1}$; $B_1, ..., B_{m+1} \in \mathscr{C} \subseteq \mathbb{R}^m$ be points such that

$$\Delta = \text{conv}\{A_1, ..., A_{m+1}\} \subset \text{conv}\{B_1, ..., B_{m+1}\} = \Delta_1$$

With $\Delta^0 \neq \varnothing$, and let $\mathscr{C}_{\Delta} = \bigcup_{j=1}^{m+1} \mathscr{C}_{Aj}$ be the cone union. Let $P = \sum_{i=1}^{n} p_i P_i$ be a convex combination of the points $P_i \in \mathscr{C}_{\Delta} \cap \Delta_1$ with the center $P \in \Delta^0$.

Then there exist two $(m+1)$-membered convex combinations $\sum_{j=1}^{m+1} \alpha_j A_j$ and $\sum_{j=1}^{m+1} \beta_j B_j$ so that

$$\sum_{j=1}^{m+1} \alpha_j A_j = \sum_{i=1}^{n} p_i P_i = \sum_{j=1}^{m+1} \beta_j B_j,$$

and consequently every convex function $f : \mathscr{C} \to \mathbb{R}$ satisfies the inequality

$$\sum_{j=1}^{m+1} \alpha_j f(A_j) = \sum_{i=1}^{n} p_i f(P_i) = \sum_{j=1}^{m+1} \beta_j f(B_j).$$

Using the integrals, we get the form as follows.

Corollary: Let $A_1, ..., A_{m+1}$; $B_1, ..., B_{m+1} \in \mathscr{C} \subseteq \mathbb{R}^m$ be points such that

$$\Delta = \text{conv}\{A_1, ..., A_{m+1}\} \subset \text{conv}\{B_1, ..., B_{m+1}\} = \Delta_1$$

With $\Delta^0 \neq \varnothing$, and let $\mathscr{C}_{\Delta} = \bigcup_{j=1}^{m+1} \mathscr{C}_{Aj}$ be the cone union. Let the barycenter P of the set $P_i \in \mathscr{C}_{\Delta} \cap \Delta_1$ belong to Δ^0 that is,

$$P\left(\frac{\int_{\mathscr{C}_{\Delta} \cap \Delta_1} x_1 \, dx_1, ..., dx_m}{\text{vol}(\mathscr{C}_{\Delta} \cap \Delta_1)}, ..., \frac{\int_{\mathscr{C}_{\Delta} \cap \Delta_1} x_m \, dx_1, ..., dx_m}{\text{vol}(\mathscr{C}_{\Delta} \cap \Delta_1)}\right)$$

$\in \Delta^0$.

Then there exist two $(m+1)$-membered convex combinations $\sum_{j=1}^{m+1} \alpha_j f(A_j)$ and $\sum_{j=1}^{m+1} \beta_j f(B_j)$ so that

$$\sum_{j=1}^{m+1}\alpha_j A_j = P = \sum_{j=1}^{m+1}\beta_j B_j,$$

and consequently every convex function $f:\mathscr{C}\to\mathbb{R}$ satisfies the inequality

$$\sum_{j=1}^{m+1}\alpha_j\, f(A_j) \le \frac{\int_{C_\Delta\cap\Delta_1} f(x_1,\dots,x_m)\,dx_1,\dots,dx_m}{\mathrm{vol}(C_\Delta\cap\Delta_1)} \le \sum_{j=1}^{m+1}\beta_j\, f(B_j).$$

If f is concave, then the reverse inequality is valid.

If the conditions of theorem are valid, then using the summarizing Jensen functional of f convex combination $\sum_{i=1}^{n} p_i P_i$, we get the functional inequality

$$J_{\alpha_1 A_1+\dots+\alpha_{m+1}A_{m+1}}(f) \le J_{p_1 P_1+\dots+p_n P_n}(f) \le J_{\beta_1 B_1+\dots+\beta_{m+1}B_{m+1}}(f).$$

If the conditions of Corollary above are satisfied, then applying another integrating Jensen functional of f for the given subset $\mathscr{V}\subseteq\mathscr{C}$ with the volume $\mathrm{vol}(\mathscr{V})>0$,

$$J_{\mathscr{V}}(f) = \frac{\int_{\mathscr{V}} f\,dx_1,\dots,dx_m}{\mathrm{vol}(\mathscr{V})} - f\left(\frac{\int_{\mathscr{V}} x_1\,dx_1,\dots,dx_m}{\mathrm{vol}(\mathscr{V})},\dots,\frac{\int_{\mathscr{V}} x_m\,dx_1,\dots,dx_m}{\mathrm{vol}(\mathscr{V})}\right),$$

we have the inequality

$$J_{\alpha_1 A_1+\dots+\alpha_{m+1}A_{m+1}}(f) \le J_{\mathscr{C}_\Delta\cap\Delta_1}(f) \le J_{\beta_1 B_1+\dots+\beta_{m+1}B_{m+1}}(f).$$

Busemann–Petty Problem

First problem, now known as the Busemann-Petty problem, states:

If K and L are origin-symmetric convex bodies in $\mathbb{R}^n$, and for each hyperplane H through the origin the volumes of their central slices satisfy

$$\mathrm{vol}_{n-1}(K\cap H) < \mathrm{vol}_{n-1}(L\cap H)$$

does it follow that the volumes of the bodies themselves satisfy

$$\text{vol}_n(K) < \text{vol}_n(L)?$$

The problem is trivially positive in $\mathbb{R}^2$. However, a surprising negative answer for $n \geq 12$ was given by Larman and Rogers [LR] in 1975. Subsequently, a series of contributions were made to reduce the dimensions to $n \geq 5$ by a number of authors. That is, the problem has a negative answer for $n \geq 5$ for a detailed description. It was proved by Gardner that the problem has a positive answer for $n = 3$. The case of $n = 4$ was considered. But the answer to this case is not correct.

A key step to the solution of the Busemann-Petty problem is the discovery of the relation of the problem to intersection bodies by Lutwak. An origin symmetric convex body K in R n is called an intersection body if its radial function ρK is the spherical Radon transform of a nonnegative measure μ on the unit sphere S n−1. The value of the radial function of K, ρK(u), in the direction u ∈ S n−1, is defined as the distance from the center of K to its boundary in that direction. When μ is a positive continuous function, K is called the intersection body of a star body. The notion of intersection body was introduced by Lutwak [Lu] who proved that the Busemann-Petty problem has a positive answer if K is an intersection body in $\mathbb{R}^n$. Based on this relation, a positive answer to the Busemann-Petty problem in $\mathbb{R}^3$ was given by Gardner who showed that all origin-symmetric convex bodies in $\mathbb{R}^3$ are intersection bodies.

Carathéodory's Theorem

The following theorem belongs to the foundations of the theory of convex sets. It was first proved by Constantin Carathéodory in 1907. The convex hull conv H was defined as the collection of convex combinations of the elements from H but we have no any restriction on the number of points of H required to make the combination. Carathéodory's theorem gives a precise answer to the question how to generate the convex hull without unnecessarily repetitions: the number of points of H in the convex combination never has to be more than n+1.

Theorem (Gauss-Lucas Theorem): Let $f(z)$ be a n-th degree polynomial with real coefficients. By fundamental theorem of algebra there are n-roots of the polynomial (they might be complex). Call the roots $\{z_1 = x_1 + iy_1, z_2 = x_2 + iy_2, ..., z_n = x_n + iy_n\}$ be the set of all the roots. Observe that each root can be represented as a point in the plane $\mathbb{R}^2$. Then the roots of the $n-1$ degree polynomial $f'(z) = \dfrac{df(z)}{dz}$ lies in the convex hull $conv(z_1, z_2, ..., z_n)$.

Proof: Without loss of generality the polynomial can be represented as

$$f(z) = (z - z_1)(z - z_2)....(z - z_n).$$

So we have

$$(z) = \sum_{k=1}^{n} \prod_{i \neq k} (z - z_i).$$

Let w be a root of $f'(z)$. Then we have $f'(w)=0$ and hence

$$f'(w)=\sum_{k=1}^{n}\prod_{i\neq k}(w-z_i)=0$$

Multiplying both sides with $\prod_{i\neq k}(\overline{w}-\overline{z}_i)$, we obatin

$$\sum_{k=1}^{n}\frac{1}{(w-z_i)}\prod_{i=1}^{n}|w-z_i|^2=0$$

which implies

$$\prod_{i=1}^{n}|w-z_i|^2\sum_{k=1}^{n}\frac{1}{(w-z_i)}=0.$$

If $\prod_{i=1}^{n}|w-z_i|^2=0$ then $w=z_i$ for some i and hence the theorem is proved. Else if

$$\sum_{k=1}^{n}\frac{1}{(w-z_k)}=0.$$

which equals

$$\sum_{k=1}^{n}\frac{\overline{w}-\overline{z_k}}{|w-z_k|^2}=0,$$

which implies

$$\overline{w}=\sum_{k=1}^{n}\frac{\dfrac{\overline{z_k}}{|w-z_k|^2}}{\displaystyle\sum_{i=1}^{n}\frac{\overline{z_k}}{|w-z_i|^2}}.$$

Taking the conjugate we get the result $w=\sum_{i=1}^{n}z_i\alpha_i$ where $\alpha_i=\dfrac{\dfrac{1}{|w-z_k|^2}}{\displaystyle\sum_{i=1}^{n}\frac{\overline{z_k}}{|w-z_i|^2}}$.

We will now look at one more theorem that looks at convex combinations.

Theorem (Caratheodory's theorem): Let $S\subset\mathbb{R}^d$. Then any point $x\in conv(S)$ can be represented as a convex combination of $d+1$ points of S.

Some observations before the proof:

1. The $(d+1)$ points depend on x in consideration.

2. $d+1$ points are necessary. Let S be three non-collinear points. Let x be the the centroid of the triangle formed by these three points. Observe that the convex combination of the three points is necessary to obtain x.

3. Observe that the $d+1$ points are from S and not conv(S).

Proof: Any point $y \in conv(S)$ can be represented as a convex combination of some m points in S. So

$$y = \sum_{i=1}^{m} \alpha_i x_i,$$

where $x_i \in S$ and $\sum_{i=1}^{m} \alpha_j = 1$. If $m \leq d+1$ we are done. Hence we will only consider the case of $m > d+1$. Let $x_i = (x_i^1, x_i^2, ..., x_i^d)$. Now consider the set of equations $\sum_{i=1}^{m} \gamma_i x_i = 0$ and $\sum_{i=1}^{m} \gamma_i = 0$. This system of equations can be put in matrix form as

$$\begin{bmatrix} x_1^1 & x_2^1 & \cdots & x_m^d \\ x_1^2 & x_2^2 & \cdots & x_m^d \\ \vdots & \vdots & \vdots & \vdots \\ x_1^d & x_2^d & \cdots & x_m^d \\ 1 & 1 & \cdots & 1 \end{bmatrix} \begin{bmatrix} \gamma_1 \\ \gamma_2 \\ \vdots \\ \gamma_m \end{bmatrix} = o_{d+1 \times 1}$$

Since the matrix is of size $(d+1) \times m$, and $d+1 < m,$, there are non-trivial solutions of this equation (look at rank-nullity theorem to observe that $\dim(\mathcal{N}(A)) \geq m-(d+1) > 0$). From now on Let $\gamma_1, ..., \gamma_m$ be one such solution. Now let $\tau > 0$. Since $\sum_{i=1}^{m} \gamma_i = 0.$, the point y can be represented as

$$y = \sum_{i=1}^{m} \alpha_i x_i - \tau \sum_{i=1}^{m} \gamma_i x_i = \sum_{i=1}^{m} (\alpha_i - \tau\gamma_i) x_i$$

Also observe that $\sum_{i=1}^{m} (\alpha_i - \tau\gamma_i) = 1$ since $\sum_{i=1}^{m} \gamma_i = 0$. Now we will choose τ such that $\alpha_i - \tau\gamma_i \geq 0$ and equals 0 for one term so that $\sum_{i=1}^{m} (\alpha_i - \tau\gamma_i) x_i$ becomes a convex combination. Since $\sum \gamma_i = 0$ implies some γ_i's are positive and others are negative. If for some i, $\gamma_i < 0$, then $\alpha - \tau\gamma_i \geq 0$. So we need to consider only the terms with $\gamma_i > 0$. For all i such that $\gamma_i > 0$, we want

$$\alpha_i - \tau\gamma_i \geq 0,$$

which implies

$$\frac{\alpha_i}{\gamma_i} - \tau \geq 0.$$

Hence if we choose $\tau = \min\{\frac{\alpha_i}{\gamma_i}, \gamma_i > 0\}$, then one term $\alpha_i - \tau\gamma_i = 0$ and every other term will be positive. So we are able to express

$$\sum_{i=1}^{m}(\alpha_i - \tau\gamma_i)x_i ,$$

where one of the coefficients is zero, \i.e., as a convex combination of $m-1$ points. We can continue this procedure till $m \le d+1$.

We will now see a simple application of Caratheodory's theorem.

Lemma: If a set $S \subset \mathbb{R}^d$ is compact, then its convex hull $conv(S)$ is also compact.

Proof: Consider a sequence $\{x_n\} \in conv(S)$. If we show that there is a convergent subsequence which converges to a point in $conv(S)$ then we are done.

By Caratheodory's theorem, each $x_i \in conv(S)$ can be represented as a convex sum of $d+1$ points in S. So

$$x_i = \sum_{k=1}^{d+1} \alpha_k^i y_k^i, \quad \sum_{k=1}^{d+1} \alpha_k^i = 1 \text{ and } y_k^i \in S$$

Now consider the following matrix of dimensions $\infty \times 2(d+1)$

$$\begin{bmatrix} y_1^1 & \alpha_1^1 & y_2^1 & \alpha_2^1 & \dots & x_{d+1}^1 & \alpha_{d+1}^1 \\ y_1^2 & \alpha_1^2 & y_2^2 & \alpha_2^2 & \dots & x_{d+1}^2 & \alpha_{d+1}^2 \\ y_1^3 & \alpha_1^3 & y_2^3 & \alpha_2^3 & \dots & x_{d+1}^3 & \alpha_{d+1}^3 \\ \vdots & \vdots & \vdots & \vdots & \vdots & \vdots & \vdots \\ y_1^n & \alpha_1^n & y_2^n & \alpha_2^n & \dots & x_{d+1}^n & \alpha_{d+1}^n \\ \vdots & \vdots & \vdots & \vdots & \vdots & \vdots & \vdots \end{bmatrix}$$

Now consider the first column $\{y_k^1\}$. Since this sequence lies in S and S is compact, there is a sub-sequence in that that converges to a point $y^1 \in S$. The indices of this subsequence are denoted by $I_1 \subset \{1, 2, 3, \dots\}$. Now consider the second column, but consider only rows corresponding to the index set I_1. Since $\alpha_1^k \in [0,1]$ a compact set, there exists a subsequene of $\{\alpha_1^k\}_{k \in I_1}$ that converges to some point α_1 in $[0,1]$. Denote the indices corresponding to this subsequence by the set $I_2 \subset I_1$. Now consider the third column corresponding to y_2^k but only corresponding to the rows I_2. Since $\{y_2^k\} \subset S$ and S is compact, we can find a subsequence which converges to $y_2 \in S$. Denote the set of the indices corresponding to this sub sequence by $I_3 \subset I_2$. Continue this till all the $2(d+1)$ are exhausted. So we will finally have an index set $I_{2(d+1)}$ corresponding to the last column.

Now consider the original sequence $\{x_n\}$. It is easy to see that, the subsequence corresponding to the index set $I_{2(d+1)}$ converges to a point

$$\sum_{i=1}^{d+1} \alpha_i y_i .$$

which belongs to the set $conv(S)$ proving the result.

References

- Bachem, Achim; Kern, Walter. Linear Programming Duality: An Introduction to Oriented Matroids. Springer Science & Business Media. p. 13. ISBN 9783642581526. Retrieved 5 April 2017
- Meyer, Robert (1970), "The validity of a family of optimization methods", SIAM Journal on Control and Optimization, 8: 41–54, MR 0312915
- Convexity: encyclopediaofmath.org, Retrieved 18 March 2018
- Kjeldsen, Tinne Hoff. "History of Convexity and Mathematical Programming" (PDF). Proceedings of the International Congress of Mathematicians (ICM 2010): 3233–3257. doi:10.1142/9789814324359_0187. Retrieved 5 April 2017
- Convex-Body: encyclopedia2.thefreedictionary.com, Retrieved 17 May 2018
- Morris, Carla C.; Stark, Robert M. Finite Mathematics: Models and Applications. John Wiley & Sons. p. 121. ISBN 9781119015383. Retrieved 5 April 2017
- Caratheodorys-theorem: wordpress.com, Retrieved 28 June 2018

6

Discrete Geometry

Discrete geometry is a branch of geometry that studies discrete geometrical objects, such as points, lines, circles, spheres, planes, etc. In order to completely understand discrete geometry, it is necessary to understand the fundamentals of polyhedral combinatorics, structural rigidity, K-set, convex lattice polytope, Helly's theorem, McMullen problem, etc. which have been extensively discussed in this chapter.

Discrete geometry is concerned with properties of finitely generated geometric objects such as polytopes and polyhedra, triangulations and polyhedral complexes, configurations of lines and, more generally, hyperplanes in Euclidean and other spaces, the theory of rigid and flexible frameworks, tilings and packings. Many problems in discrete geometry arise from questions in computational geometry related to algorithms for analyzing discrete geometric structures.

Combinatorics is the study of finite structures, many of which arise in other branches of mathematics or from problems arising in science or engineering. The study of combinatorics involves general questions of enumeration and structure, matroid theory and aspects of graph theory, partially ordered sets, set partitions and permutations and combinatorial structures such as finite geometries and designs. Techniques tend to be algebraic and topological, involving methods from commutative ring theory, algebraic topology, representation theory and Hopf algebras.

Historically, there have been connections between combinatorics, in particular enumeration theory, and questions in probability. In recent decades, there have been close connections between certain areas of combinatorics and questions arising in theoretical computer science and discrete optimization. Even more recently, there have arisen links to biology, in particular, the study of phylogenetics.

Polyhedral Combinatorics

Polyhedral combinatorics can be viewed as a technique within combinatorics and discrete geometry that uses methods from polyhedral theory in order to solve combinatorial problems and also describes the faces of convex polyhedra and higher-dimensional convex polytopes. The main idea is to transform a combinatorial problem into a polyhedral problem by, for instance, considering the convex hull of the incidence vectors of the feasible solutions of the combinatorial problem, and to employ techniques from linear and integer programming in order to solve the combinatorial problem.

Faces of Polyhedra

An inequality $\alpha x \leq \beta$, where $\alpha \neq 0$, is a valid inequality for a polyhedron $P = \{x \mid Ax \leq b\}$ *if* $\alpha x \leq \beta \ \forall x \in P$.

The inequality is a supporting hyperplane if it is valid and has a non-empty intersection with P.

A face of a polyhedron P is the intersection of P with $\{x \mid \alpha x = \beta\}$ where $\alpha x \leq \beta$ is a valid inequality for P.

We are interested in non-empty faces. Notice that a face of a polyhedron is also a polyhedron. A face of P is an extreme point or a vertex if it has dimension 0. It is a facet if the dimension of the face is dim $(P) - 1$. The face is an edge if it has dimension 1.

Another way to define a face is to say that F is a face of P if $F = \{x \in P \mid A'x = b'\}$ where $A'x \leq b'$ is a subset of the inequalities of $Ax \leq b$. In other words $F = \{x \in P \mid a_i x = b_i, i \in I\}$ where, $I \subseteq \{1, 2, \ldots, m\}$ is a subset of the rows of A.

Now we will show that these two definitions are equivalent .

Theorem 1: Let, $A \in \mathbb{R}^{m \times n}$, $b \in \mathbb{R}^n$. Let $P = \{x \mid Ax \leq b\}$ be a polyhedron. Let F be a face defined by a valid inequality $\alpha x \leq \beta$. Then $\exists I \subseteq \{1.2. \ldots, m\}$ such that, $F = \{x \in P \mid a_i x = b_i, i \in I\}$.

Proof: Let $F = \{x \mid x \in P, \alpha x = \beta\}$ where $\alpha x \leq \beta$ is a supporting hyperplane. Then, the following claim is easy to see.

Claim: F is the set of all optimal solutions to the LP

$$\max \alpha x$$

$$Ax \leq b.$$

The above LP has an optimal value β. This implies that the dual LP is feasible and has an optimum solution y^*. *Let* $I = \{i \mid y_i^* > 0\}$. Let X be the set of all optimal solutions to the primal. For any $x' \in X$, by complimentary slackness for x' *and* y^* we have that $y_i^* > 0$ implies, $a_i x' = b_i$. Therefore X is a subset of the solutions to the following system of inequalities:

$$a_i x = b_i \qquad i \in I$$

$$a_i x \leq b_i \qquad i \notin I$$

Again, by complementary slackness any x' that satisfies the above is optimal (via y^*) for the primal LP and. Therefore $F = X = \{x \in P \mid a_i xl = b_i, i \in I\}$.

Theorem: Let $P = \{x \mid Ax \leq b\}$ where $A \in \mathbb{R}^{m \times n}$ and $b \in \mathbb{R}^m$ *Let* $I \subseteq \{1, \ldots, m\}$ and $F = \{x \in P \mid a_i x = b_i, i \in I\}$. If F is non-empty, then there is a valid inequality $\alpha x \leq \beta$ such that $F = P \cap \{x \mid \alpha x = \beta\}$.

Proof: Let, $\alpha = \sum_{i \in I} a_i$ be a row vector and $\beta = max\{\alpha x \mid Ax \leq b\}$.

We claim that,

$$F = \{x \mid x \in P, \alpha x = \beta\}$$

Which, implies that F is the intersection of P with the supporting hyperplane $\alpha x \leq \beta$

Corollary:

- The number of faces of $P = \{x \mid Ax \leq b\}$ where A is a m × n matrix is at most 2^m.
- Each face is a polyhedron.
- If F is a face of P and $F' \subseteq F$ then F' is a face of P if and only if F' is a face of F.
- The intersecton of two faces is either a face or is empty.

Facets

A facet of P is an inclusion-wise maximal face distinct from P. Equivalently, a face F of P is a facet if and only if, $dim(F) = dim(P) - 1$.

Theorem: Let $P = \{x \mid Ax \leq b\} = \{x \mid A^{=}x = b^{=}, A^{+}x \leq b^{+}\}$. If no inequality of $A^{+}x \leq b^{+}$ is redundant in $Ax \leq b$, then there is a one to one correspondence between the facets of P and the inequalities in $A^{+}x \leq b^{+}$. That is, F is a facet of P if and only if $F = \{x \in P \mid a_i x = b_i\}$ for some inequality $a_i x \leq b_i$ from $A^{+}x \leq b^{+}$.

Proof: Let F be a facet of P. Then $F = \{x \in P \mid A'x = b'\}$ where $A'x \leq b'$ is a subsystem of $A^{+}x \leq b^{+}$. Take some inequality $\alpha x \leq \beta$ in $A'x \leq b'$. Then $F' = \{x \in P \mid \alpha x = \beta\}$ is a face of P and $F \subseteq F'$. Moreover, $F' \neq P$ P since no inequality in $A^{+}x \leq b^{+}$ is an implicit equality.

Let $F = \{x \in P \mid \alpha x = \beta\}$ for some inequality $\alpha x \leq \beta$ from $A^{+}x \leq b^{+}$. We claim that $dim(F) = dim(P) - 1$ which implies that F is a facet. To prove the claim it is sufficient to show that there is a point $x_0 \in P$ such that $A^{=}x_0 = b^{=}$, $\alpha x_0 = \beta$ and $A'x_0 < b'$ where $A' \leq b'$ is the inequalities in $A^{+}x \leq b^{+}$ with $\alpha x \leq \beta$ omitted. From Exercise 1, there is a point x_1 such that $\alpha x_1 = \beta$ and $A^{=}x_1 = b^{=}$ and $A^{+}x_1 \leq b^{+}$. Moreover since $\alpha x \leq \beta$ is irredundat in $Ax \leq b$, there is a point x_2 such that $A^{=}x_2 = b^{=}$ and $A'x_2 \leq b'$ and $\alpha x_2 \leq \beta$ A convex combination of x_1 and x_2 implies the existence of the desired x_0.

Corollary: Each face of P is the intersection of some of the facets of P.

A polyhedrom P has no facet if and only if P is an affine subspace.

Prove the above two corollary using theorem.

Minimal Faces and Vertices

A face is inclusion-wise minimal if it does not contain any other face. From Corollary and the fact that a face of a polyhedron is a polyhedron the next proposition follows.

Proposition: A face F of P is minimal if and only if F is an affine subspace.

Theorem: A set F is minimal face of P if and only if $\varnothing \neq F$, $F \subseteq P$ and $F = \{x \mid A'x = b'\}$ for some subsystem $A'x \leq b'$ and $Ax \leq b$.

Proof: Suppose F is a face and $F = \{x \mid A'x = b'\}$ then by Proposition, it is minimal. For the converse direction suppose F is a minimal face of P. Since F is a face, $F = \{x \mid A''x \leq b'', A'x = b'\}$ where $A''x \leq b''$ and $A'x \leq b'$ are two subsystems of $Ax \leq b$. We can assume that $A''x \leq b''$ is as small as possible and therefore, irredundant. From theorem, if $A''x \leq b''$ has any inequality then F has a facet which implies that F is not minimal. Therefore, $F = \{x \mid A'x = b'\}$.

Prove that all minimal faces of a polyhedron $\{x \mid Ax \leq b\}$ are of the form $\{x \mid A'x = b'\}$ for some subsystem $A'x \leq b'$ of $Ax \leq b$ where, rank $(A') = rank(A)$. Conclude that all minimal faces are translates of linespace (P) and have the same dimension.

A vertex or an extreme point of P is a (minimal) face of dimension 0. That is, a single point. A polyhedron is pointed if and only if it has a vertex. Note that since all minimal faces have the same dimension, if P has a vertex than all minimal faces are vertices. Since a minimal face F of P is defined by $A'x = b'$ for some subsystem $A'x \leq b'$ of $Ax \leq b$ if a vertex of P is the unique solution to $A'x = b'$ then rank $(A') = n$. We can then assume that A' has n rows. Vertices are also called basic feasible solutions.

Corollary: A polyhedron $\{x \mid Ax \leq b\}$ has a vertex only if A has rank n. The polyhedron $\{x \mid Ax \leq b\}$ is pointed if it is not empty.

Decomposition of Polyhedra

Theorem: Any polyhedron P can be written as $Q + C$ where Q is a convex hull of a finites set of vectors and $C = \{x \mid Ax \leq 0\}$ is the charcone of P.

We can give more details of the decomposition now. Given P, let $F_1, F_2, \ldots, F_h$ be its minimal faces. Choose $x_i \in F_i$ arbitrarily. Then, Püüüüüüüüüüü(§x_1 x_2 x_h) $+ C$.

In particular, if P is pointed then $x_1, x_2, \ldots, x_h$ are vertices of P and hence, $P = \text{convexhull}(vertices(P)) + C$.

We will prove the above for polytopes.

Theorem: A polytope (bounded polyhedron) is the convex hull of its vertices

Proof: First observe that a bounded polyhedron is necessarily pointed.

Let, $X = \{x_1, x_2, \ldots, x_h\}$ be the vertices of P. Clearly $\text{convexhull}(X) \subseteq P$ We prove the converse. Suppose $x^* \in P$ does no belong to convexhull(X) .

Claim: There exists a hyperplane $\alpha x = \beta$ such that $\alpha x_i < \beta \; \forall x_i \in X \; and \; \alpha x^* \geq \beta$.

Proof: One can prove this by using Farkas lemma or appeal to the general theorem that if two convex sets do not intersect then there is a separating hyperplane for them; in particular if one of the sets is bounded then there is a strict separating hyperplane. A sketch of this for the restricted case we have is as follows. Let $y \in convexhull(X)$ minimize the distance from x^* to $\text{convexhull}(X)$. The claim is that a hyperplane that is normal to the line segment joining x^* and y and passing through is the desired hyperplane. Otherwise $\text{convexhull}(X)$ intersects this hyperplane, and let y' be a point in the intersection. Since $\text{convexhull}(X)$ is convex, the line segment joining y' and y is contained in $\text{convexhull}(X)$. Now consider the right angled triangle formed by y, x^*, y'. From elementary geometry, it follows that there is a point closer to x^* than y on the line segment joining y and y', contradicting the choice of y.

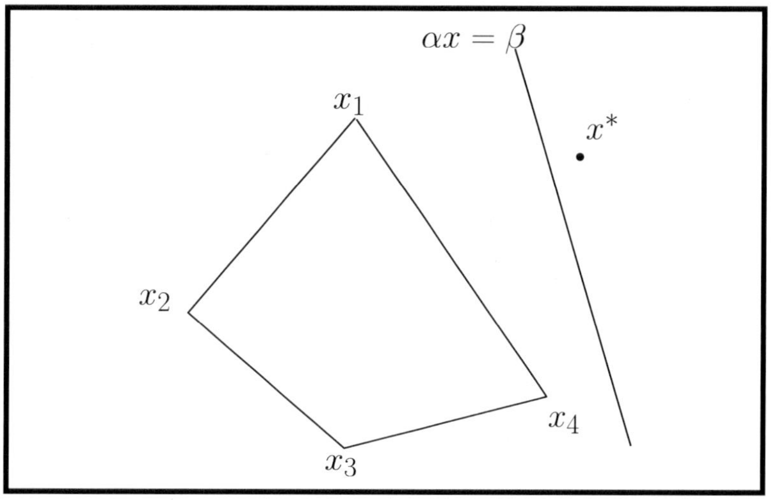

Now consider, $\max\{\alpha x \mid x \in P\}$. The set of optimal solutions to this LP is a face of F. By *Claim*, $X \cap F = \varnothing$. Since F is a face of P, it has a vertex of P since P is pointed. This contradicts that X is the set of all vertices of P.

Theorem: If $P = \{x \mid Ax \leq b\}$ is pointed then for any $c \neq 0$ the LP

$$\max\ cx$$

$$Ax \leq b$$

is either unbounded, or there is a vertex x^* such that x^* is an optimal solution.

Equalities and Inequalities

The most important relation among the coefficients of the *f*-vector of a polytope is Euler's formula $\Sigma(-1)^i f_i = 0$, where the terms of the sum range over the coefficients of the extended *f*-vector. In three dimensions, moving the two 1's at the left and right ends of the extended *f*-vector $(1, v, e, f, 1)$ to the right hand side of the equation transforms this identity into the more familiar form $v - e + f = 2$. From the fact that each facet of a three-dimensional polyhedron has at least three edges, it follows by double counting that $2e \geq 3f$, and using this inequality to eliminate *e* and *f* from Euler's formula leads to the further inequalities $e \leq 3v - 6$ and $f \leq 2v - 4$. By duality, $e \leq 3f - 6$ and $v \leq 2f - 4$. It follows from Steinitz's theorem that any 3-dimensional integer vector satisfying these equalities and inequalities is the *f*-vector of a convex polyhedron.

In higher dimensions, other relations among the numbers of faces of a polytope become important as well, including the Dehn–Sommerville equations which, expressed in terms of *h*-vectors of simplicial polytopes, take the simple form $h_k = h_{d-k}$ for all *k*. The instance of these equations with $k = 0$ is equivalent to Euler's formula but for $d > 3$ the other instances of these equations are linearly independent of each other and constrain the *h*-vectors (and therefore also the *f*-vectors) in additional ways.

Another important inequality on polytope face counts is given by the upper bound theorem, first proven by McMullen (1970), which states that a d-dimensional polytope with n vertices can have at most as many faces of any other dimension as a neighborly polytope with the same number of vertices:

$$f_{k-1} \leq \sum_{i=0}^{d/2}{}^{*} \left(\binom{d-i}{k-i} + \binom{i}{k-d+i} \right) \binom{n-d-1+i}{i},$$

Where, the asterisk means that the final term of the sum should be halved when d is even. Asymptotically, this implies that there are at most $o(n^{\lfloor d/2 \rfloor})$ faces of all dimensions.

Even in four dimensions, the set of possible f-vectors of convex polytopes does not form a convex subset of the four-dimensional integer lattice, and much remains unknown about the possible values of these vectors.

Graph-theoretic Properties

Along with investigating the numbers of faces of polytopes, researchers have studied other combinatorial properties of them, such as descriptions of the graphsobtained from the vertices and edges of polytopes.

Balinski's theorem states that the graph obtained in this way from any d-dimensional convex polytope is d-vertex-connected. In the case of three-dimensional polyhedra, this property and planarity may be used to exactly characterize the graphs of polyhedra: Steinitz's theorem states that G is the skeleton of a three-dimensional polyhedron if and only if G is a 3-vertex-connected planar graph.

A theorem (Blind & Mani-Levitska)states that one can reconstruct the face structure of a simple polytope from its graph. That is, if a given undirected graph is the skeleton of a simple polytope, there is only one polytope (up to combinatorial equivalence) for which this is true. This is in sharp contrast with (non-simple) neighborly polytopes whose graph is a complete graph; there can be many different neighborly polytopes for the same graph. Another proof of this theorem based on unique sink orientations was given by Kalai (1988), and Friedman (2009) showed how to use this theorem to derive a polynomial time algorithm for reconstructing the face lattices of simple polytopes from their graphs. However, testing whether a given graph or lattice can be realized as the face lattice of a simple polytope is equivalent (by polarity) to realization of simplicial polytopes, which was shown to be complete for the existential theory of the reals by Adiprasito & Padrol.

In the context of the simplex method for linear programming, it is important to understand the diameter of a polytope, the minimum number of edges needed to reach any vertex by a path from any other vertex. The system of linear inequalities of a linear program define facets of a polytope representing all feasible solutions to the program, and the simplex method finds the optimal solution by following a path in this polytope. Thus, the diameter provides a lower bound on the number of steps this method requires. The Hirsch conjecture, now disproved, suggested a strong bound on how large the diameter could be. Weaker (quasi-polynomial) upper bounds on the diameter are known, as well as proofs of the Hirsch conjecture for special classes of polytope.

Computational Properties

Deciding whether the number of vertices of a given polytope is bounded by some natural number k is a computationally difficult problem and complete for the complexity class.

Facets of 0-1 Polytopes

It is important in the context of cutting-plane methods for integer programming to be able to describe accurately the facets of polytopes that have vertices corresponding to the solutions of combinatorial optimization problems. Often, these problems have solutions that can be described by binary vectors, and the corresponding polytopes have vertex coordinates that are all zero or one.

As an example, consider the Birkhoff polytope, the set of $n \times n$ matrices that can be formed from convex combinations of permutation matrices. Equivalently, its vertices can be thought of as describing all perfect matchings in a complete bipartite graph, and a linear optimization problem on this polytope can be interpreted as a bipartite minimum weight perfect matching problem. The *Birkhoff–von Neumann theorem* states that this polytope can be described by two types of linear inequality or equality. First, for each matrix cell, there is a constraint that this cell has a non-negative value. And second, for each row or column of the matrix, there is a constraint that the sum of the cells in that row or column equal one. The row and column constraints define a linear subspace of dimension $n^2 - 2n + 1$ in which the Birkhoff polytope lies and the non-negativity constraints define facets of the Birkhoff polytope within that subspace.

However, the Birkhoff polytope is unusual in that a complete description of its facets is available. For many other 0-1 polytopes, there are exponentially many or superexponentially many facets, and only partial descriptions of their facets are available.

Convex Polytope

Convex polytopes are convex hulls of finite point sets in R^d. They constitute the most important class of convex sets with an enormous number of applications and connections.

Three-dimensional convex polytopes, especially the regular ones, have been fascinating people since the antiquity. Their investigation was one of the main sources of the theory of planar graphs, and thanks to this well-developed theory they are quite well understood. But convex polytopes in dimension 4 and higher are considerably more challenging, and a surprisingly deep theory, mainly of algebraic nature, was developed in attempts to understand their structure.

A strong motivation for the study of convex polytopes comes from practically significant areas such as combinatorial optimization, linear programming, and computational geometry. Let us look at a simple example illustrating how polytopes can be associated with combinatorial objects. The 3-dimensional polytope in the picture is called the permutahedron. Although it is 3-dimensional, it is most naturally defined as a subset of R^4, namely, the convex hull of the 24 vectors obtained by permuting the coordinates of the vector $(1, 2, 3, 4)$ in all possible ways. In the picture, the (visible) vertices are labeled by the corresponding permutations. Similarly, the d-dimensional permutahedron is the convex hull of the $(d+1)$ vectors in R^{d+1} arising by permuting the coordinates of $(1, 2, \ldots, d+1)$. One can observe that the edges of the polytope connect exactly pairs of permutations differing by a transposition of two adjacent numbers, and a closer examination reveals other connections between the structure of the permutahedron and properties of permutations.

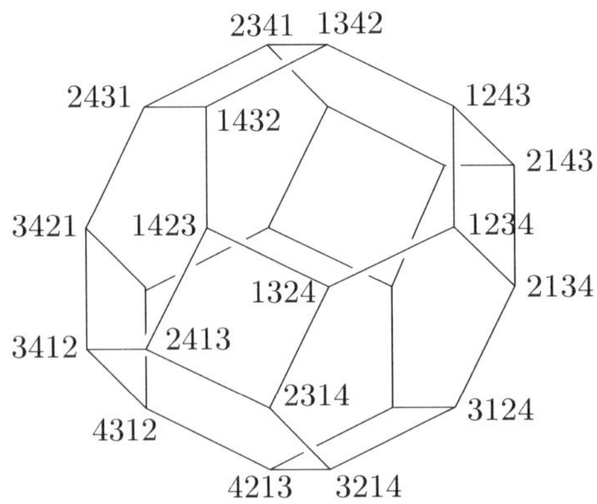

There are many other, more sophisticated, examples of convex polytopes assigned to combinatorial and geometric objects such as graphs, partially ordered sets, classes of metric spaces, or triangulations of a given point set. In many cases, such convex polytopes are a key tool for proving hard theorems about the original objects or for obtaining efficient algorithms.

Geometric Duality

First we discuss geometric duality, a simple technical tool indispensable in the study of convex polytopes and handy in many other situations. We begin with a simple motivating question.

How can we visualize the set of all lines intersecting a convex pentagon as in the picture?

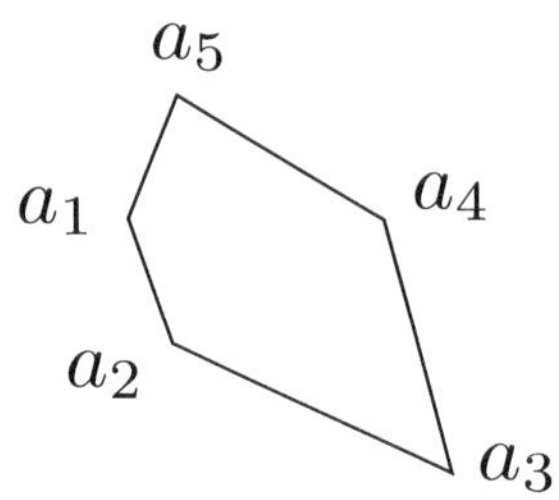

A suitable way is provided by line–point duality.

Duality Transform

The (geometric) duality transform is a mapping denoted by D_0. To a point $a \in R^d \setminus \{0\}$ it assigns the hyperplane,

$$D_0(a) = \{x \in R^d : \langle a, x\rangle = 1\}$$

And to a hyperplane h not passing through the origin, which can be uniquely written in the form $h = \{x \in R^d : \langle a, x\rangle\ 1\}$ it assigns the point $D_0(h) = a \in R^d \setminus \{0\}$.

Here is the geometric meaning of the duality transform. If a is a point at distance δ from 0, then $D_0(a)$ is the hyperplane perpendicular to the line $0a$ and intersecting that line at distance $\frac{1}{\delta}$ from o, in the direction from o towards a.

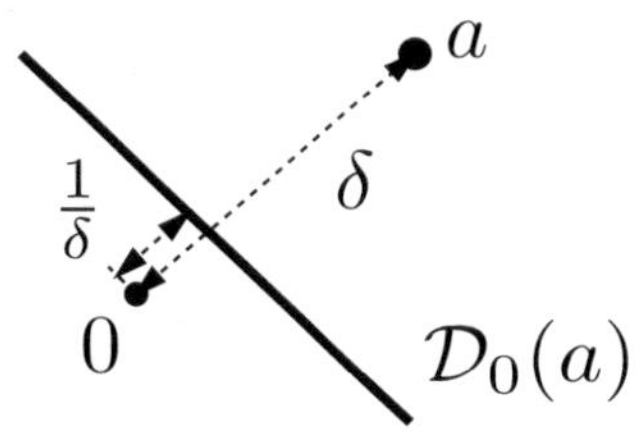

A nice interpretation of duality is obtained by working in R^{d+1} and identifying the "primal" R^d with the hyperplane $\pi = \{x \in R^{d+1} : x_{d+1} = 1\}$ and the "dual" R^d with the hyperplane. $\rho = \{x \in R^{d+1} : x_{d+1} = -1\}$. The hyperplane dual to a point $a \in \pi$ is produced as follows:

We construct the hyperplane in R^{d+1} perpendicular to $0a$ and containing 0, and we intersect it with ρ. Here is an illustration for $d = 2$,

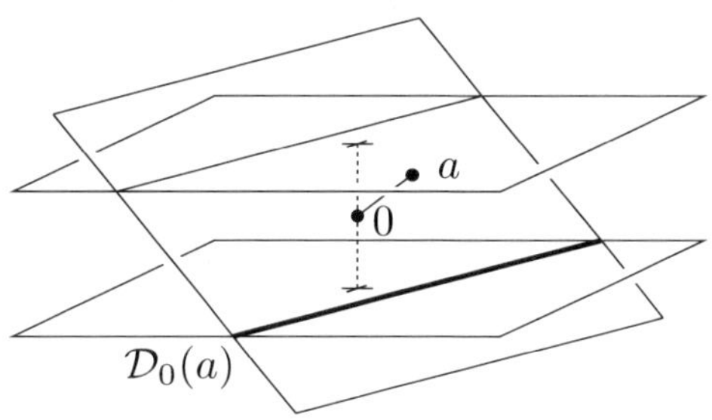

In this way, the duality D_0 can be naturally extended to k-flats in R^d , whose duals are $(d-k-1)-$ flats. Namely, given a $k-$ flat $f \subset \pi$, we consider the $(k+1)-$ flat F through 0 and f, we construct the orthogonal complement of F, and we intersect it with ρ, obtaining $D_0(f)$.

Let us consider, the pentagon drawn above and place it so that the origin lies in the interior. Let, $v_i = D_0(\ell_i)$, where ℓ_i is the line containing the side $a_i a_{i+1}$. Then the points dual to the lines intersecting the pentagon $a_1 a_2 \ldots a_5$ fill exactly the exterior of the convex pentagon $v_1 v_2 \ldots v_5$:

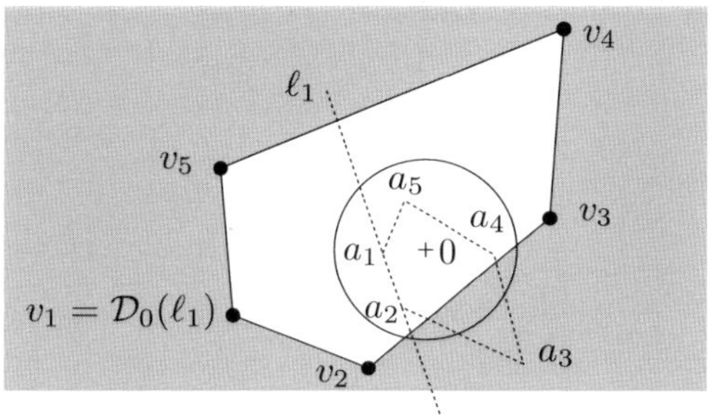

This follows easily from the properties of duality listed (of course, there is nothing special about a pentagon here). Thus, the considered set of lines can be nicely described in the dual plane. A similar passage from lines to points or back is useful in many geometric or computational problems.

Properties of the Duality Transform

Let, p be a point of R^d distinct from the origin and let h be a hyperplane in R^d not containing the origin. Let h^- stand for the closed half-space bounded by h and containing the origin, while h^+ denotes the other closed half-space bounded by h.

If $h = \{x \in R^d : \langle a, x\rangle = 1\}$, then $h^- = \{x \in R^d : \langle a, x\rangle \leq 1\}$,

Lemma (Duality Preserves Incidences)

- $p \in h$ *if and only if* $D_0(h) \in D_0(p)$.
- $p \in h^-$ *if and only if* $D_0(h) \in D_0(p)^-$.

Proof: Let, $h = \{x \in R^d : \langle a, x\rangle = 1\}$. Then $p \in h$ means $\langle a, x\rangle = 1$. Now, $D_0(h)$ is the point a, and $D_0(p)$ is the hyperplane $\{y \in R^d : \langle y, p\rangle = 1\}$, and hence $D_0(h) = a \in D_0(p)$ also means just $\langle a, x\rangle = 1$.

Part is proved similarly.

For a set $X \subseteq R^d$, we define the set dual to X, denoted by X^*, as follows:

$$X^* = \{y \in R^d : \langle x, y\rangle \leq 1 \textit{ for all } x \in X\}.$$

Another common name used for the duality is polarity; the dual set would then be called the polar set. Sometimes it is denoted by X°.

Geometrically, X^* is the intersection of all half-spaces of the form $D_0(x)^-$ with $x \in X$

Or in other words, X^* consists of the origin plus all points y such that, $X \subseteq D_0(y)^-$.

For example, if X is the pentagon $a_1a_2 \ldots a_5$ drawn above, then X^* is the pentagon $v_1v_2 \ldots v_5$.

For any set X, the set X^* is obviously closed and convex and contains the origin. Using the separation theorem, it is easily shown that for any set $X \subseteq R^d$, the set $(X^*)^*$ is the closure of conv($X \cup \{0\}$).

In particular, for a closed convex set containing the origin we have, $(X^*)^* = X$.

For a hyperplane h, the dual set h^* is different from the point $D_0(h)$.

For readers familiar with the duality of planar graphs, let us remark that it is closely related to the geometric duality applied to convex polytopes in R^3.

For example, the next drawing illustrates a planar graph and its dual graph (dashed):

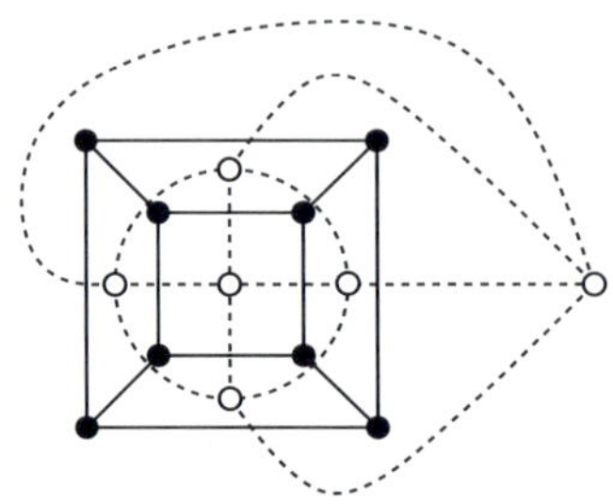

Later we will see that these are graphs of the 3-dimensional cube and of the regular octahedron, which are polytopes dual to each other in the sense defined above. A similar relation holds for all 3-dimensional polytopes and their graphs.

Other Variants of Duality

The duality transform D_0 defined above is just one of a class of geometric transforms with similar properties. For some purposes, other such transforms (dualities) are more convenient. A particularly important duality, denoted by D, corresponds to moving the origin to the "minus infinity" of the x_d-axis (the x_d-axis is considered vertical).

Another Duality

A formal definition is as follows:

A nonvertical hyperplane h can be uniquely written in the form:

$$h = \{x \in R^d : x_d = a_1x_1 + \cdots + a_{d-1}x_{d-1} - a_d\}.$$

We set $D(h) = (a_1, \ldots, a_{d-1}, a_d)$.

Conversely, the point $a = (a_1, \ldots, a_{d-1}, a_d)$ maps back to h.

A point p lies above a hyperplane h if and only if the point D(h) lies above the hyperplane D(p).

H-polytopes and V-polytopes

A convex polytope in the plane is a convex polygon. Famous examples of convex polytopes in R^3 are the Platonic solids: regular tetrahedron, cube, regular octahedron, regular dodecahedron, and regular icosahedron. A convex polytope in R^3 is a convex set bounded by finitely many convex polygons. Such a set can be regarded as a convex hull of a finite point set, or as an intersection of finitely many half-spaces. We thus define two types of convex polytopes, based on these two views:

- An *H*-polyhedron is an intersection of finitely many closed half-spaces in some R^d. An *H*-polytope is a bounded H-polyhedron.
- A *V*-polytope is the convex hull of a finite point set in R^d.

A basic theorem about convex polytopes claims that from the mathematical point of view, *H*-polytopes and *V*-polytopes are equivalent.

Theorem: Each *V*-polytope is an *H*-polytope. Each *H*-polytope is a *V*-polytope.

This is one of the theorems that may look "obvious" and whose proof needs no particularly clever idea but does require some work. In the present case, we do not intend to avoid it.

Although *H*-polytopes and *V*-polytopes are mathematically equivalent, there is an enormous difference between them from the computational point of view. That is, it matters a lot whether a convex polytope is given to us as a convex hull of a finite set or as an intersection of half-spaces. For example, given a set of n points specifying a *V*-polytope, how do we find its representation as an

H-polytope? It is not hard to come up with some algorithm, but the problem is to find an efficient algorithm that would allow one to handle large real-world problems. This algorithmic question is not yet satisfactorily solved. Moreover, in some cases the number of required halfspaces may be astronomically large compared to the number n of points.

Another, computational difference between *V*-polytopes and *H*-polytopes, we consider the maximization of a given linear function over a given polytope. For *V*-polytopes it is a trivial problem, since it suffices to substitute all points of V into the given linear function and select the maximum of the resulting values. But maximizing a linear function over the intersection of a collection of half-spaces is the basic problem of linear programming, and it is certainly nontrivial.

Terminology: The usual terminology does not distinguish *V*-polytopes and *H*-polytopes. A convex polytope means a point set in R^d that is a *V*-polytope (and thus also an H -polytope). An arbitrary, possibly unbounded, *H*-polyhedron is called a convex polyhedron. All polytopes and polyhedra considered in this chapter are convex, and so the adjective "convex" is often omitted.

The dimension of a convex polyhedron is the dimension of its affine hull. It is the smallest dimension of a Euclidean space containing a congruent copy of P. The phrase "d-dimensional polytope" is often abbreviated to "d-polytope."

Example: One of the easiest classes of polytopes is that of cubes. The d-dimensional cube as a point set is the Cartesian product $[-1, 1]^d$.

$d = 1$ $d = 2$ $d = 3$

As a *V*-polytope, the d-dimensional cube is the convex hull of the set $\{-1, 1\}^d$ (2^d points), and as an *H*-polytope, it can be described by the inequalities $-1 \leq x_i \leq 1, i = 1, 2, \ldots, d$, *i.e.*, by 2d half-spaces. We note that it is also the unit ball of the maximum norm $\|x\|_\infty = \max_i |x|$.

Another important example is the class of crosspolytopes (or generalized octahedra). The d-dimensional crosspolytope is the convex hull of the "coordinate cross," i.e. $v\{e_1, -e_1, e_2, -e_2, \ldots, e_d, -e_d\}$, where, $e_1, \ldots, e_d$ are the vectors of the standard orthonormal basis.

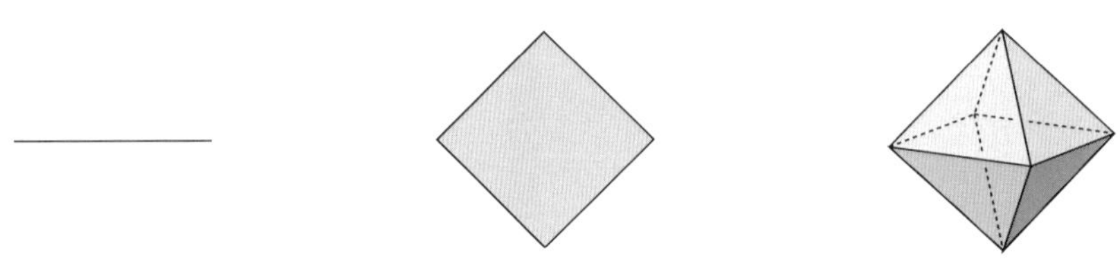

$d = 1$ $d = 2$ $d = 3$

It is also the unit ball of the $\ell_1 -$norm $\|x\|_1 = \sum_{i=1}^{d} |x_i|$. As an H-polytope, it can be expressed by the 2^d half-spaces of the form $\langle \sigma, x \rangle \leq 1$, where σ runs through all vectors in $\{-1, 1\}^d$.

The polytopes with the smallest possible number of vertices (for a given dimension) are called simplices.

Simplex

A Simplex is the convex hull of an affinely independent point set in some R^d.

A d-dimensional simplex in R^d can also be represented as an intersection of $d+1$ half-spaces, as is not difficult to check.

A regular d-dimensional simplex is the convex hull of $d+1$ points with all pairs of points having equal distances.

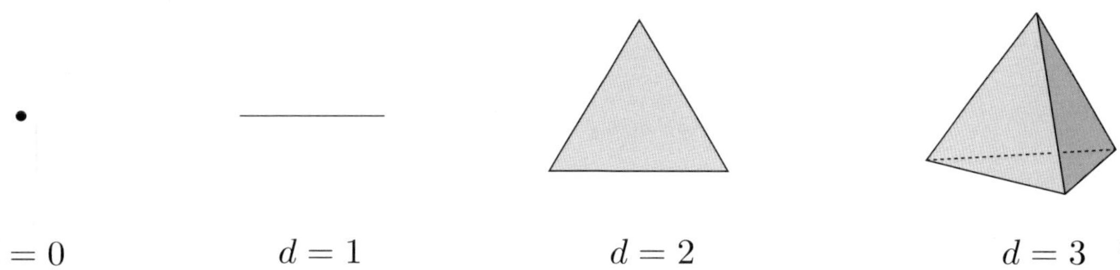

Unlike, cubes and crosspolytopes, d-dimensional regular simplices do not have a very nice coordinate representation in R^d. The simplest and most useful representation lives one dimension higher: The convex hull of the $d+1$ vectors $e_1, \ldots, e_{d+1}$ of the standard orthonormal basis in R^{d+1} is a d-dimensional regular simplex with side length $\sqrt{2}$.

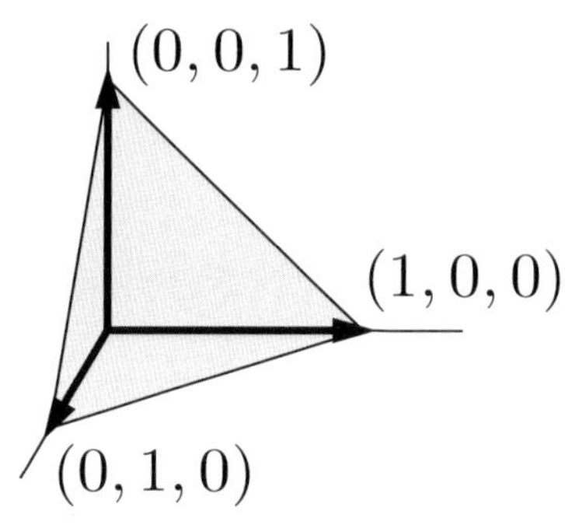

Proof of Theorem (Equivalence of H-polytopes and V-polytopes)

We first show that any *H*-polytope is also a *V*-polytope. We proceed by induction on d. The case $d = 1$ being trivial, we suppose that, $d \geq 2$.

So let, Γ be a finite collection of closed half-spaces in R^d such that $P = \bigcap \Gamma$ is nonempty and bounded. For each $\gamma \in \Gamma$, *let* $F_\gamma = P \bigcap \partial\gamma$ be the intersection of P with the bounding hyperplane of γ. Each nonempty F_γ is an H-polytope of dimension at most $d-1$ (correct?), and so it is the convex hull of a finite set $V_\gamma \subset F_\gamma$ by the inductive hypothesis.

We claim that $P = \text{conv}(V)$, where, $V = \bigcup_{\gamma\in\Gamma} V_\gamma$. Let $x \in P$ and let ℓ be a line passing through x the intersection $\ell \cap P$ is a segment; let y and z be its endpoints. There are $\alpha, \beta \in \Gamma$ such that, $y \in F_\alpha$ and $z \in F_\beta$ (if y were not on the boundary of any $\gamma \in \Gamma$, we could continue along ℓ a little further within P).

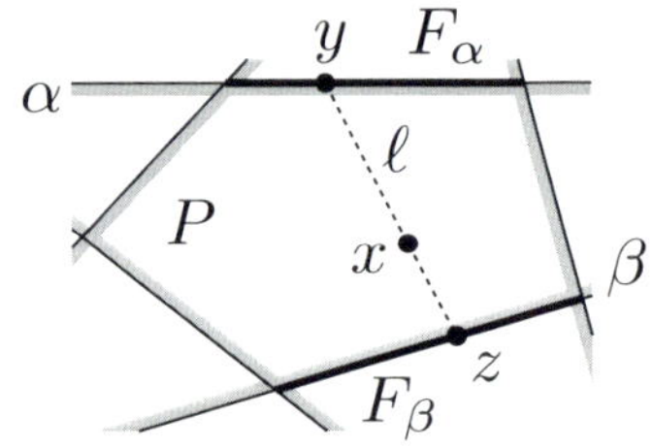

We have $y \in \text{conv}(V_\alpha)$ and $z \in \text{conv}(V_\beta)$, and thus, $x \in \text{conv}(V_\alpha \cup V_\beta) \subseteq \text{conv}(V)$.

We have proved that any *H*-polytope is a *V*-polytope, and it remains to show that a *V*-polytope can be expressed as the intersection of finitely many half-spaces. This follows easily by duality (and implicitly uses the separation theorem).

Let $P = \text{conv}(V)$ with V finite, and assume that 0 is an interior point of *P*.

Faces of a Convex Polytope

The surface of the 3-dimensional cube consists of 8 "corner" points called vertices, 12 edges, and 6 squares called facets. According to the perhaps more usual terminology in 3-dimensional geometry, the facets would be called faces. But in the theory of convex polytopes, the word face has a slightly different meaning, defined below. For the cube, not only the squares but also the vertices and the edges are all called faces of the cube.

Face: A face of a convex polytope P is defined as,

- Either P itself;
- A subset of P of the form $P \cap h$, where h is a hyperplane such that P is fully contained in one of the closed half-spaces determined by h.

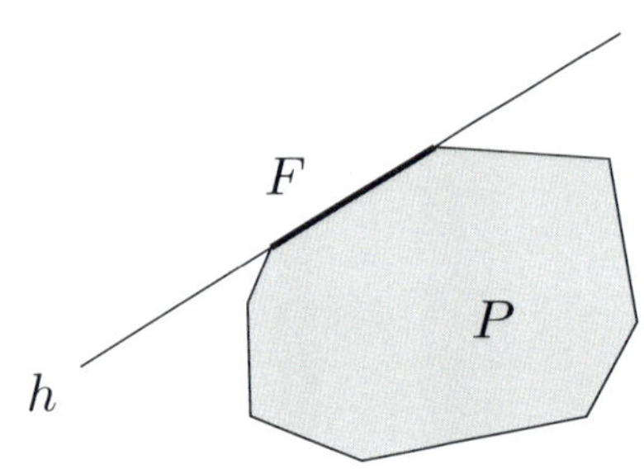

We observe that each face of P is a convex polytope. This is because P is the intersection of finitely many half-spaces and h is the intersection of two half-spaces, so the face is an H-polyhedron, and moreover, it is bounded.

If P is a polytope of dimension d, then its faces have dimensions: $-1, 0, 1, \ldots, d$.

Where, −1 is by definition, the dimension of the empty set. A face of dimension j is also called a j-face.

Names of faces: The 0-faces are called vertices, the 1-faces are called edges, and the $(d-1)$-faces of a d-dimensional polytope are called facets. The $(d-2)$–faces of a d-dimensional polytope are ridges; in the familiar 3-dimensional situation, edges = ridges. For example, the 3-dimensional cube has 28 faces in total: the empty face, 8 vertices, 12 edges, 6 facets, and the whole cube.

The following proposition shows that each V-polytope is the convex hull of its vertices, and that the faces can be described combinatorially: They are the convex hulls of certain subsets of vertices. This includes some intuitive facts such as that each edge connects two vertices.

A helpful notion is that of an extremal point of a set: For a set $X \subseteq R^d$, a point $x \in X$ is extremal if $x \neq \operatorname{conv}(X \setminus \{x\})$.

Proposition: Let $P \subset R^d$ be a (bounded) convex polytope.

- ("Vertices are extremal") The extremal points of P are exactly its vertices, and P is the convex hull of its vertices;
- ("Face of a face is a face") Let, F be face of P. The vertices of F are exactly those vertices of P that lie in F. More generally, the faces of F are exactly those faces of P that are contained in F.

The proof is not essential for our further considerations, and it is given at the end of this topic. The proposition has an appropriate analogue for polyhedra, but in order to avoid technicalities, we treat the bounded case only.

Example: A d-dimensional simplex has been defined as the convex hull of a $(d+1)-$ point affinely independent set V. It is easy to see that each subset of V determines a face of the simplex. Thus, there are $\binom{d+1}{k+1}$ faces of dimension k, k = −1, 0, . . . , d, and 2^{d+1} faces in total.

The d-dimensional crosspolytope has $V = \{e_1, -e_1, \ldots, e_d, -e_d\}$ as the vertex set. A proper subset $F \subset V$ determines a face if and only if there is no i such that both $e_i \in F$ and $-e_i \in F$. It follows that there are $3^d + 1$ faces, including the empty one and the whole crosspolytope.

The nonempty faces of the d-dimensional cube $[-1, 1]^d$ correspond to vector $v \in \{-1, 1, 0\}^d$. The face corresponding to such v has the vertex set $\{u \in \{-1, 1\}^d : u_i = v_i$ for all i with $v_i \neq 0\}$. Geometrically, the vector v is the center of gravity of its face.

The face lattice: Let $F(P)$ be the set of all faces of a (bounded) convex polytope P (including the empty face $\varnothing$ of dimension −1). We consider the partial ordering of $F(P)$ by inclusion.

Combinatorial equivalence: Two convex polytopes P and Q are called combinatorially equivalent if $F(P)$ and $F(Q)$ are isomorphic as partially ordered sets.

We are going to state some properties of the partially ordered set $F(P)$ without proofs.

It turns out that $F(P)$ is a lattice (a partially ordered set satisfying additional axioms). We recall that this means the following two conditions:

- Meets condition: For any two faces F, $G \in F(P)$, there exists a face $M \in F(P)$, called the meet of F and G, that is contained in both F and G and contains all other faces contained in both F and G.

- Joins condition: For any two faces F, $G \in F(P)$, there exists a face $J \in F(P)$, called the join of F and G, that contains both F and G and is contained in all other faces containing both F and G.

The meet of two faces is their geometric intersection $F \cap G$.

For verifying the joins and meets conditions, it may be helpful to know that for a finite partially ordered set possessing the minimum element and the maximum element, the meets condition is equivalent to the joins condition, and so it is enough to check only one of the conditions.

Here is the face lattice of a 3-dimensional pyramid:

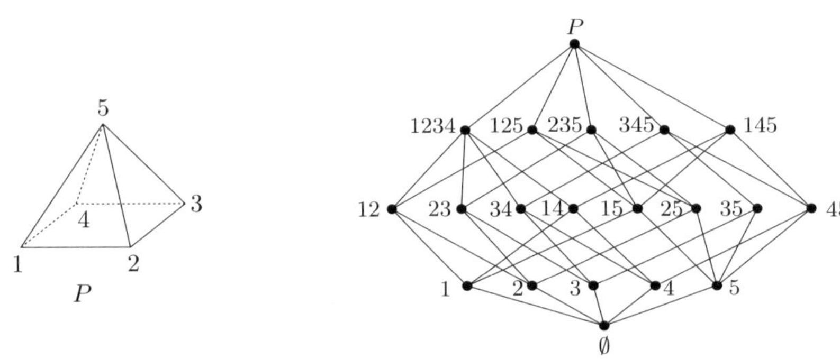

The vertices are numbered 1–5, and the faces are labeled by the vertex sets.

The face lattice is graded, meaning that every maximal chain has the same length (the rank of a face F is $dim(F)+1$).

Quite obviously, it is atomic: Every face is the join of its vertices. A little less obviously, it is coatomic; that is, every face is the meet (intersection) of the facets containing it. An important consequence is that combinatorial type of a polytope is determined by the vertex–facet incidences. More precisely, if we know the dimension and all subsets of vertices that are vertex sets of facets (but without knowing the coordinates of the vertices, of course), we can uniquely reconstruct the whole face lattice in a simple and purely combinatorial way.

Face lattices of convex polytopes have several other nice properties, but no full algebraic characterization is known, and the problem of deciding whether a given lattice is a face lattice is algorithmically difficult (even for 4-dimensional polytopes).

The face lattice can be a suitable representation of a convex polytope in a computer. Each j-face is connected by pointers to its $(j-1)-$ faces and to the $(j+1)-$ faces containing it.

On the other hand, it is a somewhat redundant representation: Recall that the vertex–facet incidences already contain the full information, and for some applications, even less data may be sufficient, say the graph of the polytope.

The dual polytope: Let P be a convex polytope containing the origin in its interior. Then the dual set P^* is also a polytope;

Proposition: For each $j = -1, 0, \ldots, d$, the j-faces of P are in a bijective correspondence with the $(d-j-1)$-faces of P^*. This correspondence also reverses inclusion; in particular, the face lattice of P^* arises by turning the face lattice of P upside down.

Among the five regular Platonic solids, the cube and the octahedron are dual to each other, the dodecahedron and the icosahedron are also dual, and the tetrahedron is dual to itself. More generally, if we have a 3-dimensional convex polytope and G is its graph, then the graph of the dual polytope is the dual graph to G, in the usual graph-theoretic sense. The dual of a d-simplex is a d-simplex, and the d-dimensional cube and the d-dimensional crosspolytope are dual to each other.

We conclude with two notions of polytopes “in general position.”

Simple and Simplicial Polytopes

A polytope P is called simplicial if each of its facets is a simplex (this happens, in particular, if the vertices of P are in general position, but general position is not necessary). A d-dimensional polytope P is called simple if each of its vertices is contained in exactly d facets.

The faces of a simplex are again simplices, and so each proper face of a sim-plicial polytope is a simplex. Among the five Platonic solids, the tetrahedron, the octahedron, and the icosahedron are simplicial; and the tetrahedron, the cube, and the dodecahedron are simple. Crosspolytopes are simplicial, and cubes are simple. An example of a polytope that is neither simplicial nor simple is the 4-sided pyramid used in the illustration of the face lattice.

The dual of a simple polytope is simplicial, and vice versa. For a simple d-dimensional polytope, a small neighborhood of each vertex looks combinatorially like a neighborhood of a vertex of the d-dimensional cube. Thus, for each vertex v of a d-dimensional simple polytope, there are d edges emanating from v, and each k-tuple of these edges uniquely determines one k-face incident to v. Consequently, v belongs to $\binom{d}{k}$ $k-faces,\ k = 0, 1, \ldots, d$.

Proof of Proposition: In “vertices are extremal”, we assume that P is the convex hull of a finite point set. Among all such sets, we fix one that is inclusion-minimal and call it V_0.

Let, V_v be the vertex set of P, and let V_e be the set of all extremal points of P. We prove that $V_0 = V_v = V_e$, which gives vertices are extremal. We have $V_e \subseteq V_0$ by the definition of an extremal point.

Next, we show that $V_v \subseteq V_e$. If $v \in V_v$ is a vertex of P, then there is a hyperplane h with $P \cap h = \{v\}$, and all of $P \setminus \{v\}$ lies in one of the open half-spaces defined by h. Hence, $P \setminus \{v\}$ is convex, which means that v is an extremal point of P, and so $V_v \subseteq V_e$.

Finally we verify, $V_0 \subseteq V_v$.

Let, $v \in V_0$; by the inclusion-minimality of V_0, we get that $v \neq C = \text{conv}(V_0 \setminus \{v\})$. Since C and $\{v\}$ are disjoint compact convex sets, they can be strictly separated by a hyperplane h. Let h_v be the hyperplane parallel to h and containing v; this h_v has all points of $V_0 \setminus \{v\}$ on one side.

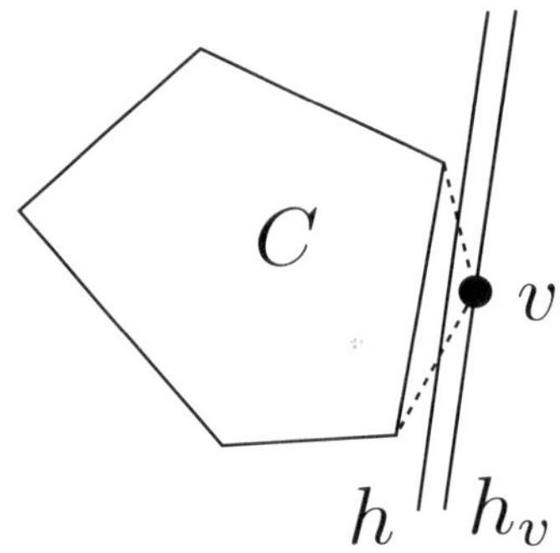

We want to show that $P \cap h_v = \{v\}$ (then v is a vertex of P, and we are done). The set $P \setminus h_v = \operatorname{conv}(V_0) \setminus h_v$, being the intersection of a convex set with an open half-space, is convex. Any segment vx, where, $x \in P \setminus h_v$, is convex as well. Since this set contains V_0 and is convex, it contains $P = \operatorname{conv}(V_0)$ and so $P \cap h_v = \{v\}$ indeed.

As for ("face of a face is a face"), it is clear that a face G of *P* contained in *F* is a face of *F* too (use the same witnessing hyperplane). For the reverse direction, we begin with the case of vertices. By a consideration similar to that at the end of the proof of ("vertices are extremal), we see that $F = \operatorname{conv}(V) \cap h = \operatorname{conv}(V \cap h)$, where h is a hyperplane defining *F*. Hence all the extremal points of *F*, which by ("vertices are extremal) are exactly the vertices of *F*, are in *V*.

Finally, let *F* be a face of *P* defined by a hyperplane h, and let $G \subset F$ be a face of F defined by a hyperplane g within h; that is, g is a $(d-2)$– dimensional affine subspace of h with $G = g \cap F$ and with all of *F* on one side. Let γ be the closed half-space bounded by h with $P \subset \gamma$. We start rotating the boundary h of γ around g in the direction such that the rotated half-space γ' still contain *F*.

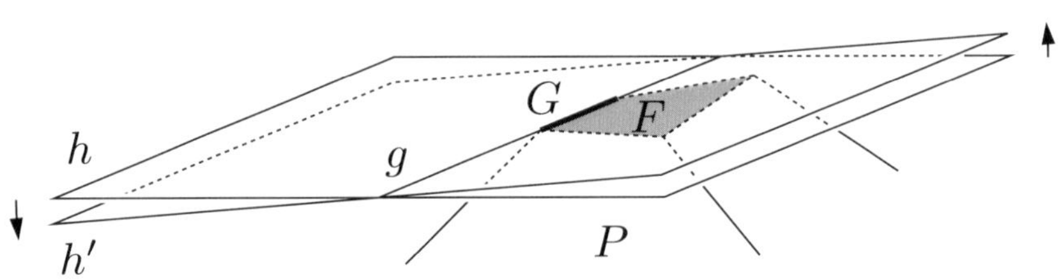

If we rotate by a sufficiently small amount, then all the vertices of *P* not lying in *F* are still in the interior of γ'. At the same time, the interior of γ' contains all the vertices of *F* not lying in *G*, while all the vertices of *G* remain on the boundary h' of γ'. So h o defines a face of *P* (since all of *P* is on one side), and this face has the same vertex set as *G*, and so it equals *G* by the first part of ("face of a face is a face") proved above.

Many Faces: The Cyclic Polytopes

A convex polytope *P* can be given to us by the list of vertices. How difficult is it to recover the full face lattice, or, more modestly, a representation of *P* as an intersection of half-spaces? The first question to ask is how large the face lattice or the collection of half-spaces can be, compared to the number of vertices. That is, what is the maximum total number of faces, or the maximum number of facets, of a convex polytope in R^d with n vertices? The dual question is, of course, the maximum number of faces or vertices of a bounded intersection of n half-spaces in R^d .

Let $f_j = f_j\ (P)$ denote the number of j-faces of a polytope P. The vector $(f_0, f_1, \ldots, f_d)$ is called the f -vector of P. We thus assume $f_0\ = n$ and we are interested in estimating the maximum value of f_{d-1} and of $\sum_{k=0}^{d} f_k$.

In dimensions 2 and 3, the situation is simple and favorable.

For $d = 2$, our polytope is a convex polygon with n vertices and n edges, and so, $f_0 = f_1 = n, f2 = 1$. The f-vector is even determined uniquely.

A 3-dimensional polytope can be regarded as a drawing of a planar graph, in our case with n vertices. By well-known results for planar graphs, we have $f_1\ \leq\ 3n-6\ and\ f_2\ \leq\ 2n-4$. Equalities hold if and only if the polytope is simplicial (all facets are triangles).

In both cases the total number of faces is linear in n. But as the dimension grows, polytopes become much more complicated. First of all, even the total number of faces of the most innocent convex polytope, the d-dimensional simplex, is exponential in d. But here we consider d fixed and relatively small, and we investigate the dependence on the number of vertices n.

Still, as we will see, for every $n\ \geq\ 5$ there is a 4-dimensional convex polytope with n vertices and with every two vertices connected by an edge, i.e. with $\binom{n}{2}$ edges. This looks counterintuitive, but our intuition is based on the 3-dimensional case. In any fixed dimension d, the number of facets can be of order $n^{[d/2]}$, which is rather disappointing for someone wishing to handle convex polytopes efficiently. On the other hand, complete desperation is perhaps not appropriate. Certainly not all polytopes exhibit this very bad behavior. For example, it is known that if we choose n points uniformly at random in the unit ball B^d , then the expected number of faces of their convex hull is only $o(n)$, for every fixed d.

It turns out that the number of faces for a given dimension and number of vertices is the largest possible for so-called cyclic polytopes, to be introduced next. First we define a very useful curve in R^d.

Moment curve: The curve $\gamma = \{(t, t^2\ , \ldots, t^d\ ) : t\ \in\ R\}$ in R^d is called the moment curve.

Cyclic polytope: The convex hull of finitely many points on the moment curve is called a cyclic polytope.

Theorem: The number of facets of a d-dimensional cyclic polytope with n vertice $(n\ \geq\ d+1)$ is:

$$\binom{n-\lfloor d/2\rfloor}{\lfloor d/2\rfloor}+\binom{n-\lfloor d/2\rfloor-1}{\lfloor d/2\rfloor-1}\ \textit{for d even, and}$$

$$2\binom{n-\lfloor d/2\rfloor-1}{\lfloor d/2\rfloor}\ \textit{for d odd.}$$

For fixed d, this has the order of magnitude $n^{\lfloor d/2\rfloor}$.

The Upper Bound Theorem

The upper bound theorem, one of the earlier major achievements of the theory of convex polytopes, claims that the cyclic polytope has the largest possible number of faces.

Theorem (Upper bound theorem): Among all d-dimensional convex polytopes with n vertices, the cyclic polytope maximizes the number of faces of each dimension.

Proposition (Asymptotic upper bound theorem): A d-dimensional convex polytope with n vertices has at most $2\binom{n}{\lfloor d/2 \rfloor}$ facets and no more than $2^{d+1}\binom{n}{\lfloor d/2 \rfloor}$ faces in total. For d fixed, both quantities thus have the order of magnitude $n^{\lfloor d/2 \rfloor}$.

First we establish this proposition for simplicial polytopes, in the following form

Proposition: Let P be a d-dimensional simplicial polytope.

Then,

(a) $f_0(P) + f_1(P) + \cdots + f_d(P) \leq 2^d f_{d-1}(P)$, and

(b) $f_{d-1}(P) \leq 2 f_{\lfloor d/2 \rfloor - 1}(P)$.

This implies Proposition(Asymptotic upper bound theorem) for simplicial polytopes, since the number of $(\lfloor d/2 \rfloor - 1)$-faces is certainly no bigger than $\binom{n}{\lfloor d/2 \rfloor}$, the number of all $\lfloor d/2 \rfloor -$ tuples of vertic

Proof of Proposition: We pass to the dual polytope P^*, which is simple. Now we need to prove $\sum_{k=0}^{d} f_k(P^*) \leq 2^d f_0(P^*)$ and $f_0(P^*) \leq 2 f_{\lfloor d/2 \rfloor}(P^*)$.

Each face of P^* has at least one vertex, and every vertex of a simple d-polytope is incident to 2^d faces, which gives the first inequality.

We now bound the number of vertices in terms of the number of $\lceil d/2 \rceil -$ faces. This is the heart of the proof, and it shows where the mysterious exponent $\lfloor d/2 \rfloor$ comes from.

Let us, rotate the polytope P^* so that no two vertices share the x_d-coordinate (i.e., no two vertices have the same vertical level).

Consider a vertex v with the d edges emanating from it. By the pigeonhole principle, there are at least $\lceil d/2 \rceil$ edges directed upwards or at least $\lceil d/2 \rceil$ edges directed downwards. In the former case, every dd/2-tuple of edges going up determines a $\lceil d/2 \rceil -$ face for which v is the lowest vertex. In the latter case, every $\lceil d/2 \rceil -$ tuple of edges going down determines a $\lceil d/2 \rceil -$ face for which v is the highest vertex. Here is an illustration, unfortunately for the not too interesting 3-dimensional case, showing a situation with 2 edges going up and the corresponding 2-dimensional face having v as the lowest vertex:

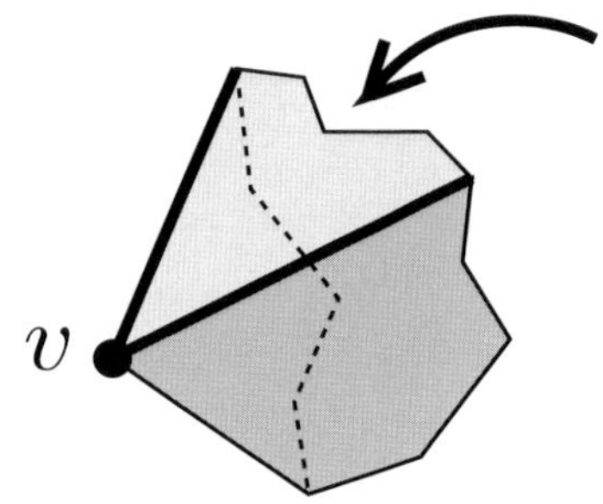

We have exhibited at least one $\lceil d/2 \rceil-$ face for which v is the lowest vertex or the highest vertex. Since the lowest vertex and the highest vertex are unique for each face, the number of vertices is no more than twice the number of $\lceil d/2 \rceil-$ faces.

Warning: For simple polytopes, the total combinatorial complexity is proportional to the number of vertices, and for simplicial polytopes it is proportional to the number of facets (considering the dimension fixed, that is). For polytopes that are neither simple nor simplicial, the number of faces of intermediate dimensions can have larger order of magnitude than both the number of facets and the number of vertices.

Nonsimplicial polytopes: To prove the asymptotic upper bound theorem, it remains to deal with nonsimplicial polytopes. This is done by a perturbation argument, similar to numerous other results. Where, general position is convenient for the proof but where we want to show that the result holds in degenerate cases as well.

Lemma: For any d-dimensional convex polytope P there exists a d-dimensional simplicial polytope, Q with $f_0(P) = f_0(Q)$ and $f_k(Q) \geq f_k(P)$ for all $k = 1, 2, \ldots, d$.

Proof: The basic idea is very simple. Move (perturb) every vertex of P by a very small amount, in such a way that the vertices are in general position, and show that each k-face of P gives rise to at least one k-face of the perturbed polytope. There are several ways of doing this proof.

We process the vertices one by one. Let V be the vertex set of P and let, $v \in V$. The operation of ε-pushing v is as follows: We choose a point v' lying in the interior of P, at distance at most ε from v, and on no hyperplane determined by the points of V, and we set $V' = (V \setminus \{v\}) \cup \{v'\}$. If we successively -push each vertex v of the polytope, the resulting vertex set is in general position and we have a simple polytope.

It remains to show that for any polytope P with vertex set V and any $v \in V$, there is an $\varepsilon > 0$ such that ε-pushing v does not decrease the number of faces.

Let, $U \subset V$ be the vertex set of a k-face of P, $0 \leq k \leq d-1$, and let V o arise from V by ε-pushing v. If $v \notin U$, then no doubt, U determines a face of $\text{conv}(V')$, and so we assume that $v \in V$. First suppose that v lies in the affine hull of $U \setminus \{v\}$; we claim that then $U \setminus \{v\}$; determines a k-face of $\text{conv}(V')$.

A subset $U \subset V$ is the vertex set of a face of $\text{conv}(V)$ if and only if the affine hull of U is disjoint from $\text{conv}(V \setminus U)$. We leave a detailed argument to the reader (one must use the fact that v is pushed inside).

If v lies outside of the affine hull $U\setminus\{v\}$; then we want to show that $U'=(U\setminus\{v\})\cup\{v'\}$ determines a k-face of $\operatorname{conv}(V')$. The affine hull of U is disjoint from the compact set $\operatorname{conv}(V\setminus U)$. If we move v continuously by a sufficiently small amount, the affine hull of U moves continuously and so there is an $\varepsilon > 0$ such that, if we move v within ε from its original position, the considered affine hull and $\operatorname{conv}(V\setminus U)$ remain disjoint.

Voronoi Diagrams

Consider a finite set $P\subset R^d$. For each point $p \in P$, we define a region $reg(p)$, which is the "sphere of influence" of the point p. It consists of the points $x\in R^d$ for which p is the closest point among the points of P.

Formally,

$$reg(p) = \{x\in R^d : \operatorname{dist}(x, p)\le \operatorname{dist}(x, q) \text{ for all } q \in P\},$$

Where, dist (x, y) denotes the Euclidean distance of the points x and y. The Voronoi diagram of P is the set of all regions $reg(p)$ for $p\in P$. (More precisely, it is the cell complex induced by these regions; that is, every intersection of a subset of the regions is a face of the Voronoi diagram.) Here an example of the Voronoi diagram of a point set in the plane:

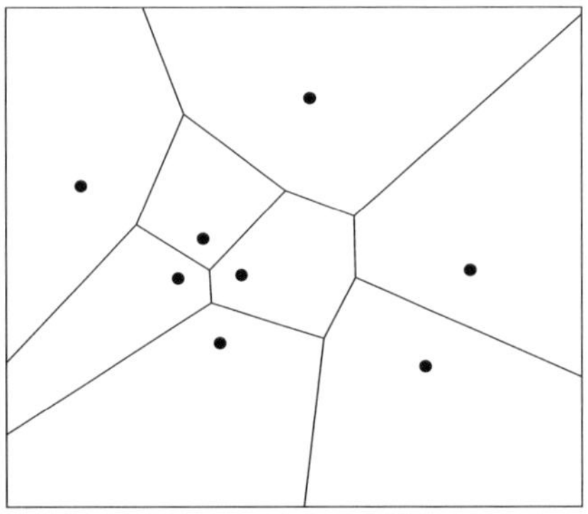

(Of course, the Voronoi diagram is clipped by a rectangle so that it fits into a finite page). The points of P are traditionally called the sites in the context of Voronoi diagrams.

Observation: Each region $reg(p)$ is a convex polyhedron with at most $|P|-1$ facets.

Indeed,

$$reg(p) = \bigcap_{q\in P\setminus\{p\}} \{x: \operatorname{dist}(x, p) \le \operatorname{dist}(x, q)\}$$

Is an intersection of $|P|-1$ half-spaces.

For $d=2$, a Voronoi diagram of n points is a subdivision of the plane into n convex polygons (some of them are unbounded). It can be regarded as a drawing of a planar graph (with one vertex at the infinity, say), and hence, it has a linear combinatorial complexity: n regions, $O(n)$ vertices, and $O(n)$ edges.

Examples of applications: Voronoi diagrams have been reinvented and used in various branches of science. Sometimes the connections are surprising. For instance, in archaeology, Voronoi diagrams help study cultural influences. Here we mention a few applications, mostly algorithmic.

- "Post office problem" or nearest neighbor searching: Given a point set P in the plane, we want to construct a data structure that finds the point of P nearest to a given query point x as quickly as possible. This problem arises directly in some practical situations or, more significantly, as a subroutine in more complicated problems. The query can be answered by determining the region of the Voronoi diagram of P containing x. For this problem (point location in a subdivision of the plane), efficient data structures are known.
- Robot motion planning: Consider a disk-shaped robot in the plane. It should pass among a set P of point obstacles, getting from a given start position to a given target position and touching none of the obstacles.

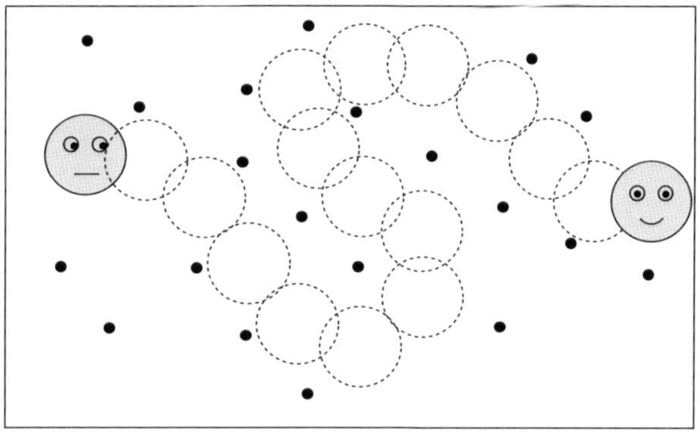

If such a passage is possible at all, the robot can always walk along the edges of the Voronoi diagram of P, except for the initial and final segments of the tour. This allows one to reduce the robot motion problem to a graph search problem: We define a subgraph of the Voronoi diagram consisting of the edges that are passable for the robot.

- A nice triangulation (Delaunay triangulation): Let $P \subset R^2$ be a finite point set. In many applications one needs to construct a triangulation of P (that is, to subdivide conv(P) into triangles with vertices at the points of P) in such a way that the triangles are not too skinny. Of course, for some sets, some skinny triangles are necessary, but we want to avoid them as much as possible. One particular triangulation that is usually very good, and provably optimal with respect to several natural criteria, is obtained as the dual graph to the Voronoi diagram of P. Two points of P are connected by an edge if and only if their Voronoi regions share an edge.

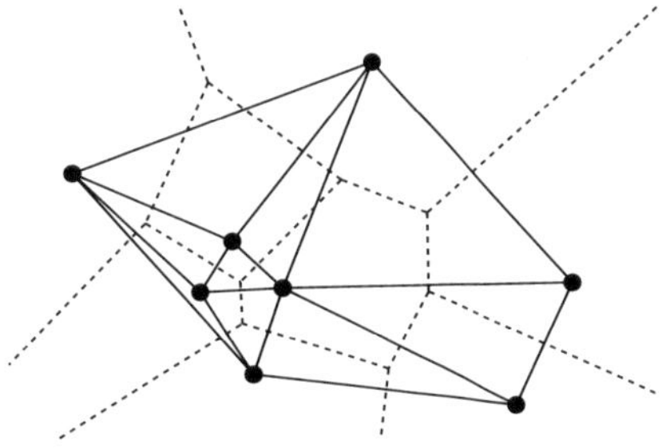

If no 4 points of P lie on a common circle then this indeed defines a triangulation, called the *Delaunay triangulation* of P; The definition extends to points sets in R^d in a straightforward manner.

- Interpolation: Suppose that $f: R^2 \to R$ is some smooth function whose values are known to us only at the points of a finite set $P \subset R^2$. We would like to interpolate f over the whole polygon conv(P). Of course, we cannot really tell what f looks like outside P, but still we want a reasonable interpolation rule that provides a nice smooth function with the given values at P. Multidimensional interpolation is an extensive semiempirical discipline which we do not seriously consider here; we explain only one elegant method based on Voronoi diagrams. To compute the interpolated value at a point $x \in \text{conv}(P)$, we construct the Voronoi diagram of P, and we overlay it with the Voronoi diagram of $P \cup \{x\}$.

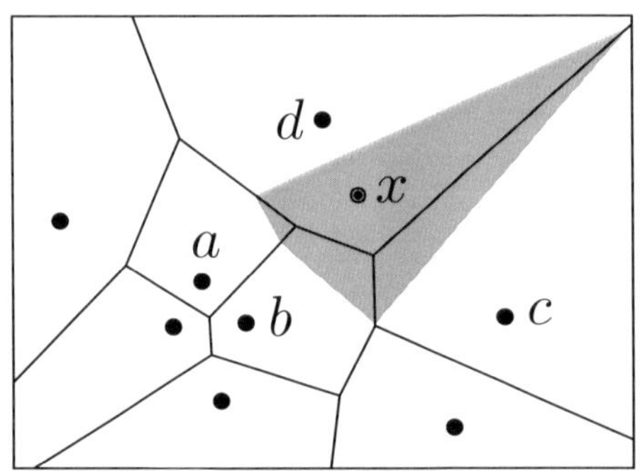

The region of the new point x cuts off portions of the regions of some of the old points. Let w_p be the area of the part of $reg(p)$ in the Voronoi diagram of P that belongs to $reg(x)$ after inserting x.

The interpolated value $f(x)$ is:

$$f(x) = \sum_{p \in P} \frac{w_p}{\sum_{q \in P} w_q} f(p).$$

An analogous method can be used in higher dimensions, too.

Relation of Voronoi Diagrams to Convex Polyhedra

We now show that Voronoi diagrams in R^d correspond to certain convex polyhedra in R^{d+1}. First we define the unit paraboloid in R^{d+1}:

$$U = \{x \in R^{d+1} : x_{d+1} = x_1^2 + x_2^2 + \cdots + x_d^2\}.$$

For $d = 1$, U is a parabola in the plane.

In the sequel, let us imagine the space as the hyperplane $x_{d+1} = 0$ in R^{d+1}. For a point $p = (p_1, \ldots, p_d) \in R^d$, let $e(p)$ denote the hyperplane in R^{d+1} with equation

$$x_{d+1} = 2p_1x_1 + 2p_2x_2 + \cdots + 2p_dx_d - p_1^2 - p_2^2 - \cdots - p_d^2.$$

Geometrically, $e(p)$ is the hyperplane tangent to the paraboloid U at the point $u(p)=(p_1, p_2, \ldots, p_d, p_1^2 + \cdots + p_d^2)$ lying vertically p. It is perhaps easier to remember this geometric definition of $e(p)$ and derive its equation by differentiation when needed. On the other hand, in the forthcoming proof we start out from the equation of $e(p)$, and as a by-product, we will see that $e(p)$ is the tangent to U at $u(p)$ as claimed.

Proposition: Let $p, x \in R^d$ be points and let $u(x)$ be the point of U vertically above x. Then $u(x)$ lies above the hyperplane $e(p)$ or on it, and the vertical distance of $u(x)$ to $e(p)$ is δ^2, where, $\delta = dist(x, p)$.

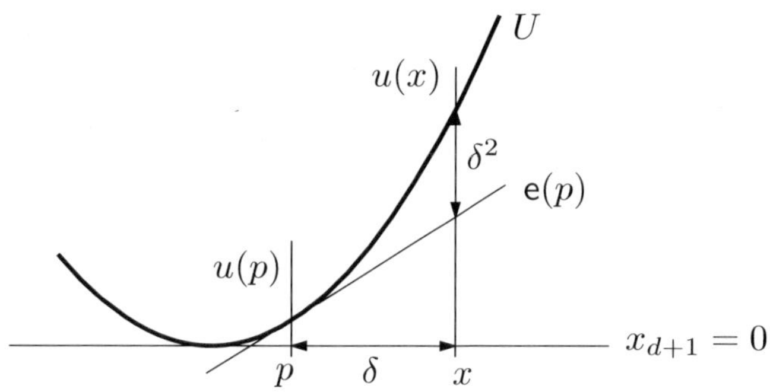

Proof: We just substitute into the equations of U and of $e(p)$. The x_{d+1}- coordinate of $u(x)$ is $x_1^2 + \cdots + x_d^2$,while the x_{d+1}-coordinate of the point of $e(p)$ above x is $2p_1x_1 + \cdots + 2p_dx_d - p_1^2 - \cdots - p_d^2$.

The difference is $(x_1 - p_1)^2 + \cdots + (x_d - p_d)^2 = \delta^2$.

Let $E(p)$ denote the half-space lying above the hyperplane $e(p)$. Consider an n -point set $P \subset R^d$.

By Proposition $x \in reg(p)$ holds if and only if $e(p)$ is vertically closest to U at x among all $e(p)$, $q \in$ P.

Here is what we have derived:

Corollary: The Voronoi diagram of P is the vertical projection of the facets of the polyhedron $\bigcap_{p\in P} E(p)$ onto the hyperplane $x_{d+1} = 0$.

Here is an illustration for a planar Voronoi diagram:

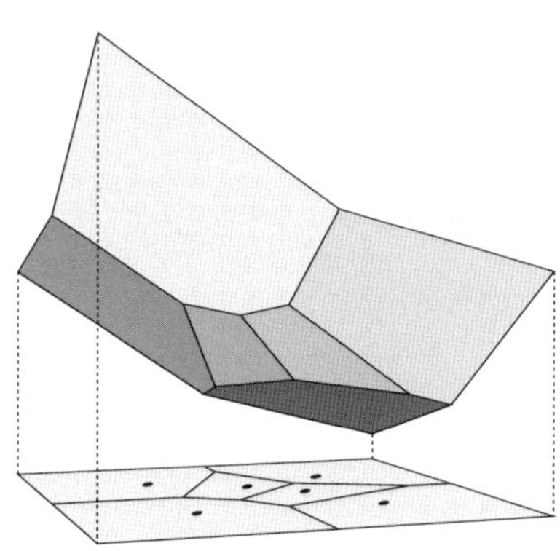

Corollary: The maximum total number of faces of all regions of the Voronoi diagram of an n-point set in R^d is $O\left(n^{\lceil d/2 \rceil}\right)$.

Proof: We know that the combinatorial complexity of the Voronoi diagram equals the combinatorial complexity of an *H*-polyhedron with at most n facets in R^{d+1}. By intersecting this *H*-polyhedron with a large simplex we can obtain a bounded polytope with at most $n+d+2$ facets, and we have not decreased the number of faces compared to the original *H*-polyhedron. Then the dual version of the asymptotic upper bound theorem implies that the total number of faces is $O\left(n^{\lceil d/2 \rceil}\right)$. Since, $\lfloor (d+1)/2 \rfloor = \lceil d/2 \rceil$.

The convex polyhedra in R^{d+1} obtained from Voronoi diagrams in R^d by the above construction are rather special, and so a lower bound for the combinatorial complexity of convex polytopes cannot be automatically transferred to Voronoi diagrams. But it turns out that the number of vertices of a Voronoi diagram on n points in R^d can really be of order $n^{\lceil d/2 \rceil}$.

Let us remark that the trick used for transforming Voronoi diagrams to convex polyhedra is an example of a more general technique, called linearization or Veronese mapping, this method sometimes allows us to convert a problem about algebraic curves or surfaces of bounded degree to a problem about k-flats in a suitable higherdimensional space.

The Farthest-point Voronoi Diagram

The projection of the *H*-polyhedron $\bigcap_{p \in P} E(p)^{op}$, where γ^{op} denotes the half-space opposite to γ, forms the farthest-point Voronoi diagram, in which each point $p \in P$ is assigned the regions of points for which it is the farthest point. It can be shown that all nonempty regions of this diagram are unbounded and they correspond precisely to the points appearing on the surface of $\text{conv}(P)$.

K-set Geometry

The *k-set problem* is a classic open problem in Combinatorial Geometry, which is considered similar in spirit to incidence problems and to distance problems. While in recent years algebraic techniques led to significant progress in incidence and distance problems, so far none of the new methods were applied to the *k*-set problem.

Let $\mathcal{P}$ be a set of n points in $\mathbb{R}^2$, and let $1 \le k \le n/2$ be an integer. A *k*-set of $\mathcal{P}$ is a subset $\mathcal{P}' \subset \mathcal{P}$ of k points such that there exists a line that *separates* $\mathcal{P}'$ from $\mathcal{P} \setminus \mathcal{P}'$. That is, $\mathcal{P}'$ is on one open half-plane defined by the line, while $\mathcal{P} \setminus \mathcal{P}'$ is on the other half-plane. The following figure depicts two 5-sets of a set of 12 points.

The k-set problem asks for the maximum number of k-sets that a set of n-points in $\mathbb{R}^2$ can have. As a first silly example, note that the maximum number of 1-sets is exactly n. It is not difficult to show that we may assume that no three points are collinear (specifically, perturbing the point set cannot significantly decrease the number of k-sets).

The number of subsets of k points of $\mathcal{P}$ is $\Theta(n^k)$, so this is a straightforward upper bound for the maximum number of k sets. However, we can easily obtain a significantly stronger upper bound. Consider a k-set $\mathcal{P}' \subset \mathcal{P}$ that is separated by a line ℓ. We translate ℓ until it is incident to a point $\mathcal{P} \in \mathcal{P}'$ and then rotate ℓ around p until it is incident to a second point of $\mathcal{P}$ (not necessarily in $\mathcal{P}'$). For example, see the figure below. Thus, to find all of the k-sets of $\mathcal{P}$ it suffices to consider every line that is incident to two points of $\mathcal{P}$. Since there are $O(n^2)$ such lines, this is also an upper bound for the number of k-sets.

The current best upper bound for the maximum number of k-sets is $O(nk^{1/3})$. This bound was derived by Dey about 20 years ago. Similarly to the pre-algebraic era in incidence and distance problems, this bound was obtained by using the crossing lemma. In the other direction, Tóth constructed a set of n points and $ne^{\Omega(\sqrt{\log k})}$ k-sets.

A Dual problem: The k-set problem has an interesting dual formulation. Consider only the k-sets $\mathcal{P}' \subset \mathcal{P}$ that are separated by a line ℓ such that $\mathcal{P}'$ is above the line (the case of k-sets below the separating line can be handled symmetrically).

Given a point, $a = (a_x, a_y) \in P$, we define the *dual line* a^* by the equation $y = a_x x - a_y$. Given a line $\ell \subset \mathbb{R}^2$ defined by $y = bx - c$, we define the *dual point* as $\ell^* = (b, c)$. Note that a is incident to ℓ if and only if ℓ^* is incident to a^*. Indeed, both cases occur if and only if $c = ba_x - a_y$. By replacing the equality with $\geq$, we get that a is above ℓ if and only if ℓ^* is above a^*.

Consider the set of lines,

$$\mathcal{L} = \{p^* : p \in \mathcal{P}\}$$

Let ℓ be a line that has a k-set $\mathcal{P}' \subset \mathcal{P}$ above it and $\mathcal{P} \setminus \mathcal{P}'$ below it. Let $\mathcal{P}^*$ be the set of lines dual to the points of $\mathcal{P}'$. Then the point ℓ^* is above the lines of $\mathcal{P}^*$ and below the lines of $\mathcal{L} \setminus \mathcal{P}^*$. That is, a point of $\mathbb{R}^2$ is dual to a line that separates a k-set if and only if it has exactly k lines of $\mathcal{L}$.

As stated above, it suffices to consider k-sets where the separating line is incident to at least two points of $\mathcal{P}$. The dual statement says that it suffices to bound the points that are incident to at least two lines of $\mathcal{L}$. That is, the k-set problem was reduced to the following:

Corollary: *Let $\mathcal{P}$ be a set of m points in the plane and let Π be a set of n open half-planes, both in $\mathbb{R}^2$. Let k be a positive integer (which may depend on m and n) such that the incidence graph of $\mathcal{P} \times \Pi$ contains no copy of $K_{2,k}$. Then,*

$$I(P,\Pi) = O(m^{2/3}n^{2/3}k^{1/3} + n + mk).$$

We can reduce the *k*-set problem to a similar incidence problem. Recall that we are given a set $\mathcal{P}$ of n points in $\mathbb{R}^2$ and can think of every *k*-set as an open half-plane that contains exactly k points. The intersection of two such half-planes contains at most $k-1$ common points of $\mathcal{P}$. That is, the number of k-sets is upper bounded by the number of *k*-rich half-planes when assuming that the incidence graph contains no copy of $K_{k,2}$. By a duality argument as described above, we get to the problem of bounding the number of k-rich points when there is no $K_{2,k}$ in the incidence graph.

Combinatorial Bounds

It is of importance in the analysis of geometric algorithms to bound the number of *k*-sets of a planar possint set, or equivalently the number of *k*-levels of a planar line arrangement, a problem first studied by Lovász and Erdős. The best known upper bound for this problem is $O\left(nk^{1/3}\right)$, as was shown by Tamal Dey using the crossing number inequality of Ajtai, Chvátal, Newborn, and Szemerédi. However, the best known lower bound is far from Dey's upper bound: it is $\Omega\left(n\,exp\left(c\left(logk\right)^{1/2}\right)\right)$ for some constant c, as shown by Toth.

In three dimensions, the best upper bound known is $O\left(nk^{3/2}\right)$, and the best lower bound known is $\Omega(nk\,exp(c\left(logk\right)^{1/2}))$. For points in three dimensions that are in convex position, that is, are the vertices of some convex polytope, the number of k-sets is $\Theta((n-k)k)$, which follows from arguments used for bounding the complexity of k-th order Voronoi diagrams.

For the case when $k = n/2$ (halving lines), the maximum number of combinatorially distinct lines through two points of *S* that bisect the remaining points when k = 1, 2, ... is 1,3,6,9,13,18,22... (sequence A076523 in the OEIS).

Bounds have also been proven on the number of $\le k-sets$, where a $\le k-sets$ is a *j*-set for some $j \le k$. In two dimensions, the maximum number of $\le k-sets$ is exactly *nk*, while in *d* dimensions the bound is .

Construction Algorithms

Edelsbrunner and Welzl first studied the problem of constructing all *k*-sets of an input point set, or dually of constructing the *k*-level of an arrangement. The *k*-level version of their algorithm can be viewed as a plane sweep algorithm that constructs the level in left-to-right order. Viewed in terms of *k*-sets of point sets, their algorithm maintains a dynamic convex hull for the points on each side of a separating line, repeatedly finds a bitangent of these two hulls, and moves each of the two points of tangency to the opposite hull. Chan surveys subsequent results on this problem, and shows that it can be solved in time proportional to Dey's $O(nk^{1/3})$ bound on the complexity of the *k*-level.

Agarwal and Matoušek describe algorithms for efficiently constructing an approximate level; that is, a curve that passes between the (k - d)-level and the (k + d)-level for some small approximation parameter d. They show that such an approximation can be found, consisting of a number of line segments that depends only on n/d and not on n or k.

Matroid Generalizations

The planar k-level problem can be generalized to one of parametric optimization in a matroid: one is given a matroid in which each element is weighted by a linear function of a parameter λ, and must find the minimum weight basis of the matroid for each possible value of λ. If one graphs the weight functions as lines in a plane, the k-level of the arrangement of these lines graphs as a function of λ the weight of the largest element in an optimal basis in a uniform matroid, and Dey showed that his $O(nk^{1/3})$ bound on the complexity of the k-level could be generalized to count the number of distinct optimal bases of any matroid with n elements and rank k.

For instance, the same $O(nk^{1/3})$ upper bound holds for counting the number of different minimum spanning trees formed in a graph with n edges and k vertices, when the edges have weights that vary linearly with a parameter λ. This parametric minimum spanning tree problem has been studied by various authors and can be used to solve other bicriterion spanning tree optimization problems.

However, the best known lower bound for the parametric minimum spanning tree problem is $\Omega(n\alpha(k))$, where α is the inverse Ackermann function, an even weaker bound than that for the k-set problem. For more general matroids, Dey's $O(nk^{1/3})$ upper bound has a matching lower bound.

Heilbronn Triangle Problem

Hans Heilbronn, the Jewish German mathematician, raised the following question in the early 1950's: Given n points to be situated in the unit square, what is the largest possible smallest triangle area defined by them? Originally, the question was posed as locating the points on a closed convex region, not necessarily the unit square, but it was later proved that as long as the containing shape is convex and of unit area, the optimal values for all shapes differ only by a multiplicative constant factor, irrespective of the shape.

To state the problem more formally, we will call $H(S)$ Heilbronn's value for a set of point S, and define it as the minimal triangle area out of all triangles determined by points of S. We will define $H(n)$ as Heilbronn's optimal value for any set S of n points in the unit square. Despite the simplicity of its formulation, no general solution method is yet known. Furthermore, no optimal solutions are known even for small numbers of points.

Let $P_uP_2,\ldots,P_n$ be a distribution of n points (where, $n \geq 3$) in a closed convex region K of unit area such that the minimum of the areas of the triangles $P_jP_jP_k$, assumes its maximum possible value $\Delta(K;n)$.

We write $\Delta(n) = \Delta(D;n)$ if D is a disc.

Heilbronn conjectured over 30 years ago that $\Delta(n) << n^{-2}$, where $A << B$ means $A \leq cB$ with an absolute constant c.

It was proved by Erdos that,

$$\Delta(n) >> n^{-2},$$

So, that Heilbronn's conjecture, if true, would be best possible.

In the direction of Heilbronn's conjecture it is trivial to show that $(n) << n^{-1}$ (by selecting $n-2$ triangles of type $P_iP_jP_k$ with disjoint interiors). The first nontrivial estimate,

$$\Delta(n) << n^{-1} (\log\log n)^{-1/2}$$

was proved by $K.\ F.$ Roth in 1950. This was improved 20 years later by W. M. Schmidt to

$$\Delta(n) << n^{-1} (\log n)^{-1/2}$$

making use of a different method.

Soon after this, using an entirely new method, Roth proved the first result of the type,

$$\Delta(n) << n^{-\mu} \quad (\mu > 1)$$

With the value,

$$\mu = 2-\left(2/y\sqrt{5}\right)-\varepsilon = 1-105... -\varepsilon.$$

This method refined to yield $\Delta(n) << n^{-\mu} \quad (\mu > 1)$ with the value,

$$\mu = (17-\sqrt{65})/8 - \varepsilon = 1-1\cdot17... -\varepsilon.$$

Very recently using combinatorial methods, the authors disproved Heilbronn's conjecture by showing that:

$$\Delta(n) >> n^{-2} (\log n).$$

The aim is to refine the method of Roth to yield $\Delta(n) << n^{-\mu} \quad (\mu > 1)$ the value

$$\mu = (8/7)-\varepsilon = 1\cdot142. ..-\varepsilon$$

In Using the notation $A <<< B$ for

$$A \le exp\left(c\sqrt{\log n}\right)B$$

(where c is an explicit absolute positive constant) we shall actually prove the following theorem:

Theorem: $\Delta(n) >> n^{-8/7}.$

We shall prove the theorem in the equivalent form,

$$\Delta(S^* ; n) >> n^{-8/7}$$

Where S^* is a square (of unit area). It is easy to show, using linear transformations, that if the region K is a triangle T, then the value of $\Delta(T;\, n)$ is independent of the shape of the triangle, and further any K contains a triangle of area $1/4$ and is contained in a triangle of area 4. Thus, one gets, for any K,

$$\Delta(K;n) << \Delta(n) << \Delta(K;n).$$

This means that, substituting $\Delta(n)$ by the general $\Delta(K;n)$, the present knowledge in Heilbronn's triangle problem can be formulated by the inequality,

$$n^{-2}\,(\log n) << \Delta(K;n) <<< n^{-8/7}$$

Variants of the Problem

Some additional aspects of the problem are discussed in the literature. One of them resolves Heilbronn's value in the expected case—when the n points are chosen independently,

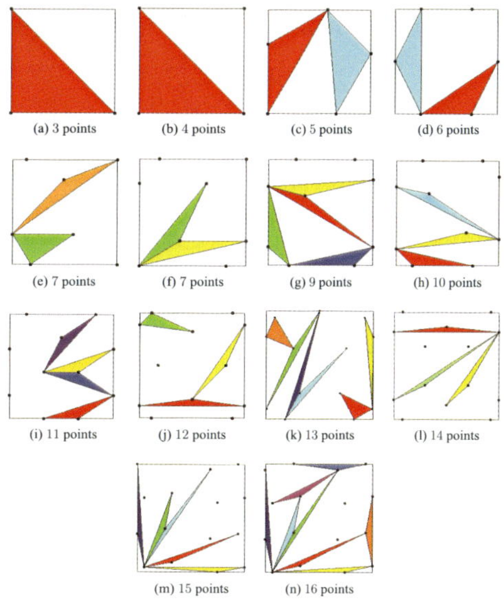

Figure: Best known configurations for 3 to 16 points.
The highlighted triangles are some of the minimal triangles in the configurations.

Table: Heilbronn's values of all best known configurations for 3 to 16 points.

Number of points	Heilbronn's value	Number of points	Heilbronn's value
3	0.5	10	0.0465...
4	0.5	11	0.0370...
5	0.1924...	12	0.0325...
6	0.125	13	0.0266...
7	0.0838...	14	0.0243...
8	0.0723...	15	0.0207...
9	0.0548...	16	0.0205...

At a uniform-probability random location in the unit square. Jiang, Li, and Vitányi showed that Heilbronn's value is $\Theta\left(\frac{1}{n^3}\right)$ in the expected case.

Another variant is a generalization of Heilbronn's problem to higher dimensions, which was first investigated by Barequet. When talking about the d-dimensional problem, the target is to maximize the volume of the minimal simplex defined by $d + 1$ out of the n points located in the d-dimensional unit cube.

The last variant presented here is the on-line version of Heilbronn's problem. In this problem, the number of points is not known in advance. The points are located in the unit square (or cube in higher dimensions) one after another, until this sequence stops without prior notice. This problem was investigated by Barequet in 3 and 4 dimensions, and was later generalized by Barequet and Shaikhet to the d-dimensional case.

Maximizing a Multiextremal Function

As explained above, the function $H(n)$ is defined as the global maximum of the function $H(S)$ over all sets of n points. The problem of finding this global maximum is very hard because $H(S)$ is a multiextremal extremely-flat function, taking $2n$ variables—two variables (x and y coordinates) for each point in the set.

Helly's Theorem

Helly's theorem is a result from combinatorial geometry that explains how convex sets may intersect each other. The theorem is often given in greater generality, though for our considerations, we will mainly apply it to the plane.

Theorem: (Eduard Helly) For a finite collection of convex subsets $X_1, X_2, \ldots, X_n \subset R^d$, where $n > d$, if the intersection of every $d + 1$ of these sets is nonempty, then

$$\bigcap_{j=1}^{n} X_j \neq \varnothing.$$

Proof:

By mathematical induction, Let proposition P_n be Helly's Theorem in the case of n subsets in $\mathbb{R}^d$. Since, $n > d$, we can use P_{d+1} as our base case. P_{d+1} is clearly true, because if the intersection of $d + 1$ of them are non-empty, then the intersection of all of them are non-empty.

Lemma A: (Johann Radon) any set with $d + 2$ points in $\mathbb{R}^d$ can be partitioned into 2 disjoint, non-empty sets such that the convex hulls of these sets have a non-empty intersection.

Proof:

Let $X = \{p_1, p_2, \ldots, p_{d+2}\} \subset \mathbb{R}^d$, let $X' = \{p'_1, p'_2, \ldots, p'_{d+2}\} \subset \mathbb{R}^{d+1}$, let $\vec{pi}$ be the position vector of the point pi and let the vector $\vec{pi}$ be such that,

$$\vec{p}_i = \begin{bmatrix} a_1 \\ a_2 \\ \vdots \\ a_d \end{bmatrix} \text{ then } \vec{p}_i{}' = \begin{bmatrix} a_1 \\ a_2 \\ \vdots \\ a_d \\ 1 \end{bmatrix}.$$

Since $p_i' \in \mathbb{R}^{d+1}$, the position vectors of the points in X' cannot be mutually linearly independent because $|X'| = d+2$. This means that there exists some non-trivial solution for the equation,

$$\sum_{i=1}^{d+2} \alpha_i \vec{p}_i = 0.$$

Which, implies that there exists some non-trivial solution to $\alpha_1, \alpha_2, \ldots, \alpha_n$ *s.t.* the two equations,

$$\sum_{i=1}^{d+2} \alpha_i \vec{p}_i = 0 \quad \text{and} \quad \sum_{i=1}^{d+2} \alpha_i = 0$$

are satisfied.

Let $I = \{1 \le i \le d+2 \mid \alpha_i > 0\}$, let $J = \{1 \le j \le d+2 \mid \alpha_j \le 0\}$, let $K = \{p_i \mid i \in I\}$ *and* let $L = \{p_i \mid i \in J\}$.

It is easy to see that $K, L \neq \varnothing$ because if either of them is empty, then either

$\sum_{i=1}^{d+2} \alpha_i = 0$ is not satisfied or the solution to $\sum_{i=1}^{d+2} \alpha_i \vec{p}_i = 0$ is trivial.

This means that the point p with position vector $\vec{\mathrm{p}} = \dfrac{\sum_{i \in I} \alpha_i \vec{p}_i}{\sum_{i \in I} \alpha_i}$ exists, and is in the convex hull of K because it is a convex combination of the position vectors of the points in K, i.e.

$$\forall i \in I \frac{\quad}{\sum_{i \in I}} \ge \quad \text{and} \quad \sum_{i \in I} \left(\frac{\alpha_i}{\sum_{i \in I} \alpha_i}\right) = 1.$$

Also,

$$\begin{aligned} \vec{\mathrm{p}} &= \frac{\sum_{i \in I} \alpha_i \vec{p}_i}{\sum_{i \in I} \alpha_i} \\ &= \frac{\sum_{i=1}^{d+2} \alpha_i \vec{\mathrm{p}}_i - \sum_{i \in J} \alpha_i \vec{p}_i}{\sum_{i=1}^{d+2} \alpha_i - \sum_{i \in J} \alpha_i} \\ &= \frac{0 - \sum_{i \in J} \alpha_i \vec{p}_i}{0 - \sum_{i \in J} \alpha_i} \\ &= \frac{\sum_{i \in J} \alpha_i \vec{p}_i}{\sum_{i \in J} \alpha_i} \end{aligned}$$

This shows that p is also in the convex hull of L because the position vector of p is a convex combination of the position vectors of the points in L. Hence, $p \in K \cap L$, so the convex hulls of K and L have a non-empty intersection.

Lemma B: The intersection of any 2 convex sets is a convex set.

Proof:

By contradiction, Let A, B be convex sets, and assume that $A \cap B = C$ is not convex. This implies that there exists two points $a, b \in C$, with position vectors $\vec{a}, \vec{b}$ respectively, such that for some $\alpha \in [0, 1]$, the point p with position vector $\vec{p} = \alpha\vec{a} + (1 - \alpha)\vec{b}$ is not in C. Since $p \notin C$, we can assume without loss of generality that $p \notin A$, which means that A is not convex because $a, b \in C$ implies $a, b \in A$, contradiction.

Although we already have a base case, we shall now consider P_{d+2}, which will later be used in conjunction with the inductive hypothesis to prove the inductive step.

Choose a common point p_i of all sets X_j, where $j \neq i$. We assume that $p_i \notin X_i$, since otherwise we are done. Let $A = \{p_1, \ldots, p_{d+2}\}$.

By Lemma A, there exists a nontrivial, disjoint partition A_1, A_2 of A such that the convex hulls of A_1 and A_2 intersect at some point p. Also, observe that $\forall i \in [d + 2]$, the only point that is not in X_i but is in A is pi . Note that since $p_i \in A$ and $A_1 \cup A_2 = A$, we can assume without loss of generality that $p_i \in A_1$. This means that $p_i \notin A_2$, so $A_2 \subset X_i$. Since X_i is convex, it has to contain the convex hull of A_2 , and in particular, the point p.

Hence, p is common to all the X_i's, and so P_{d+2} is true.

Now, we are ready to prove the inductive step. Assume there exists some $k \in \mathbb{N}$ with $k > d$ such that P_k is true.

Consider, P_{k+1}, and let $Y_i = X_i \cap X_{k+1}$.

$$\bigcap_{R \subset [k], \|R\| = d+1} Y_i = \left[\bigcap_R X_i\right] \cap X_{k+1}$$

$$\neq \varnothing \qquad \because P_{d+2} \textit{ is true}.$$

By Lemma B, $\forall i \in [k + 1]$, Y_i is also convex. Since the Y_i 's are convex and every $d + 1$ of them have a nonempty intersection, by the inductive hypothesis, $\bigcap_{i=1}^{k} Y_i \neq \varnothing$.

Which, implies that $\bigcap_{i=1}^{k+1} X_i \neq \varnothing$. Hence, P_k is true implies that P_{k+1} is true, and this proves the theorem.

Now that we have proven Helly's theorem for a finite number of convex sets in $\mathbb{R}^d$, we will try to extend this theorem to an infinite number of convex sets. However, we have to add an additional restriction of compactness in place of removing the finiteness restriction on the number of sets.

Theorem: For any infinite collection of convex, compact subsets $X_1, X_2, \cdots \in \mathbb{R}^d$, if the intersection of every $d + 1$ of these sets is nonempty, then:

$$\bigcap_{j=\infty} X_j \neq \varnothing.$$

Before we attempt to prove this theorem, let us demonstrate that the restriction of compactness is necessary. This will be done by creating two counter-examples; in the first, we will show that restricting ourselves only to closed, convex sets is insufficient, while in the second, we will show that restricting ourselves to bounded, convex sets is also insufficient.

False Assertion: For any infinite collection of convex, closed subsets $X_1, X_2, \cdots \in \mathbb{R}^d$, if the intersection of every $d + 1$ of these sets is nonempty, then

$$\bigcap_{j=1}^{\infty} A_i \neq \varnothing.$$

Which, is contrary to Proposition.

False Assertion: For any infinite collection of convex, bounded subsets $X_1, X_2, \cdots \in \mathbb{R}^d$, if the intersection of every $d + 1$ of these sets is nonempty, then

$$\bigcap_{j=\infty}^{\infty} X_j \neq \varnothing\cdot$$

This counter-example will again be in the case where d = 1. Consider the sets $A_i = \left(0, \frac{1}{i}\right)$, where $i \in N$. The A_i 's are clearly bounded, and any two Ai 's have a non-empty intersection because $A_i \subset A_j$ *if* $i > j$. Again however,

$$\bigcap_{i=1}^{\infty} A_i \neq \varnothing$$

Now, we will prove Helly's theorem for an infinite number of compact, convex sets.

Proof:

In Theorem, we proved that for any n convex sets $A_1, \ldots, A_n$ with every $d + 1$ of them having a nonempty intersection, there exists some point x_n s.t.

$$x_n \in \bigcap_{i=1}^{n} A_i.$$

Consider the infinite set $Z_i = \{x_i, x_{i+1}, \ldots\}$. For all $j \in N, j > i, x_j \in A_i$ because x_j is a point in the intersection of a set of sets, one of which is A_i. This means that for all $i \in N$, $Z_i \subset A_i$. Since we restricted that A_i to be compact, A_i is bounded, and so its subset Z_i also has to be bounded. We also know that Z_i is infinite, which implies that Z_i has at least one limit point, because any infinite bounded set has a limit point.

Observe that if $i > j$ with $i, j \in N$, then $Z_i \subset Z_j$, which means that the set of limit points of is a subset of the set of limit points of Z_j. Since, every Z_i has at least one limit point, this means that for every $i > j$, Z_i shares at least one limit point with Z_j. Also, for all $k < j$ with

$k \in \mathbb{N}$, $Z_j \subset Z_k$, and this implies that all the limit points of Z_j are also limit points of Z_k, which means that for every $k < j$, Z_k and Z_j share a limit point. Since this is true for all $j \in \mathbb{N}$, all the Z_i 's share a limit point, q.

Now, $Z_i \subset A_i$, the limit points of Z_i are also limit points of A_i. This means that q is a limit point of all the A_i 's because it is a limit point of all the Z_i's. Moreover, since all the A_i 's are closed, for every $i \in \mathbb{N}$, $q \in A_i$.

So,

$$\bigcap_{i=1}^{\infty} A_i \neq \varnothing .$$

Tverberg's Theorem

A set S of points in d-dimensional Euclidean space, whose cardinality exceeds $(d + 1)(r - 1)$, r a positive integer, can be partitioned into r subsets $S_1, S_2, \ldots, S_r$, whose convex hulls have non-empty intersection:

$$S = \bigcup_{i=1}^{r} S_i, \qquad for \quad i \neq j, S_i \cap s_j = \varnothing \qquad \bigcap_{i=1}^{r} \text{conv}(S_i) \neq \varnothing$$

With, $d = r = 3$, Tverberg's theorem requires at least $1 + (3-1)(3+1) = 9$ points.

A set of 8 points which cannot be partitioned into three subsets satisfying the conclusion of Tverberg's theorem, by adding any single distinct point we can apply Tverberg's theorem. Here the point $i = (2, 3, 0)$ has been added; the 9 points have been partitioned into 3 subsets of 3 points (whose convex hulls are triangles), intersecting in the point $(1/2, 3/2, 1/2)$.

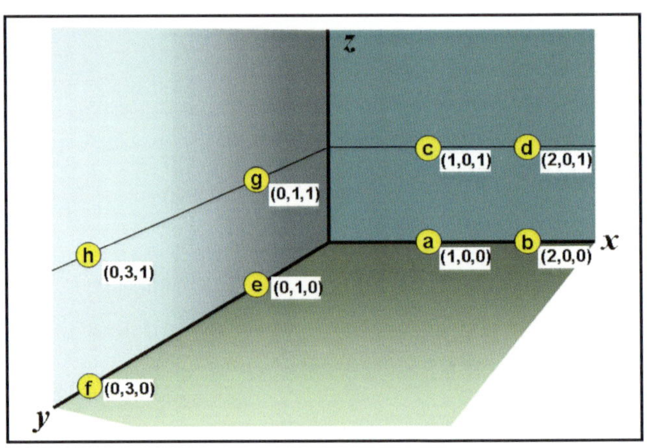

There is an important 1979 conjecture of Gerard Sierksma which says that there are at least $((r-1)!)^d$ partitions satisfying Tverberg's theorem. There are a total of six different 3-triangle partitions of the nine integer points on the right; as an exercise you may like to try to find a further two partitions to reach Sierksma's lower bound of: $((3 - 1)!) = 8$.

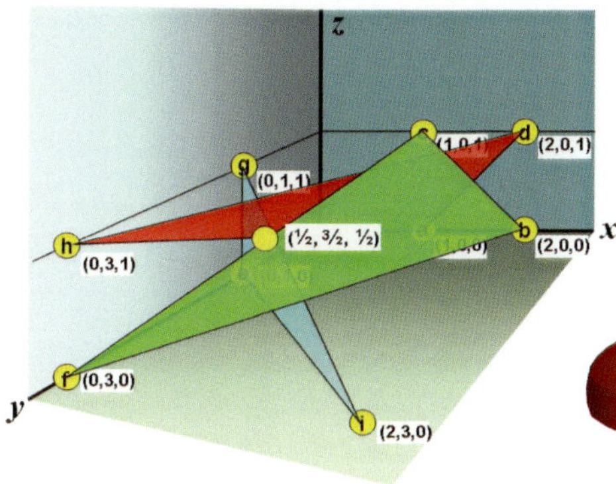

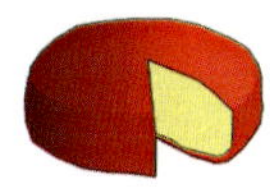

G. Sierskma's conjecture is known as the Dutch Cheese Problem because he offered a Web link: Dutch cheese for its solution.

Conjecture

$$T(d,\, r,\, k) \;=\; T(d,\, r) \text{ for all } 2 \le k \le r\,.$$

The meaning of Reay's conjecture is: If $n < T(d,\, r)$ then there exists a set $X \subset \mathbb{R}^d, |X| = n,$ such that for every r-partition of X there are two parts whose convex hulls are disjoint.

We don't really believe this is true. To press our point, consider the case $d = r = 1000$. By Tverberg's theorem, a million points in $\mathbb{R}^{1000}$ can be partitioned into one thousand parts whose convex hulls have a common point. Is there a set of $999{,}999$ points in $\mathbb{R}^{1000}$ that cannot be partitioned into 1000 parts whose convex hulls intersect just pairwise.

Theorem: For every dimension $d \ge 2$ and for every $r\left(\ge \left[\frac{d+3}{2}\right]\right)$:

$$T\,(d, r, \left[\frac{d+3}{2}\right]) = Y(d\,, r) = (d+1)(r-1)+1.$$

Properties of the spanning set: $P = \{p_0,\, p_1,\, \ldots,\, p_d\}$.

Proposition: Every point $x \in \mathbb{R}^d$ has a representation,

$$x = \xi_0 p_0 + \xi_1 p_1 + \cdots + \xi_d p_d$$

Where, $min\{\xi_0,\, \xi_1,\, \ldots,\, \xi_d\} = 0$. This representation is unique.

Proof: The vectors $p_0,\, p_1,\, \ldots,\, p_d$ span $\mathbb{R}^d$ linearly. In fact, each d of them form a linear basis of $\mathbb{R}^d$. Let $x = \sum_{i=0}^{d} \alpha_i p_i$ be some fixed representation of x in terms of P. The only linear dependences among $p_0,\, p_1,\, \ldots,\, p_d$ are $\sum_{i=0}^{d} \lambda p_i = 0\,, \lambda \in \mathbb{R}$. Therefore the most general representation of x in terms of P is $\sum_{i=0}^{d} (\alpha_i - \lambda)\, p_i \;\; , \lambda \in \mathbb{R}$. To obtain a representation with the smallest coefficient equal 0, we must choose $\lambda = \min\{\alpha_i \,:\, i \in D\}$.

We call the non-negative representation of x (in terms of P). The support of x, (with respect to P) is defined by,

$$supp\ x = \{i \in D : \xi_i > 0\}.$$

Simple properties of supp x:

- $\varnothing \subseteq$ supp x D.
- supp $x = \varnothing$ iff $x = \underline{0}$.
- supp $p_i = \{i\}$.
- supp $\lambda x =$ supp x for $\lambda > 0$.
- $\text{supp}(x + y) \subseteq \text{supp } x \cup \text{supp } y$, with equality iff supp $x \cup$ supp $y \neq D$.
- If $x \neq 0$, then supp $x \cup \text{supp}(-x) = D$.

Recall that our set X consists of r–1 distinct points on each ray R_i $(i \in D)$. For a subset $C \subseteq X$, define $I(C) = \{i \in D : C \cap R_i \neq \varnothing\}$. Now make the following observations:

Proposition: If $C \subseteq X$ and $x \in$ conv C, then $supp\ x \subseteq I(C)$. (This is obviously true also when $I(C) = D$).

When I is a subset of D, we shall denote by $R(I)$ the union $\cup\{R_i : i \in I\}$.

Proposition: Suppose, $C \subseteq X$, and $x \in$ conv C. If $I(C)$ 6= D then $x \in \text{conv}\{C \cap R(supp\ x)\}$.

Proof: Suppose $x = \sum_{v=1}^{n} \gamma_v c_v$, where $c_v \in C, \gamma_v > 0, \sum_{v=1}^{n} \gamma_v = 1$. If $c_v = \lambda_v p_i \in R_i$, $\lambda_v > 0$, then p_i will appear with a positive coefficient in the non-negative representation of x in terms of P, and therefore $i \in supp\ x$. Note that we have used the fact that $I(C) \neq D$.

For points $a = \alpha p_i \in R_i$, $b = \beta p_i \in R_i$, $(\alpha, \beta > 0)$ we say that a is lower than b (or b is higher than a) on R_i if $\alpha < \beta$ (or, equivalently, if $\|a\| < \|b\|$).

Proposition: Suppose, $I \subseteq D$. Let C, C' be two finite subsets of $R(I)(= \cup\{R_i : i \in I\})$. If, for each $i \in I$, every point of $C \cap R_i$ is lower (on R_i) than every point of $C' \cap R_i$, then conv $C \cap$ conv $C' = \varnothing$.

Proof: Assume, w.l.o.g., that $|I| = d$. (We do not assume that $C \cap R_i \neq \varnothing$ and $C' \cap R_i \neq \varnothing$ for all $i \in I$).

For each, $i \in I$ choose a point $s_i = \sigma_i p_i \in R_i$ that is higher (on R_i) than every point of $C \cap R_i$ and lower than every point of $C' \cap R_i$. The d points s_i $(i \in I)$ are linearly independent, and their affine hull $H = \text{aff}\{s_i : i \in I\} \subset \mathbb{R}^d$ is a hyperplane that does not pass through the origin. Denote by H_-, H_+ the two open half spaces determined by H, and assum $\underline{0} \in H_-$. From our assumptions it follows that $C \subset H_-$ and $C' \subset H_+$.

Hence, conv $C \cap$ conv $C' = \varnothing$.

Proposition: Suppose, $U \subseteq D$. Let $C_1, C_2, \ldots, C_n$ $(n \geq 2)$ be subsets of X.

Assume,

- $\bigcap_{V=1}^{n} I(C_V) \subseteq U$.
- $I(C_V) \subseteq D$ *for* $\nu = 1, 2$.

For each, $i \in U$, each point of $C_1 \cap R_i$ is lower (on R_i) than every point of $C_2 \cap R_i$.

Then, $\bigcap_{V=1}^{n} \operatorname{conv} C_V = \varnothing$.

Proof: Assume, on the contrary, that $\bigcap_{V=1}^{n} \operatorname{conv} C_V \neq \varnothing$, and suppose that $x \in \bigcap_{V=1}^{n} \operatorname{conv} C_V$. By Proposition we conclude that supp $x \subseteq \bigcap_{V=1}^{n} I(C_V) \subseteq U$. Applying Proposition C_1 and C_2, we find that,

$$x \in \operatorname{conv}(C_\nu \cap R(U)) \text{ for } \nu = 1, 2.$$

Now invoke proposition $C = C_1 \cap R(U)$, $C' = C_2 \cap R(U)$, and $I = U$, to conclude that $\operatorname{conv}(C_1 \cap R(U)) \cap \operatorname{conv}(C_2 \cap R(U)) = \varnothing$, which contradicts.

Proof of Theorem

For the proof we will use the following (counter) example: let $p_0, p_1, \ldots, p_d \in \mathbb{R}^d$ be the vertices of a d-simplex centered at the origin, i.e., $\sum_{i=0}^{d} p_i = \underline{0}$ and each d of the points $p_0, p_1, \ldots, p_d$ are linearly independent. Let $D = \{0, 1, \ldots, d\}$ and for $i \in D$ define $R_i = \{\lambda p_i : \lambda > 0\}$ (the open ray emanating from o through p_i).

On each ray R_i we choose $r - 1$ distinct points. The chosen points form a set $X \subset \mathbb{R}^d$, $|X| = (d+1)(r-1) = T(d, r) - 1$. We show that in every partition of X into r parts $(X = C_1 \cup \cdots \cup C_r)$ there is $j \leq \left\lfloor \frac{d+3}{2} \right\rfloor$ and there are parts $C_{i_1}, \ldots, C_{i_j}$, whose convex hulls have empty intersection.

This will show that, $T(d, r, k) = T(d, r)$ $\frac{d+3}{2} \leq k \leq r$. We start with some preliminaries concerning the "positive basis" $P = \{p_0, p_1, \ldots, p_d\}$ of $\mathbb{R}^d$.

Let $X \subset R$ d be the set described at the beginning of this topic ($r - 1$ points on each of the rays $R_0, R_1, \ldots, R_d$), and let $C_1, \ldots, C_r$ be an arbitrary partition of X into r disjoint sets. Our aim is to apply Proposition to some n of the parts C_i, with n as small as possible. We shall be able to do this with some $n \leq \left\lfloor \frac{d+3}{2} \right\rfloor$.

Assume the parts C_i are ordered in such a way that,

- $|C_1| \leq |C_i|$ *for* $i = 2, 3, \ldots, r$.
- $|C_2 \cap R(I(C_1))| \leq |C_i \cap R(I(C_1))|$ *for* $i = 3, 4, \ldots, r$.

From condition ($|C_1| \le |C_i|$ *for* $i = 2, 3, \ldots, r.$) we have,

$$|I(C_1)| \le |C_1| \le \left[\frac{1}{r}|X|\right]\left[\frac{r-1}{r}(d+1)\right],$$

And therefore $|I(C_1)| \le |C_1| \le d.$

Condition ($|C_2 \cap R(I(C_1))| \le |C_i \cap R(I(C_1))|$ *for* $i = 3, 4, \ldots, r$) yields:

$$|C_2 \cap R(I(C_1))| \le \frac{1}{r-1}\sum_{i=2}^{r}|C_i \cap R(I(C_1))| = \frac{1}{r-1}\left|\bigcup_{i=2}^{r} C_i \cap R(I(C_1))\right|$$

$$\frac{1}{r-1}\,|(X\ C_1) \cap R(I(C_1))| \le \frac{r-2}{r-1}|I(C1)|$$

And therefore,

$$|C_2 \cap R(I(C_1))| < |I(C_1)|$$

This, in turn, implies $|I(C_2)| \le d$

Next, we define the set U to be plugged into Proposition.

For $i = 1, 2$ we divide $I(C_i)$ into two disjoint sets:

$$S_i = \{j \in D : |C_i \cap R_j| = 1\}$$

$$M_i = \{j \in D : |C_i \cap R_j| > 1\}$$

And get:

$$|C_i| \ge |I(C_i)| + |M_i|.$$

Furthermore, for every subset J of D:

$$|C_i \cap R(J)| \ge |I(C_i) \cap J| + |M_i \cap J|.$$

Assume $|I(C_1)| = d - a\ (a \ge 0)$. From ($|C_i| \ge |I(C_i)| + |M_i|$).

We obtain,

$$|M_1| \le |C_1| - |I(C_1)| \le d - (d - a) = a$$

And from,

$$|C_2 \cap R(I(C_1))| \le d - a - 1.$$

The set $S_1 \cap S_2$ can be divided into two disjoint subsets:

$$U_1 = \{j \in S_1 \cap S_2 : C1 \textit{ is lower than } C_2 \textit{ on } R_j\},$$

$$U_2 = \{j \in S_1 \cap S_2 : C_2 \textit{ is lower than } C_1 \textit{ on } R_j\}.$$

If $|U_1| \geq |U_2|$ we define, $U = U_1$, otherwise we define $U = U_2$. In any case, $|U| \geq \frac{1}{2}|S_1 \cap S_2|$.

We choose for each index $i \in I(C_1) \cap I(C_2) \setminus U$ a set C_q $(3 \leq q \leq r)$ that does not meet Ri, and call it $C(i)$. (Such a set does exist, since $|(X \setminus (C_1 \cup C_2)) \cap R_i| \leq r - 3$). Note that the sets $C(i)$ $(i \in I(C_1) \cap I(C_2) \setminus U)$ are not necessarily distinct.

Under these conditions, the sets $C_1, C_2, \{C(i) : i \in I(C_1) \cap I(C_2) \setminus U\}$ satisfy the assumptions of Proposition $n \leq 2 + \left\lceil \frac{d-1}{2} \right\rceil = \left\lceil \frac{d+3}{2} \right\rceil$, and therefore $\operatorname{conv} C_1 \cap \operatorname{conv} C_2 \cap \bigcap_{i \in I(C_1) \cap I(C_2) \setminus U} \operatorname{conv} C(i) = \varnothing$.

To sum it up, we have shown that for every $d \geq 2$ and $r \geq \left\lceil \frac{d+3}{2} \right\rceil$ there is a set X of $(d+1)(r-1)$ points in $\mathbb{R}^d$, such that in r-partition of X, there are $\left\lceil \frac{d+3}{2} \right\rceil$ parts whose convex hulls have empty intersection.

References

- Cook, William; Seymour, Paul D. (1989), Polyhedral Combinatorics, DIMACS Series in Discrete Mathematics and Theoretical Computer Science, American Mathematical Society, ISBN 978-0-8218-6591-0
- Balinski, Michel L. (1961), "On the graph structure of convex polyhedra in n-space", Pacific Journal of Mathematics, 11: 431–434, doi:10.2140/pjm.1961.11.431
- Combinatorics-and-discrete-geometry: math.cornell.edu, Retrieved 21 March 2018
- Alon, N.; Győri, E. (1986). "The number of small semi-spaces of a finite set of points in the plane". Journal of Combinatorial Theory, Series A. 41: 154–157. doi:10.1016/0097-3165(86)90122-6
- Kalai, Gil; Ziegler, Günter M. (2000), Polytopes: Combinatorics and Computation, DMV Seminar, 29, Birkhäuser, ISBN 978-3-7643-6351-2
- The-k-set-problem: adamsheffer.wordpress.com, Retrieved 29 June 2018
- Blind, Roswitha; Mani-Levitska, Peter (1987), "Puzzles and polytope isomorphisms", Aequationes Mathematicae, 34 (2-3): 287–297, doi:10.1007/BF01830678, MR 0921106

7

Other Branches of Geometry

Geometry is a vast subject that branches out into a number of significant sub-disciplines. Some of the prominent branches discussed in this chapter include Spherical geometry, Incidence geometry, Non-Euclidean geometry, Projective geometry and Taxicab geometry.

Spherical Geometry

Spherical geometry is the study of geometric objects located on the surface of a sphere. Spherical geometry works similarly to Euclidean geometry in that there still exist points, lines, and angles. For instance, a "line" between two points on a sphere is actually a great circle of the sphere, which is also the projection of a line in three-dimensional space onto the sphere.

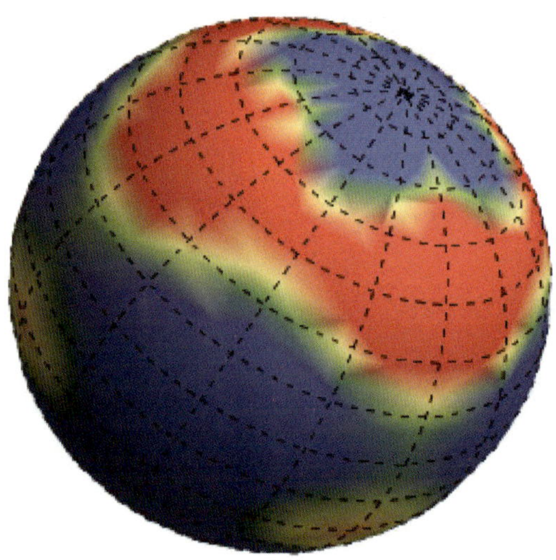

Spherical body

Properties of Spherical Geometry

In spherical geometry, there are no parallel lines. In fact, there are no straight lines. Therefore, in spherical geometry, a great circle is comparable to a line. There are no straight lines in spherical geometry. Instead, the shortest distance from one point to the next lying on a sphere is along the arc of a great circle. On a sphere, the angle between two curved arcs is measured by the angle formed from the intersection of the lines lying tangent to the two arcs.

When three curved arcs intersect one another, a spherical triangle is formed. A spherical triangle is any three-sided region enclosed by sides that are the arcs of great circles. Spherical triangles can have angle sums that range between one hundred eighty degrees and five hundred forty degrees. Spherical triangles can have angle sums larger than the usual one hundred eighty degrees found in a triangle because lines connecting points have a slight curve to them. Even so, spherical triangles can have ninety-degree angles just like triangles in the plane.

Differing from Euclidean geometry, two spherical triangles are not only similar, but congruent if they share the same angles. If one of the angles of a spherical triangle is a right angle, the triangle is known as a spherical right triangle, and a Spherical Pythagorean Theorem exists. In a spherical right triangle, let C denote the length of the side opposite the right angle. Let A and B denote the lengths of the other two sides. Let R denote the radius of the sphere.

Then, we can use the following formula:

$$\cos(C/R) = \cos(A/R)*\cos(B/R)$$

To calculate the area of a spherical triangle on a unit sphere, one must sum the angles of the spherical triangle (in radians) and subtract pi. For example, say a spherical triangle had two right angles and one forty-five degree angle. To find the area of the spherical triangle, restate the angles given in degrees to angles in radians. Thus, we are working with a spherical triangle with two pi/2 angles and one pi/4 angle. Add the three angles together (pi/2 + pi/2 + pi/4). Thus, the angles total 5pi/4. Subtract pi from 5pi/4 to find the area of the spherical triangle. Therefore, we have a total area of pi/4 units squared.

It has been established that a great circle is formed from a plane that intersects a sphere through its center. In analyzing two great circles that lie in a sphere, the two planes that form the great circles must intersect in a line, which in turn intersects the sphere at two distinct points. Therefore, two great circles intersect each other at two antipodal points. The planes also divide the sphere into four parts. These regions are known as lunes, which comes from the Latin word "luna," meaning moon. A lune is a two-sided polygon lying on a sphere.

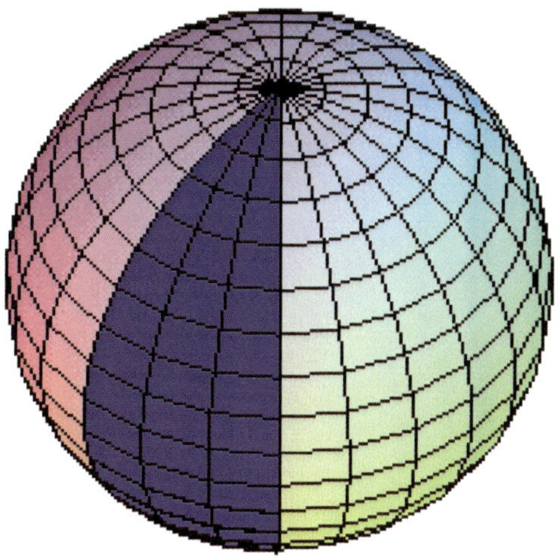

To find the area of a lune, one must know the lunar angle and the formula $2R^2$ (lunar angle), where R represents the radius of the sphere. If the lunar angle is not given, find p(pi)/q, where, p represents the union of the lunes and q is the number of equal lunes drawn in a hemisphere.

Knowledge about a lune's area can be helpful in solving the area of the sphere. For example, say the area of a lune is thirty-two centimeters squared, and we want to find the area of the sphere where the lune is present. The area of a sphere is 4piR^2, where R represents the radius. Since we know the formula for the area of a lune, we can solve for the radius, the unknown variable. Then we can use the radius in the formula for the area of the sphere to solve for the area of the sphere. After a few calculations, we can see that the radius is four centimeters. We then can square this and multiply the new result by 4pi. Thus, we have the sphere's area equal to 64pi centimeters squared.

The shortest distance between two points on a sphere is along the arc of a great circle. In spherical geometry, there exists a formula to find this distance. Given two points A and B, first locate the angle (measured in radians) that is created between the two points and the center of the sphere. Then, multiply by the radius (R) of the sphere, giving the formula d(A,B)= Ra, where a represents the angle formed between points A and B. For example, to find the distance between two points A and B with an angle of pi/4 between them on a sphere having a radius of 4, we multiply pi/4 by the radius length of 4 units to get a spherical distance of pi.

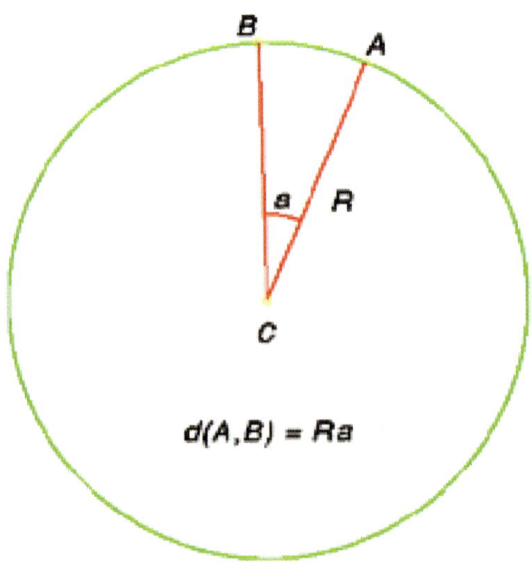

The minimal distance between two points is known as a geodesic and is the spherical analog of a line segment: an arc of a great circle. The distance between two points is therefore $R\varphi$, where R is the radius of the sphere and φ is the measure (in radians) of the central angle subtended by the radii to the two points.

Example:

What is the minimal distance on the sphere, centered at the origin and of radius 2, between points $\left(1,1,\sqrt{2}\right)$ and $\left(-1,1,\sqrt{2}\right)$?

The dot product of the vectors to the two points is $R^2\cos\varphi = 1\cdot(-1)+1\cdot1+\sqrt{2}\cdot\sqrt{2}=2$, implying $\cos\varphi=\frac{1}{2}$.

Therefore, $\varphi=\frac{\pi}{3}$, and the distance along the sphere is $R\varphi=\frac{2\pi}{3}$.

Example: The equator has a latitude of 0^0. Find its azimuthal position as measured from the North Pole.

The red arc represents the arc subtended by the azimuth.

In order to get from the North Pole to the equator, one must travel $\frac{1}{4}$ of the way around the circumference of the earth, so the azimuthal position of the equator is $\frac{1}{4}$ of 360° or 90°.

So, since latitude is measured off the equator, the convention is actually off of a true azimuthal measurement.

Relation to Euclid's Postulates

Spherical geometry obeys two of Euclid's postulates: the second postulate ("to produce [extend] a finite straight line continuously in a straight line") and the fourth postulate ("that all right angles are equal to one another"). However, it violates the other three: contrary to the first postulate, there is not a unique shortest route between any two points (antipodal points such as the north and south poles on a spherical globe are counterexamples); contrary to the third postulate, a sphere does not contain circles of arbitrarily great radius; and contrary to the fifth (parallel) postulate, there is no point through which a line can be drawn that never intersects a given line.

A statement that is equivalent to the parallel postulate is that there exists a triangle whose angles add up to 180°. Since spherical geometry violates the parallel postulate, there exists no such triangle on the surface of a sphere. The sum of the angles of a triangle on a sphere is $180°(1 + 4f)$, where, f is the fraction of the sphere's surface that is enclosed by the triangle. For any positive value of f, this exceeds 180°.

Incidence Geometry

Incidence geometry is essentially geometry based on the first postulate in Euclid's.

- Incidence Axiom 1 (IA1): For every pair of points P and Q there exists a unique line l incident with P and Q.

- Incidence Axiom 2 (IA2): For every line l there exist at least two distinct points incident with l.
- Incidence Axiom 3 (IA3): There exist three distinct points A, B, C not simultaneously incident with a common line l.

Proposition: If l, m are distinct lines and are not parallel, then l and m meet at a unique point.

Proof: The lines l, m must intersect since they are not parallel. Suppose they meet at two distinct points P, Q. Then there are two lines l, m through points P, Q. This is contradictory to Incidence Axiom 1. So l and m meet at a unique point.

Proposition: There exist three distinct lines not through any common point.

Proof: Let A, B, C be three distinct points not collinear. Since any two distinct points, determine a unique line, the three line $\overleftrightarrow{AB}$, $\overleftrightarrow{AC}$, $\overleftrightarrow{BC}$ must be distinct. We claim that the three lines have no common point. Suppose they have a common point P. We must have $P \neq A$. (Otherwise, $P = A$ implies that A lies on the line $\overleftrightarrow{BC}$ since P is a common point of $\overleftrightarrow{AB}$, $\overleftrightarrow{AC}$, $\overleftrightarrow{BC}$. Thus A, B, C are collinear, a contradiction.) Then two lines $\overleftrightarrow{AB}$, $\overleftrightarrow{AC}$ pass through points P, A. This is contradictory to Incidence Axiom 1.

Proposition: For every line there is at least one point not lying on it.

Proof: Let A, B, C be three distinct points not collinear by Incidence Axiom 3. Suppose there is a line l which has no point outside l, i.e., l contains every point. Then l contains all A, B, C. This means that A, B, C are collinear, a contradction.

Proposition: For every point there is at least one line not passing through it.

Proof: Let l, m, n be three distinct lines not concurrent by Proposition2. Given an arbitrary point P; then one of l, m, n does not pass through P.

Proposition: For every point P there exist at least two lines through P.

Proof: Let A, B, C be three points not collinear. If P is outside the line $\overleftrightarrow{AB}$, then the lines $\overleftrightarrow{AP}$, $\overleftrightarrow{BP}$ pass through point P and must be distinct. Otherwise, $\overleftrightarrow{AP} = \overleftrightarrow{BP}$ implies that A, B, P are collinear; so P is on the line $\overleftrightarrow{AB}$, contradicting to that P is outside $\overleftrightarrow{AB}$. If P is on the line $\overleftrightarrow{AB}$, then $\overleftrightarrow{AB}$, $\overleftrightarrow{CP}$ are two distinct lines passing through P.

Abstract Geometry

Suppose P is a set and $\mathcal{L}$ is a set of non-empty subsets of P such that,

a. For every A ∈ P and B ∈ P there exists $l \in \mathcal{L}$ such that A ∈ l and B ∈ l,

b. For every $l \in \mathcal{L}$, l has at least two elements from P.

We call $\{P, \mathcal{L}\}$ an abstract geometry, the elements of P points, and the elements of $\mathcal{L}$ lines. Moreover, if $P \in P$, $l \in \mathcal{L}$, and $P \in l$, we say P lies on l, or that l passes through P.

Note: The conditions in the definition say that every pair of points lies on a line and that every line passes through at least two points.

Example, Recall that the unit-sphere in $\mathbb{R}^3$ is the set

$$S^2 = \{(x, y, z) : x \in \mathbb{R}, y \in \mathbb{R}, z \in \mathbb{R}, x^2 + y^2 + z^2 = 1\}$$

And, for any given real numbers a, b, c, and d, where not all of a, b, and c are 0, the set:

$$\{(x, y, z) : x \in \mathbb{R}, y \in \mathbb{R}, z \in \mathbb{R}, ax + by + cz = d\}$$

is a plane in $\mathbb{R}^3$.

If d = 0, the plane passes through the origin. Given a plane passing through the origin, the set $S^2 \cap A$ is a great circle of S^2. Given P = (x_1, y_1, z_1) and Q = (x_2, y_2, z_2) on S^2, we will show that P and Q both lie on some great circle. To do so, we need to find real numbers a, b, and c, not all 0, such that,

$$ax_1 + by_1 + cz_1 = 0$$

and

$$ax_2 + by_2 + cz_2 = 0$$

Since this is a homogeneous system in three unknowns, we know it is possible to solve for a, b, and c, not all 0. Thus if we let $\mathcal{L}_R$ be the set of great circles of S^2, then $\{S^2, \mathcal{L}_R\}$ is an abstract geometry, called the Riemann sphere.

Note, that on the Riemann sphere, the points P = (0, 0, 1) (the "north pole") and Q =(0, 0, −1) (the "south pole") lie both on the great circle determined by the plane with equation x = 0 and the great circle determined by the plane y = 0 in this geometry, two distinct points do not determine a unique line.

The definition given an abstract geometry {P, $\mathcal{L}$}, a set of points S ⊂ P is collinear if there exists a line $l \in \mathcal{L}$ such that, $S \subset \mathcal{L}$. A set of points which is not collinear is non-collinear.

An abstract geometry {P, $\mathcal{L}$} is an incidence geometry if distinct points P and Q in P lie on a unique line, and there exists a set of three non-collinear points.

Example, given a real number a, let

$$L_a = \{(x, y) : x \in \mathbb{R}, y \in \mathbb{R}, x = a\}$$

Given real numbers m and b, let

$$L_{m,b} = \{(x, y) : x \in \mathbb{R}, y \in \mathbb{R}, y = mx + b\}$$

Let

$$\mathcal{L}_E = \{L_a : a \in \mathbb{R}\} \cup \{L_{m,b} : m \in \mathbb{R}, b \in \mathbb{R}\}$$

We will show that $\{\mathbb{R}^2, \mathcal{L}_E\}$ is incidence geometry.

Let P = (x_1, y_1) and Q = (x_2, y_2) be distinct points in $\mathbb{R}^2$. If $x_1 = x_2$, let a $= x_1$ Then $P \in L_a$ and $Q \in L_a$.

Clearly, there is no $b \neq a$ for which $P \in L_b$. Suppose there exist real numbers m and b for which ($P \in L_{m,b}$ and $Q \in L_{m,b}$).

Then,

$$y_1 = mx_1 + b = ax + b$$

And,

$$y_2 = mx_2 + b = ax + b$$

But then,

$$P = (x_1, y_1) = (a, ax + b) = (a, y_2) = (x_2, y_2) = Q$$

Contradicting our assumption that P and Q are distinct points.

Now suppose $x_1 \neq x_2$. We need to find m and b such that P and Q both lie on $L_{m,b}$. That is, we want,

$$y_1 = mx_1 + b$$

And

$$y_2 = mx_2 + b$$

Subtracting, we have

$$y_2 - y_1 = m(x_2 + x_1)$$

From which we have, since $x_1 - x_2 \neq 0$,

$$m = \frac{y_2 - y_1}{x_2 - x_1}$$

We then have,

$$b = y_1 - mx_1 = y_2 - mx_2$$

Note, that since the solutions are unique, P and Q lie on a unique line $L_{m,b}$.

Hence,

$\varepsilon = \{\mathbb{R}^2, \mathcal{L}_E\}$ is an incidence geometry, which we call the Cartesian Plane.

We call the lines L_a vertical lines and the lines $L_{m,b}$ non-vertical lines.

Example: Let, $\mathbb{H}=\{(x,y):(x,y)\in\mathbb{R}^2, y>0\}$. Given any real number a, let $= a^L = \{(x,y):(x,y)\in\mathbb{H}, x=a\}$, and given any real numbers c and r > 0 let

$$ {}_cL_r = \{(x,y):(x,y)\in\mathbb{H}, (x-c)^2+y^2=r^2\} $$

Let

$$ \mathcal{L}_H = \{a_L : a\in\mathbb{R}\}\cup\{{}_cL_r : c\in\mathbb{R}, r>0\} $$

We will show that $\{\mathbb{H},\mathcal{L}_H\}$ is a incidence geometry.

Let P = (x_1, x_2) and Q = (x_2, x_2) be distinct points in $\mathbb{H}$. If $x_1 = x_2$, let $a = x_1$. Then P and Q both lie on ${}_aL$.

Suppose P and Q also lie on ${}_cL_r$. Then we must have

$$ (x_1-c)^2 + y_1^2 = r^2 $$

And

$$ (x_2-c)^2 + y_2^2 = r^2 $$

Hence, since $a = x_1 = x_2$

$$ y_1^2 = r^2 - (x_1-c)^2 = r^2 - (a-c)^2 = r^2 - (x_2-c)^2 = y_2^2 $$

Since $y_1 > 0$ and $y_2 > 0$, it follows that $y_1 = y_2$, and P = Q, contradicting our assumption that P and Q are distinct points in $\mathbb{H}$

Now suppose $x_1 \neq x_2$. We need to find c and r > 0 such that.

$$ (x_1-c)^2 + y_1^2 = r^2 $$

And

$$ (x_2-c)^2 + y_2^2 = r^2 $$

Subtracting and expanding gives us,

$$ x_1^2 - 2x_1c - x_2^2 + 2x_2c = y_2^2 - y_1^2 $$

From which it follows that

$$ c = \frac{y_2^2 - y_1^2 + x_2^2 - x_1^2}{2(x_2 - x_1)} $$

We then have

$$ r = \sqrt{(x_1-c)^2 + y_1^2} = \sqrt{(x_2-c)^2 + y_2^2} $$

Thus P and Q lie on a unique line ${}_cL_r$.

Hence,

$H = \{\mathbb{H}, \mathcal{L}_H\}$ is an incidence geometry, which we call the Poincaré Plane.

We call the lines ${}_aL$ type I lines and the lines ${}_cL_r$ type II lines.

Notation: Given distinct points P and Q in an incidence geometry, we let $\overleftrightarrow{PQ}$ denote the unique line on which both P and Q lie.

Theorem: If l_1 and l_2 are lines in an incidence geometry and $l_1 \cap l_2$ has two or more distinct points, then $l_1 = l_2$.

Proof: If P and Q are distinct points in $l_1 \cap l_2$, then

$$l_1 = \overleftrightarrow{PQ} = l_2$$

Definition: We say lines l_1 and l_2 in an abstract geometry are parallel, denoted $l_1 \| l_2$, if either l_1 = l_2 or $l_1 \cap l_2 = \emptyset$.

Theorem, in an incidence geometry, two distinct lines are either parallel or they intersect in exactly one point.

Incidence Structures

An *incidence structure* (*P*, *L*, I) consists of a set *P* whose elements are called *points*, a disjoint set *L* whose elements are called *lines* and an *incidence relation,* I between them, that is, a subset of *P* × *L* whose elements are called *flags*. If (*A*, *l*) is a flag, we say that *A* is *incident with l* or that *l* is incident with *A* (the relation is symmetric), and write *A* I *l*. Intuitively, a point and line are in this relation if and only if the point is *on* the line. Given a point *B* and a line *m* which do not form a flag, that is, the point is not on the line, the pair (*B*, *m*) is called an *anti-flag*.

Distance in an Incidence Structure

There is no natural concept of distance (a metric) in an incidence structure. However, a combinatorial metric does exist in the corresponding incidence graph (Levi graph), namely the length of the shortest path between two vertices in this bipartite graph. The distance between two objects of an incidence structure – two points, two lines or a point and a line – can be defined to be the distance between the corresponding vertices in the incidence graph of the incidence structure.

Another way to define a distance again uses a graph-theoretic notion in a related structure, this time the *collinearity graph* of the incidence structure. The vertices of the collinearity graph are the points of the incidence structure and two points are joined if there exists a line incident with both points. The distance between two points of the incidence structure can then be defined as their distance in the collinearity graph.

When distance is considered in an incidence structure, it is necessary to mention how it is being defined.

Partial Linear Spaces

Incidence structures that are most studied are those that satisfy some additional properties (axioms), such as projective planes, affine planes, generalized polygons, partial geometries and near polygons. Very general incidence structures can be obtained by imposing "mild" conditions, such as, A partial linear space is an incidence structure for which the following axioms are true:

- Every pair of distinct points determines at most one line.
- Every line contains at least two distinct points.

In a partial linear space, it is also true that every pair of distinct lines meet in at most one point.

Further constraints are provided by the regularity conditions:

- RLk: Each line is incident with the same number of points. If finite this number is often denoted by k.
- RPr: Each point is incident with the same number of lines. If finite this number is often denoted by r.

The second axiom of a partial linear space implies that $k > 1$. Neither regularity condition implies the other, so it has to be assumed that $r > 1$.

A finite partial linear space satisfying both regularity conditions with $k, r > 1$ is called a *tactical configuration*. Some authors refer to these simply as *configurations*, or *projective configurations*. If a tactical configuration has n points and m lines, then, by double counting the flags, the relationship $nr = mk$ is established. A common notation refers to (n_r, m_k)-*configurations*. In the special case where $n = m$ (and hence, $r = k$) the notation (n_k, n_k) is often simply written as (n_k).

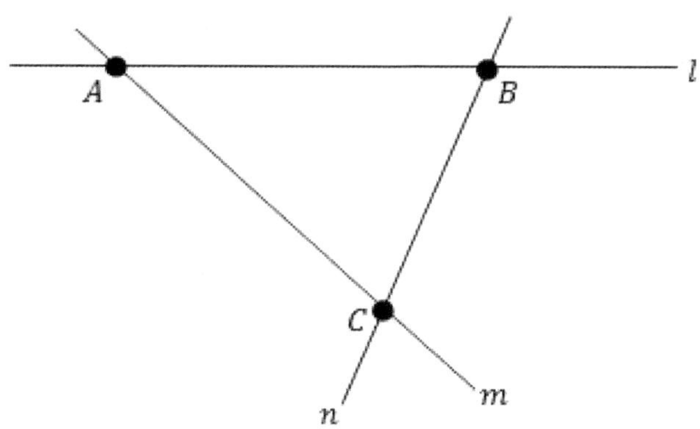

Figure: Simplest non-trivial linear space

- A linear space is a partial linear space such that: Every pair of distinct points determines exactly one line.
- Some authors add a "non-degeneracy" (or "non-triviality") axiom to the definition of a (partial) linear space, such as: There exist at least two distinct lines.

This is used to rule out some very small example, that would normally be exceptions to general statements made about the incidence structures. An alternative to adding the axiom is to refer to incidence structures that do not satisfy the axiom as being *trivial* and those that do as *non-trivial*.

Each non-trivial linear space contains at least three points and three lines, so the simplest non-trivial linear space that can exist is a triangle.

A linear space having at least three points on every line is a Sylvester–gallai design.

Fundamental Geometric Examples

Some of the basic concepts and terminology arises from geometric examples, particularly projective planes and affine planes.

Projective Planes

A *projective plane* is a linear space in which:

- Every pair of distinct lines meet in exactly one point, and that satisfies the non-degeneracy condition:
- There exist four points, no three of which are collinear.

There is a bijection between P and L in a projective plane. If P is a finite set, the projective plane is referred to as a *finite* projective plane. The order of a finite projective plane is $n = k - 1$, that is, one less than the number of points on a line. All known projective planes have orders that are prime powers. A projective plane of order n is an $((n^2 + n + 1)_{n+1})$ configuration.

The smallest projective plane has order two and is known as the *Fano plane*.

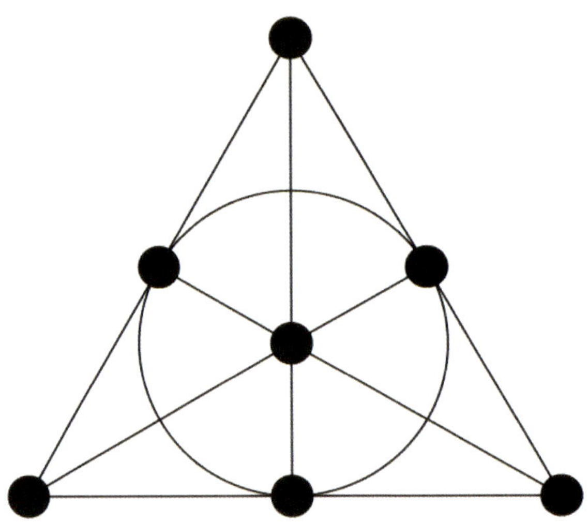

Projective plane of order 2
the Fano plane

Fano Plane

This famous incidence geometry was developed by the Italian mathematician Gino Fano. In his work on proving the independence of the set of axioms for projective n-space that he developed, he produced a finite three-dimensional space with 15 points, 35 lines and 15 planes, in which each line had only three points on it. The planes in this space consisted of seven points and seven lines and are now known as Fano planes.

The Fano plane cannot be represented in the Euclidean plane using only points and straight line segments (i.e. it is not realizable). This is a consequence of the Sylvester–Gallai theorem, according to which every realizable incidence geometry must include an *ordinary line*, a line containing only two points. The Fano plane has no such line (that is, it is a Sylvester–Gallai configuration), so it is not realizable.

A complete quadrangle consists of four points, no three of which are collinear. In the Fano plane, the three points not on a complete quadrangle are the diagonal points of that quadrangle and are collinear. This contradicts the *Fano axiom*, often used as an axiom for the Euclidean plane, which states that the three diagonal points of a complete quadrangle are never collinear.

Affine Planes

An *affine plane* is a linear space satisfying:

- For any point A and line l not incident with it (an anti-flag) there is exactly one line m incident with A (that is, $A \text{ I } m$), that does not meet l (known as Playfair's axiom).

And satisfying the non-degeneracy condition:

- There exists a triangle, i.e. three non-collinear points.

The lines l and m in the statement of Playfair's axiom are said to be *parallel*. Every affine plane can be uniquely extended to a projective plane. The *order* of a finite affine plane is k, the number of points on a line. An affine plane of order n is an $((n^2)_{n+1}, (n^2 + n)_n)$ configuration.

Hesse Configuration

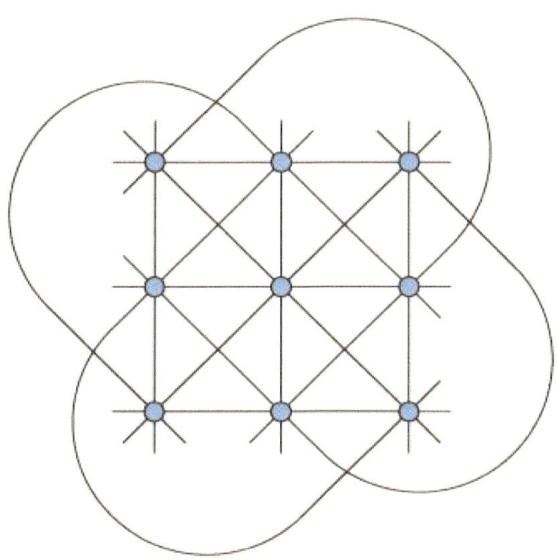

Figure: Affine plane of order 3 (Hesse configuration)

The affine plane of order three is a $(9_4, 12_3)$ configuration. When embedded in some ambient space it is called the Hesse configuration. It is not realizable in the Euclidean plane but is realizable in the complex projective plane as the nine inflection points of an elliptic curve with the 12 lines incident with triples of these.

The 12 lines can be partitioned into four classes of three lines apiece where, in each class the lines are mutually disjoint. These classes are called *parallel classes* of lines. Adding four new points, each being added to all the lines of a single parallel class (so all of these lines now intersect), and one new line containing just these four new points produces the projective plane of order three, a (13_4) configuration. Conversely, starting with the projective plane of order three (it is unique) and removing any single line and all the points on that line produces this affine plane of order three (it is also unique).

Removing one point and the four lines that pass through that point (but not the other points on them) produces the (8_3) Möbius–Kantor configuration.

Partial Geometries

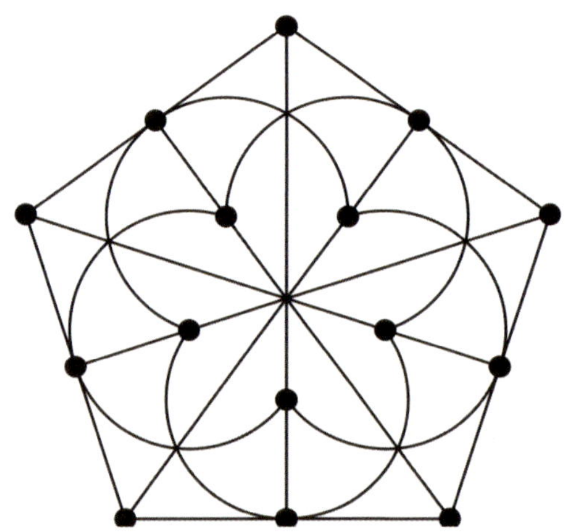

Partial geometry pg(2,2,1)

Given an integer $\alpha \geq 1$, a tactical configuration satisfying:

> For every anti-flag (B, m) there are α flags (A, l) such that B I l and A I m, is called a *partial geometry*.
>
> If there are $s + 1$ points on a line and $t + 1$ lines through a point, the notation for a partial geometry is pg(s, t, α).

If $\alpha = 1$ these partial geometries are generalized quadrangles.

If $\alpha = s + 1$ these are called Steiner systems.

Generalized Polygons

For $n > 2$, a generalized n-gon is a partial linear space whose incidence graph Γ has the property:

- The girth of Γ (length of the shortest cycle) is twice the diameter of Γ (the largest distance between two vertices, n in this case).

- A *generalized 2-gon* is an incidence structure, which is not a partial linear space, consisting of at least two points and two lines with every point being incident with every line. The incidence graph of a generalized 2-gon is a complete bipartite graph.
- A generalized n-gon contains no ordinary m-gon for $2 \le m < n$ and for every pair of objects (two points, two lines or a point and a line) there is an ordinary n-gon that contains them both.
- Generalized 3-gons are projective planes. Generalized 4-gons are called generalized quadrangles. By the Feit-Higman theorem the only finite generalized n-gons with at least three points per line and three lines per point have $n = 2, 3, 4, 6$ or 8.

Near Polygons

For a non-negative integer d a near $2d$-gon is an incidence structure such that:

- The maximum distance (as measured in the collinearity graph) between two points is d;
- For every point X and line l there is a unique point on l that is closest to X.

A near 0-gon is a point, while a near 2-gon is a line. The collinearity graph of a near 2-gon is a complete graph. A near 4-gon is a generalized quadrangle (possibly degenerate). Every finite generalized polygon except the projective planes is a near polygon. Any connected bipartite graph is a near polygon and any near polygon with precisely two points per line is a connected bipartite graph. Also, all dual polar spaces are near polygons.

Many near polygons are related to finite simple groups like the Mathieu groups and the Janko group J2. Moreover, the generalized $2d$-gons, which are related to Groups of Lie type, are special cases of near $2d$-gons.

Möbius Planes

An abstract Möbius plane (or inversive plane) is an incidence structure where, to avoid possible confusion with the terminology of the classical case, the lines are referred to as *cycles* or *blocks*.

Specifically, a Möbius plane is an incidence structure of points and cycles such that:

- Every triple of distinct points is incident with precisely one cycle;
- For any flag (P, z) and any point Q not incident with z there is a unique cycle z^* with $P \mathrel{I} z^*$, $Q \mathrel{I} z^*$ and $z \cap z^* = \{P\}$. (The cycles are said to *touch* at P);
- Every cycle has at least three points and there exists at least one cycle.

The incidence structure obtained at any point P of a Möbius plane by taking as points all the points other than P and as lines only those cycles that contain P (with P removed), is an affine plane. This structure is called the *residual* at P in design theory.

A finite Möbius plane of *order* m is a tactical configuration with $k = m + 1$ points per cycle that is a 3-design, specifically a 3-$(m^2 + 1, m + 1, 1)$ block design.

Incidence Theorems in the Euclidean Plane

The Sylvester-Gallai Theorem

A question raised by J.J. Sylvester in 1893 and finally settled by Tibor Gallai concerned incidences of a finite set of points in the Euclidean plane.

Theorem (Sylvester-Gallai): A finite set of points in the Euclidean plane is either collinear or there exists a line incident with exactly two of the points.

A line containing exactly two of the points is called an *ordinary line* in this context. Sylvester was probably led to the question while pondering about the embeddability of the Hesse configuration.

The De Bruijn–Erdős Theorem

A related result is the de Bruijn–Erdős theorem. Nicolaas Govert de Bruijn and Paul Erdős proved the result in the more general setting of projective planes, but it still holds in the Euclidean plane. The theorem is:

> In a projective plane, every non-collinear set of n points determines at least n distinct lines.

As the authors pointed out, since their proof was combinatorial, the result holds in a larger setting, in fact in any incidence geometry in which there is a unique line through every pair of distinct points. They also mention that the Euclidean plane version can be proved from the Sylvester-Gallai theorem using induction.

The Szemerédi–Trotter Theorem

A bound on the number of flags determined by a finite set of points and the lines they determine is given by:

Theorem (Szemerédi–Trotter): Given n points and m lines in the plane, the number of flags (incident point-line pairs) is:

$$O\left(n^{\frac{2}{3}}m^{\frac{2}{3}}+n+m\right),$$

And this bound cannot be improved, except in terms of the implicit constants.

This result can be used to prove Beck's theorem.

Beck's Theorem

Beck's theorem says that finite collections of points in the plane fall into one of two extremes; one where a large fraction of points lie on a single line, and one where a large number of lines are needed to connect all the points.

The theorem asserts the existence of positive constants C, K such that given any n points in the plane, at least one of the following statements is true:

- There is a line that contains at least $\frac{n}{C}$ of the points.
- There exist at least $\frac{n^2}{K}$ lines, each of which contains at least two of the points.

In Beck's original argument, C is 100 and K is an unspecified constant; it is not known what the optimal values of C and K are.

Non-Euclidean Geometry

Non-Euclidean geometry is any geometry that is not the same as Euclidean geometry. Although the term is frequently used to refer only to hyperbolic geometry, common usage includes those few geometries (hyperbolic and spherical) that differ from but are very close to Euclidean geometry.

The non-Euclidean geometries developed along two different historical threads. The first thread started with the search to understand the movement of stars and planets in the apparently hemispherical sky. For example, Euclid (flourished c. 300 BC) wrote about spherical geometry in his astronomical work phenomena. In addition to looking to the heavens, the ancients attempted to understand the shape of the Earth and to use this understanding to solve problems in navigation over long distances (and later for large-scale surveying). These activities are aspects of spherical geometry.

The second thread started with the fifth ("parallel") postulate in Euclid's Elements:

If a straight line falling on two straight lines makes the interior angles on the same side less than two right angles, the two straight lines, if produced indefinitely, will meet on that side on which the angles are less than the two right angles.

For 2,000 years following Euclid, mathematicians attempted either to prove the postulate as a theorem (based on the other postulates) or to modify it in various ways. These attempts culminated when the Russian Nikolay Lobachevsky and the Hungarian János Bolyai independently published a description of a geometry that, except for the parallel postulate, satisfied all of Euclid's postulates and common notions. It is this geometry that is called hyperbolic geometry.

The Parallel Postulate

The fifth postulate in Euclid's Elements can be rephrased as

> Given a line and a point not on it, exactly one line parallel to the given line can be drawn through the point.

The postulate is not true in 3D but in 2D it seems to be a valid statement. Considering the importance of postulates however, a seemingly valid statement is not good enough.

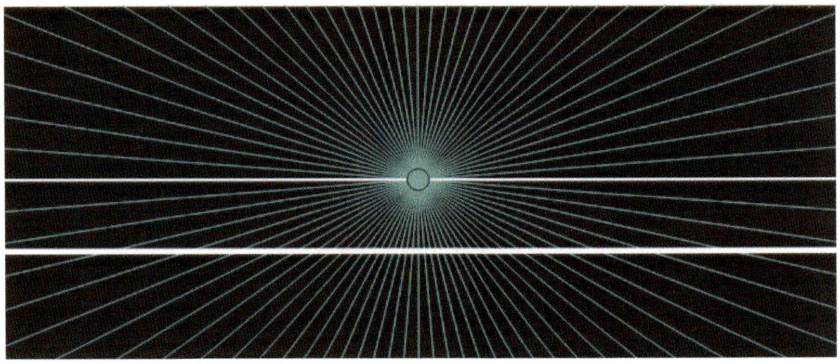

Exactly one line through the green point is parallel to the thick white line.

A postulate (or axiom) is a statement that acts as a starting point for a theory. Since a postulate is a starting point it cannot be proven using previous result. As a statement that cannot be proven, a postulate should be self-evident. The Elements of Euclid is built upon five postulates. For over 2000 years the fifth postulate has been considered to be less intuitive than the other postulates, and not sufficiently self-evident. Many attempts have been made to prove the fifth postulate using the other four postulates. All these attempts have failed. In the 19th century it was shown that the fifth postulate is independent of the other postulates. It is possible to build a theory of geometry where the fifth postulate is not true. Such geometries are called non-Euclidean. Furthermore, it can be shown that non-Euclidean theories can be just as consistent as Euclidean theory. A theory is consistent if it does not contain any contradiction.

The converse statement to the parallel postulate, as stated above, is that there are either no lines, or at least two lines "parallel to the given line through the point". It can be shown that if there is at least two lines, there are in fact infinitely many lines "parallel to".

If the parallel postulate is replaced by:

> Given a line and a point not on it, no lines parallel to the given line can be drawn through the point.

You get an elliptic geometry.

If the parallel postulate is replaced by:

> Given a line and a point not on it, infinitely many lines parallel to the given line can be drawn through the point.

you get a hyperbolic geometry.

Distances, Angles and Lines

The theorems in the Elements of Euclid are built on a number of postulates and a number of definitions. Definition states that two lines are parallel if they never meet. There is nothing in the definition indicating that the distance between two parallel lines is the same everywhere. The parallel postulate is seemingly obvious only if you assume that parallel lines look like railroad tracks. If you redefine what you mean by a line, you may have that two parallel lines either converge towards each other or diverge from one another.

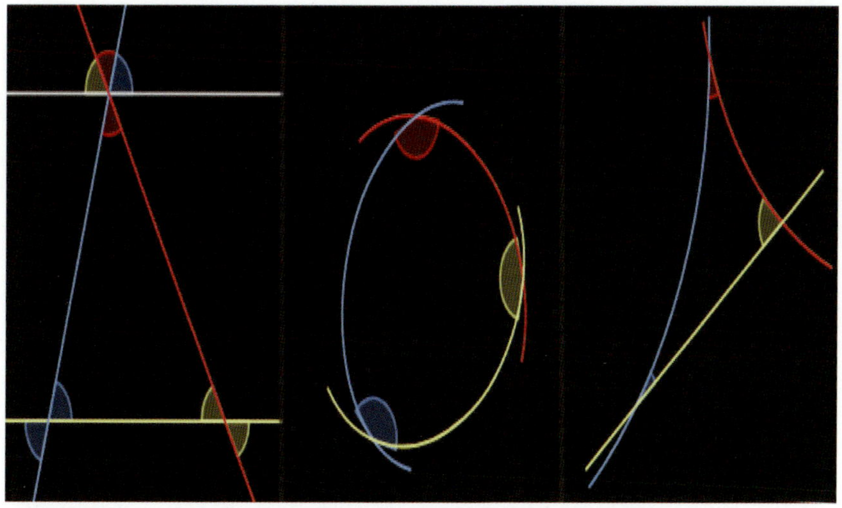

Euclidean, elliptical, and hyperbolic triangle.

In non-Euclidean geometry, the concept corresponding to a line is a curve called a geodesic. In non-Euclidean geometry a shortest path between two points is along such a geodesic, or "non-Euclidean line".

All theorems in Euclidean geometry that use the fifth postulate, will be altered when you rephrase the parallel postulate. As an example; in Euclidean geometry the sum of the interior angles of a triangle is 180°, in non-Euclidean geometry this is not the case.

Elliptic Geometry

Elliptic Geometry is a type of non-Euclidean geometry, that is, a geometric theory based on axioms whose requirements are to a considerable degree different from the requirements of the axioms of Euclidean geometry. The basic objects, or elements, of three-dimensional elliptic geometry are points, lines, and planes; the basic concepts of elliptic geometry are the concepts of incidence (a point is on a line, a line is in a plane), order (for example, the order of points on a line or the order of lines passing through a given point in a given plane), and congruence.

The requirements of the axioms of elliptic geometry concerning membership and order completely coincide with the requirements of the axioms of projective geometry. Accordingly, such propositions as the following are found in elliptic geometry: through any two points there passes one line; any two planes intersect in one line; any two coplanar lines intersect in one point; and the points on a line are arranged in a cyclic order, as are lines lying in the same plane and passing through the same point.

The requirements of the axioms of elliptic geometry concerning congruence are similar to the corresponding requirements of the axioms of Euclidean geometry; in any case, they provide for motions of figures in the elliptic plane and in elliptic space that are just as free as in the plane and space of Euclidean geometry.

The metric properties of the elliptic plane coincide "in the small" with the metric properties of the ordinary sphere. More precisely, for any point of the elliptic plane there exists a part of the plane, containing this point, that is isometric to some part of the sphere; the radius R of this sphere is the same for all planes in the given elliptic space. The number K = 1/R2 is called the curvature of the elliptic space. The smaller is K, the closer are the properties of this space to those of Euclidean space.

The properties of the elliptic plane "in the large" differ from the properties of the sphere as a whole. For example, in the elliptic plane, two lines intersect in one point; on the sphere, two great circles, which play the role of lines in spherical geometry, intersect in two points. A line in a plane does not separate the plane—that is, if the line a is in the plane α, then any two points of α not on a can be joined by a line segment that does not intersect a.

Two Dimensions

Elliptic Plane

The elliptic plane is the real projective plane provided with a metric: Kepler and Desargues used the gnomonic projection to relate a plane σ to points on a hemisphere tangent to it. With O the center of the hemisphere, a point *P* in σ determines a line *OP* intersecting the hemisphere, and any line $L \subset \sigma$ determines a plane *OL* which intersects the hemisphere in half of a great circle. The hemisphere is bounded by a plane through O and parallel to σ. No ordinary line of σ corresponds to this plane; instead a line at infinity is appended to σ. As any line in this extension of σ corresponds to a plane through O, and since any pair of such planes intersects in a line through O, one can conclude that any pair of lines in the extension.

Intersect: the point of intersection lies where the plane intersection meets σ or the line at infinity. Thus the axiom of projective geometry, requiring all pairs of lines in a plane to intersect, is confirmed.

Given *P* and *Q* in σ, the elliptic distance between them is the measure of the angle *POQ*, usually taken in radians. Arthur Cayley initiated the study of elliptic geometry when he wrote "On the definition of distance". This venture into abstraction in geometry was followed by Felix Klein and Bernhard Riemann leading to non-Euclidean geometry and Riemannian geometry.

Comparison with Euclidean Geometry

In Euclidean geometry, a figure can be scaled up or scaled down indefinitely, and the resulting figures are similar, i.e., they have the same angles and the same internal proportions. In elliptic geometry this is not the case. For example, in the spherical model we can see that the distance between any two points must be strictly less than half the circumference of the sphere (because antipodal points are identified). A line segment therefore cannot be scaled up indefinitely. A geometer measuring the geometrical properties of the space he or she inhabits can detect, via measurements, that there is a certain distance scale that is a property of the space. On scales much smaller than this one, the space is approximately flat, geometry is approximately Euclidean, and figures can be scaled up and down while remaining approximately similar.

A great deal of Euclidean geometry carries over directly to elliptic geometry. For example, the first and fourth of Euclid's postulates, that there is a unique line between any two points and that all right angles are equal, hold in elliptic geometry. Postulate 3, that one can construct a circle with any given center and radius, fails if "any radius" is taken to mean "any real number", but holds if it is taken to mean "the length of any given line segment". Therefore any result in Euclidean geometry that follows from these three postulates will hold in elliptic geometry, such as proposition 1 from book I of the *Elements*, which states that given any line segment, an equilateral triangle can be constructed with the segment as its base.

Elliptic geometry is also like Euclidean geometry in that space is continuous, homogeneous, isotropic, and without boundaries. Isotropy is guaranteed by the fourth postulate, that all right angles are equal. For an example of homogeneity, note that Euclid's proposition1, implies that the same equilateral triangle can be constructed at any location, not just in locations that are special in some way. The lack of boundaries follows from the second postulate, extensibility of a line segment.

One way in which elliptic geometry differs from Euclidean geometry is that the sum of the interior angles of a triangle is greater than 180 degrees. In the spherical model, for example, a triangle can be constructed with vertices at the locations where the three positive Cartesian coordinate axes intersect the sphere, and all three of its internal angles are 90 degrees, summing to 270 degrees. For sufficiently small triangles, the excess over 180 degrees can be made arbitrarily small.

The Pythagorean theorem fails in elliptic geometry. In the 90°–90°–90° triangle described above, all three sides have the same length, and consequently do not satisfy $a^2 + b^2 = c^2$. The Pythagorean result is recovered in the limit of small triangles.

The ratio of a circle's circumference to its area is smaller than in Euclidean geometry. In general, area and volume do not scale as the second and third powers of linear dimensions.

Elliptic Space

Elliptic space can be constructed in a way similar to the construction of three-dimensional vector space: with equivalence classes. One uses directed arcs on great circles of the sphere. As directed line segments are equipollent when they are parallel, of the same length, and similarly oriented, so directed arcs found on great circles are equipollent when they are of the same length, orientation, and great circle. These relations of equipollence produce 3D vector space and elliptic space, respectively.

Access to elliptic space structure is provided through the vector algebra of William Rowan Hamilton: he envisioned a sphere as a domain of square roots of minus one. Then Euler's formula $\exp(\theta r) = \cos\theta + r\sin\theta$ (where r is on the sphere) represents the great circle in the plane perpendicular to r. Opposite point r and $-r$ correspond to oppositely directed circles. An arc between θ and φ is equipollent with one between 0 and $\varphi - \theta$. In elliptic space, arc length is less than π, so arcs may be parameterized with θ in $[0, \pi)$ or $(-\pi/2, \pi/2]$.

For $z = \exp(\theta r)$, $z^* = \exp(-\theta r) \Rightarrow zz^* = 1$. It is said that the modulus or norm of z is one (Hamilton called it the tensor of z). But since r ranges over a sphere in 3-space, $\exp(\theta\ r)$ ranges over a sphere in 4-space, now called the 3-sphere, as its surface has three dimensions. Hamilton called his algebra quaternions and it quickly became a useful and celebrated tool of mathematics. Its space of four dimensions is evolved in polar co-ordinates $t\exp(\theta r)$, with t in the positive real numbers.

When doing trigonometry on Earth or the celestial sphere, the sides of the triangles are great circle arcs. The first success of quaternions was a rendering of spherical trigonometry to algebra. Hamilton called a quaternion of norm one a versor, and these are the points of elliptic space.

With r fixed, the versors

$$e^{ar}, \quad 0 \le a < \pi$$

Form an *elliptic line*, the distance from e^{ar} to 1 is a. For an arbitrary versor u, the distance will be that θ for which $\cos\theta = (u + u^*)/2$ since this is the formula for the scalar part of any quaternion.

An *elliptic motion* is described by the quaternion mapping:

$$q \mapsto uqv$$

Where, u and v are fixed versors.

Distances between points are the same as between image points of an elliptic motion. In the case that u and v are quaternion conjugates of one another, the motion is a spatial rotation, and their vector part is the axis of rotation. In the case $u = 1$ the elliptic motion is called a right *Clifford translation*, or a *parataxy*. The case $v = 1$ corresponds to left Clifford translation.

Elliptic lines through versor u may be of the form:

$$\{ue^{ar} : 0 \le a < \pi\} \text{ or } \{e^{ar}u : 0 \le a < \pi\} \text{ for a fixed } r.$$

They are the right and left Clifford translations of u along an elliptic line. The *elliptic space* is formed by from S^3 by identifying antipodal points.

Elliptic space has special structures called Clifford parallels and Clifford surfaces.

The versor points of elliptic space are mapped by the Cayley transform to $\mathbb{R}^3$ for an alternative representation of the space.

Higher-dimensional Spaces

Hyperspherical Model

The hyperspherical model is the generalization of the spherical model to higher dimensions. The points of n-dimensional elliptic space are the pairs of unit vectors $(x, -x)$ in R^{n+1}, that is, pairs of opposite points on the surface of the unit ball in $(n+1)$-dimensional space (the n-dimensional hypersphere). Lines in this model are great circles, i.e., intersections of the hypersphere with flat hypersurfaces of dimension n passing through the origin.

Projective Elliptic Geometry

In the projective model of elliptic geometry, the points of n-dimensional real projective space are used as points of the model. This models an abstract elliptic geometry that is also known as projective geometry.

The points of n-dimensional projective space can be identified with lines through the origin in

$(n+1)$-dimensional space, and can be represented non-uniquely by nonzero vectors in R^{n+1}, with the understanding that u and λu, for any non-zero scalar λ, represent the same point. Distance is defined using the metric,

$$d(u,v) = \arccos\left(\frac{|u \cdot v|}{\|u\|\|v\|}\right);$$

That is, the distance between two points is the angle between their corresponding lines in R^{n+1}. The distance formula is homogeneous in each variable, with $d(\lambda u, \mu v) = d(u,v)$ if λ and μ are non-zero scalars, so it does define a distance on the points of projective space.

A notable property of the projective elliptic geometry is that for even dimensions, such as the plane, the geometry is non-orientable. It erases the distinction between clockwise and counter-clockwise rotation by identifying them.

Stereographic Model

A model representing the same space as the hyperspherical model can be obtained by means of stereographic projection. Let E^n represent $R^n \cup \{\infty\}$, that is, n-dimensional real space extended by a single point at infinity. We may define a metric, the *chordal metric*, on E^n by

$$\delta(u,v) = \frac{2\|u-v\|}{\sqrt{(1+\|u\|^2)(1+\|v\|^2)}}$$

Where, u and v are any two vectors in R^n and $\|\cdot\|$ is the usual Euclidean norm. We also define,

$$\delta(u,\infty) = \delta(\infty,u) = \frac{2}{\sqrt{1+\|u\|^2}}.$$

The result is a metric space on E^n, which represents the distance along a chord of the corresponding points on the hyperspherical model, to which it maps bijectively by stereographic projection. We obtain a model of spherical geometry if we use the metric,

$$d(u,v) = 2\arcsin\left(\frac{\delta(u,v)}{2}\right).$$

Elliptic geometry is obtained from this by identifying the point u and $-u$, and taking the distance from v to this pair to be the minimum of the distances from v to each of these two points.

Self-consistency

Because spherical elliptic geometry can be modeled as, for example, a spherical subspace of a Euclidean space, it follows that if Euclidean geometry is self-consistent, so is spherical elliptic geometry. Therefore it is not possible to prove the parallel postulate based on the other four postulates of Euclidean geometry.

Tarski proved that elementary Euclidean geometry is complete: there is an algorithm which, for every proposition, can show it to be either true or false. (This does not violate Gödel's theorem, because Euclidean geometry cannot describe a sufficient amount of arithmetic for the theorem to apply). It therefore follows that elementary elliptic geometry is also self-consistent and complete.

Hyperbolic Geometry

Hyperbolic geometry is a type of non-Euclidean geometry that arose historically when mathematicians tried to simplify the axioms of Euclidean geometry, and instead discovered unexpectedly that changing one of the axioms to its negation actually produced a consistent theory. Later, physicists discovered practical applications of these ideas to the theory of special relativity.

Hyperbolic geometry also inspired the art of M. C. Escher, and has various theoretical applications as well, including geometric group theory and the theory of modular forms.

Background and Origins

The first four axioms of Euclidean geometry, laid out in Euclid's Elements, are essentially self-evident:

(1) Any two points can be connected by a line;

(2) Any line segment can be extended indefinitely;

(3) Given a line segment, a circle can be drawn with center at one of the endpoints and radius equal to the length of the segment;

(4) Any two right angles are congruent.

These four axioms define what is sometimes called absolute geometry. Euclid's first 28 propositions used only these four axioms, but he was forced to add a fifth axiom which was much less obvious than the first four:

(5) The parallel postulate: If a line segment intersects two straight lines forming two interior angles on the same side that sum to less than two right angles, then the two lines, if extended indefinitely, meet on that side on which the angles sum to less than two right angles.

This statement is a bit complicated and wordy; many modern accounts of Euclidean geometry use a different fifth axiom known as Playfair's axiom.

(5') Playfair's axiom: Given a line and a point not on, there is at most one line that can be drawn through that is parallel to.

In fact, axioms (1)-(4) imply that such a line always exists, by dropping a perpendicular to through and then dropping a perpendicular to the perpendicular at P.

It can be shown that $(1), (2), (3), (4), (5) \Leftrightarrow (1), (2), (3), (4), (5')$ logically.

Many mathematicians, going back to the ancient greeks, attempted to show that (5) or (5') were actually consequences of (1)-(4). One method that many of them used was proof by contradiction; start with the axioms (1), (2), (3), (4), and the negation of (5), and try to produce something that is "wrong." A lot of progress was made along these lines, and the properties of the subsequent geometry that was produced were quite strange and different from the propositions of Euclid's geometry. But there was nothing logically inconsistent about them. Eventually, 19-century mathematicians (starting with the Russian mathematician Nikolai Ivanovich Lobachevsky) shifted from trying to prove that non-Euclidean geometries were impossible, and began instead to explore the consequences of changing the fifth axiom.

In fact, the fifth axiom cannot be proved (or disproved) from the first four, a fact that was established in the late century by the Italian mathematician Eugenio Beltrami and others.

Results in Hyperbolic Geometry

Consider the geometry obtained from replacing Playfair's axiom by its negation:

> Given a line L and P a point not on L, there are at least two distinct lines that can be drawn through P that are parallel to L.

Here are some consequences of these axioms:

- The interior angles of a triangle sum to less than 180°.
- The interior angles of a quadrilateral sum to less than 360°.
- There are no rectangles.
- Given a line l and a point p not on l, there are infinitely many lines through p that are parallel to l.
- Similar triangles are congruent.

Proofs: The proof of angle180^{o} and 360° is lengthy and is omitted for now.

Clearly (180°) implies on (360°), since a quadrilateral can be split into two triangles by drawing a diagonal.

Since a rectangle is a quadrilateral with all right angles.

A perpendicular from Q to l; say it meets l at Q. (It is convenient to use lower-case letters for lines, and upper-case letters for points.) Then draw a perpendicular line m to PQ with right angle at P. It's impossible for l and m to meet, because if they met at a point then PQX would be a triangle with two right angles, which would contradict. So m is parallel to l:

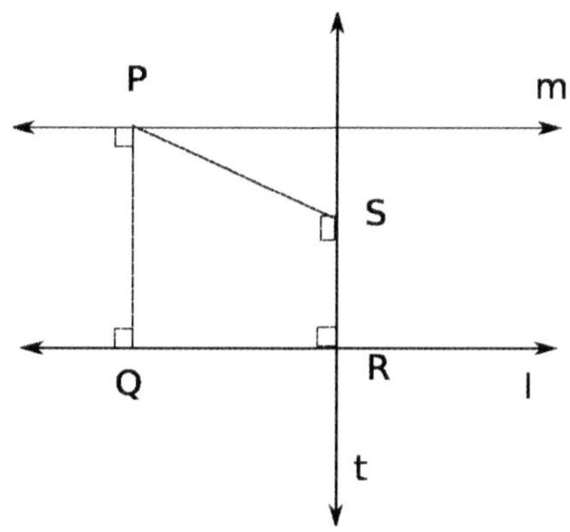

Pick a different point R on L, draw a line n perpendicular to l at R, and drop a perpendicular PS to n. Now PS is parallel to l by the same two-right-angles argument as above, but PS is not the same line as m; if it were, then $PQRS$ would be a rectangle. For any on the line, we thus get a new line PS that is parallel to l, and a similar argument shows that different choices of R lead to different lines.

Suppose the triangles are as pictured. Find a triangle $AB''C''$ congruent to the smaller one lying inside the bigger one,

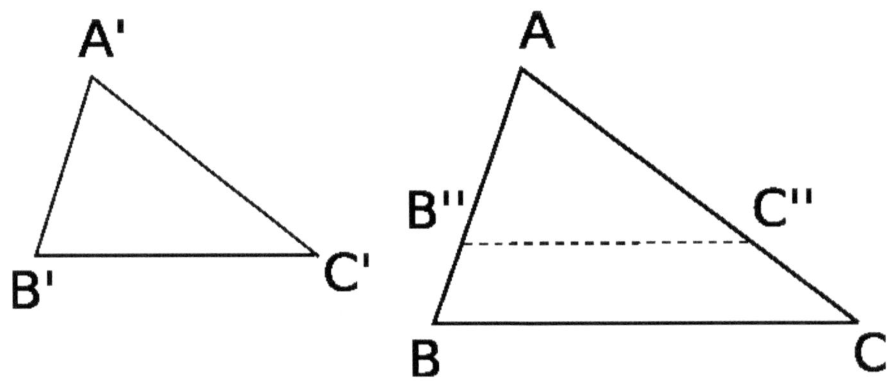

But then it is straightforward to see that the two angles on the left side of the quadrilateral sum to, and similarly for the two angles on the right side, so the angles sum to, which contradicts.

Models

There are two common models used to picture lines and angles in plane (two-dimensional) hyperbolic geometry. They are both due to Poincare.

Poincare half-plane model: The points are the complex numbers in the set $\mathbb{H} = \{z \in \mathbb{C} : \mathrm{Im}(z) > 0\}$. This is called the upper half-plane; in Cartesian coordinates it consists of points (x, y) such that y is positive.

Lines in this model are of two types: straight vertical lines, and half-circles whose centers lie on the -axis. Angles are measured as one would expect (the angle between the two curves at the point of intersection), but distances are trickier. Generally speaking, one thinks of distances between points near the -axis as "blowing up"; one way to represent this is that the infinitesimal unit of distance satisfies the formula:

$$(ds)^2 = \frac{(dx)^2 + (dy)^2}{y^2}$$

Rather than the usual $(ds)^2 = (dx)^2 + (dy)^2$ in standard analysis. There is a formula for the distance between two points z and w that uses the inverse hyperbolic trigonometric functions, similar to the one in the Poincare disk model, but it is unwieldy to work with. (One way to think of this is that it is the price one pays for keeping angles at their normal values; there is another model due to Beltrami and Klein with a nicer distance function, and lines which are straight, but angle measures in this model are distorted).

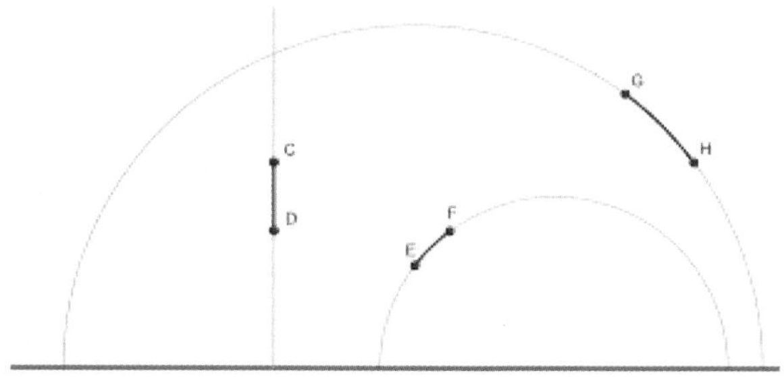

CD and EF are parallel, EF and GH are parallel, but CD and GH are not.

Poincare disk model: Using a conformal mapping that takes the -axis to the unit circle gives a model of hyperbolic geometry contained inside the unit disk. In this model, lines are either diameters of the disk or the intersection of a circle with the disk, where is perpendicular to the unit circle at its two points of intersection. Angles continue to be measured as expected.

The distance formula for two complex numbers z, w inside the disk becomes,

$$d(z,w) = \operatorname{arccosh}\left(1 + 2\frac{|z-w|^2}{(1-|z|^2)\left(1-|w|^2\right)}\right),$$

Where, $(x) = In\left(x + \sqrt{x^2 - 1}\right)$ is the inverse function of the hyperbolic cosine.

Note that as $|z|$ approaches, the distance between z and another point goes to ∞.

This is a tiling of the hyperbolic plane by congruent triangles.

Applications

Modular forms are fundamental objects in modern number theory; they were famously used in Wiles' proof of Fermat's last theorem. They are functions defined on the upper half-plane which are invariant under isometry, which is a (hyperbolic-) distance-preserving map from the upper half-plane to itself. So properties of hyperbolic geometry become important when studying modular forms.

There is an explicit connection between special relativity and hyperbolic geometry via Minkowski space, which is a common setting for both of them.

Hyperbolic geometry graphs have been suggested as a promising model for social networks where the hyperbolicity appears through a competition between similarity and popularity of an individual.

Taxicab Geometry

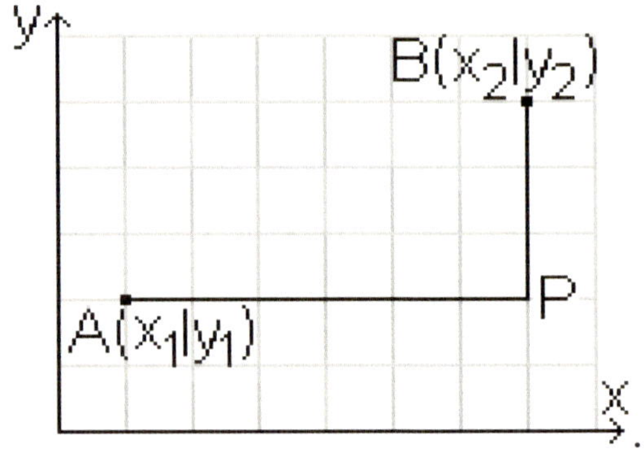

Taxicab geometry is a form of geometry, where the distance between two points A and B is not the length of the line segment AB as in the Euclidean geometry, but the sum of the absolute differences of their coordinates.

Formula: AP+PB $= |x_2 - x_1| + |y_2 - y_1|$

The points have only integer coordinates in the simple version and their connections are grid lines.

Example: If we have the point P (2, 3) and point Q (7, 6) we can find the distance first using taxicab geometry.

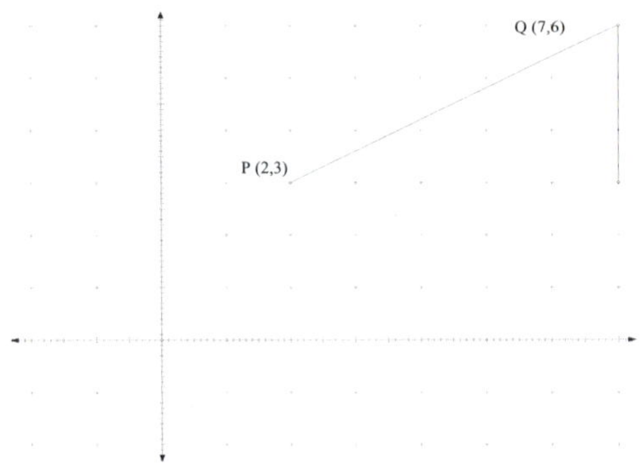

$$Taxicab\, d_T(P,Q)$$
$$= d_T[(x_1, y_1),(x_2, y_2)] = |X_1 - X_2| + |y_1 - y_2|$$
$$d_T[(2,3),(7,6)]$$
$$= |7-2| + |6-3|$$
$$= |5| + |3|$$
$$= 5 + 3 = 8$$

Using P, Q as your points, plug them in. Subtract the x and y's and then takes the absolute value. Add them together and get 8 for your solution.

Now, how about using the Euclidean approach?

Euclidean $d_E(A,B)$

$$= d_E[(x_1,y_1),(x_2,y_2)] = \sqrt{(x_1-x_2)^2+(y_1+y_2)^2}$$
$$= d_E[(2,3),(7,6)] = \sqrt{(2-7)^2+(3-6)^2}$$
$$= \sqrt{(-5)^2+(-3)^2}$$
$$= \sqrt{25+9}$$
$$= \sqrt{34} = 5.83$$

We would have to use 8 if we have to stay on roads, but if "as the crow flies" it would only be (5.83).

Properties

Taxicab distance depends on the rotation of the coordinate system, but does not depend on its reflection about a coordinate axis or its translation. Taxicab geometry satisfies all of Hilbert's axioms (a formalization of Euclidean geometry) except for the side-angle-side axiom, as two triangles with equally "long" two sides and an identical angle between them are typically not congruent unless the mentioned sides happen to be parallel.

Circles

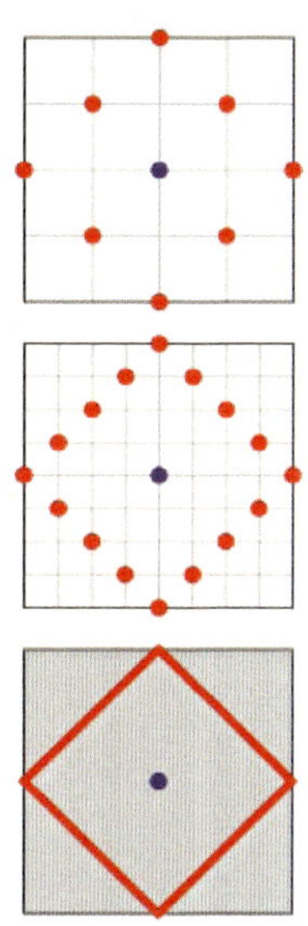

Circles in discrete and continuous taxicab geometry

A circle is a set of points with a fixed distance, called the *radius*, from a point called the *center*. In taxicab geometry, distance is determined by a different metric than in Euclidean geometry and the shape of circles changes as well. Taxicab circles are squares with sides oriented at a 45° angle to the coordinate axes. The image to the right shows why this is true, by showing in red the set of all points with a fixed distance from a center, shown in blue. As the size of the city blocks diminishes, the points become more numerous and become a rotated square in continuous taxicab geometry. While each side would have length $\sqrt{2}r$ using a Euclidean metric, where r is the circle's radius, its length in taxicab geometry is $2r$. Thus, a circle's circumference is $8r$. Thus, the value of a geometric analog to π is 4 in this geometry. The formula for the unit circle in taxicab geometry is $|x|+|y|=1$ in Cartesian coordinates and

$$r = \frac{1}{|\sin\theta|+|\cos\theta|}$$

in polar coordinates.

A circle of radius 1 (using this distance) is the von Neumann neighborhood of its center.

A circle of radius r for the Chebyshev distance (L_∞ metric) on a plane is also a square with side length $2r$ parallel to the coordinate axes, so planar Chebyshev distance can be viewed as equivalent by rotation and scaling to planar taxicab distance. However, this equivalence between L_1 and L_∞ metrics does not generalize to higher dimensions.

Whenever each pair in a collection of these circles has a nonempty intersection, there exists an intersection point for the whole collection; therefore, the Manhattan distance forms an injective metric space.

Applications

Measures of Distances in Chess

In chess, the distance between squares on the chessboard for rooks is measured in taxicab distance; kings and queens use Chebyshev distance, and bishops use the taxicab distance (between squares of the same color) on the chessboard rotated 45 degrees, i.e., with its diagonals as coordinate axes. To reach from one square to another, only kings require the number of moves equal to their respective distance; rooks, queens and bishops require one or two moves (on an empty board, and assuming that the move is possible at all in the bishop's case).

Compressed Sensing

In solving an underdetermined system of linear equations, the regularisation term for the parameter vector is expressed in terms of the $\ell 1$-norm (taxicab geometry) of the vector. This approach appears in the signal recovery framework called compressed sensing.

Differences of Frequency Distributions

Taxicab geometry can be used to assess the differences in discrete frequency distributions. For example, in RNA splicing positional distributions of hexamers, which plot the probability of each hexamer appearing at each given nucleotide near a splice site, can be compared with L1-distance. Each position distribution can be represented as a vector where each entry represents the likelihood of the hexamer starting at a certain nucleotide. A large L1-distance between the two vectors indicates a significant difference in the nature of the distributions while a small distance denotes similarly shaped distributions. This is equivalent to measuring the area between the two distribution curves because the area of each segment is the absolute difference between the two curves' likelihoods at that point. When summed together for all segments, it provides the same measure as L1-distance.

Projective Geometry

Projective geometry is an extension (or a simplification, depending on point of view) of Euclidean geometry, in which there is no concept of distance or angle measure. Intuitively, projective geometry can be understood as only having points and lines; in other words, while Euclidean geometry can be informally viewed as the study of straightedge and compass constructions, projective geometry can be viewed as the study of straightedge only constructions.

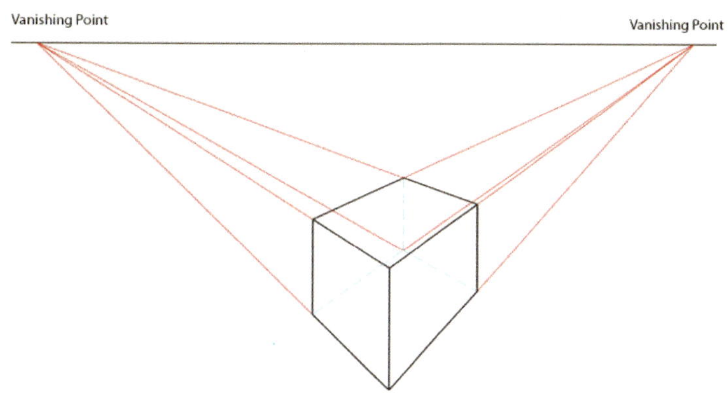

A cube drawn in perspective drawing, which, motivated projective geometry

Surprisingly, though originally conceived in the context of art, projective geometry has significant application to Euclidean geometry. For instance, results such as Desargues, 'Pappus', and Pascal's theorems are consequences of projective geometry, but are of much interest in "normal" geometry as well; this is due to the fact that the Euclidean plane can be naturally extended into a projective one. Projective geometry is also useful in avoiding edge cases of particular configurations, particularly the case of parallel lines (as in projective geometry, there are no parallel lines).

Formal Definitions

A projective plane is defined by a set of points, a set of lines, and a property of incidence satisfying three properties:

- For any two points, there is exactly one line incident with both of them.
- For any two lines, there is exactly one point incident with both of them.
- There exist four points such that no line is incident with more than two of them.

The final condition is not strictly necessary; it exists only to exclude degenerate cases from consideration. This is useful, however, in order to avoid having to manually check several degenerate cases over and over again; therefore, it is generally assumed as a matter of convenience.

The second condition has an important consequence: there are no parallel lines in the projective plane.

Additionally, the first and second conditions are very similar, differing only in the swapping of "points" and "lines". This forms the motivation behind defining the property of incidence: it shows the duality between points and lines.

The simplest (and most commonly used) example of a projective plane is the extended Euclidean plane, which is the ordinary Euclidean plane equipped with two additional properties:

- Every set of parallel lines is equipped with an additional point that is incident to any of these lines. That point is one of the points at infinity.
- The line at infinity is incident to every point at infinity, and is incident to only those points.

It can be verified that this results in a projective plane. Because of this projective plane's similarity to the Euclidean plane (and thus properties of this projective plane are generally true in the Euclidean plane as well), this is often referred to as the projective plane.

Projective planes need not be infinite. For instance, the projective plane of order 3 contains 13 lines and 13 points. More generally, the projective plane of order n contains, n^2+n+1 points and lines, so that every point is incident to lines and every line is incident to $n+1$ points. It is worth noting that this provides the solution to the following combinatorics question:

Example, what is the maximum number of -element sets such that any two sets share exactly one element, and there is no element that appears in every set?

Each n-element set can be interpreted as the set of lines with which a point is incident.

Duality

Since the projective plane is defined in terms of a set of points and a set of lines, equivalent definitions come from interchanging the two. Formally speaking, a duality p is a mapping of points to lines and lines to points, with the property that if P is incident to l, then l^p is incident to P^p. The simplest example of this was the first two properties in the definition:

- For any two points, there is exactly one line incident with both of them.
- For any two lines, there is exactly one point incident with both of them.

More generally, this means any property of a projective plane is equivalent to the dual of that property in the dual projective plane. The extended Euclidean plane is self-dual, meaning the dual of the extended Euclidean plane is the extended Euclidean plane itself, so any property of the extended Euclidean plane can be transferred into an additional property by duality. For example:

> Ceva's theorem and Menelaus' theorem are dual, and thus "equivalent" (meaning that proving one is sufficient to show the other), in the extended Euclidean plane. Doing away with degenerate cases (where the triangle in either theorem is degenerate), this means that Ceva's and Menelaus' are also equivalent in the "regular" Euclidean plane.

Though there are projective planes that are not self-dual, they are somewhat difficult to construct, and the most natural extensions of the extended Euclidean plane to an arbitrary field are also self-dual.

Polarities

A polarity of a projective plane is a duality that is an involution, meaning that the duality is its own inverse (or, equivalently, applying the duality twice results in the original plane). An important example stems from the concept of inversion, defined as follows: in the Euclidean plane, fix a circle O with radius r. For every point, P define the image of P to be the point on OP such that, $OP \cdot OQ = r^2$. The mapping is an $P \rightarrow Q$ inversion of the plane.

Now define a duality as follows: for every line l in the plane not passing through O, let P be the perpendicular from O to P, and let Q be the P image of under inversion of the plane. Then map l to Q. Duality also requires a mapping of points to lines, so the analogous definition is made: for any point $P \neq 0$ in the plane, is mapped to the line through Q (the image P) of that is, perpendicular to OP. Since this needs to be extended to the projective plane (the extended Euclidean plane), two more conditions are necessary:

- O is mapped to the line at infinity.
- A line through with slope is mapped to the point at infinity associated to the class of parallel lines with slope $-\frac{1}{m}$.

This mapping, from P to the polar of P and from to l the pole* of l, define a polarity (this is relatively easy to verify). There are many properties of this particular polarity, but the most important and most general fact is:

P lies on the polar of Q if and only if Q lies on the polar of P. Equivalently, lies on the pole of l if and only if the pole of lies on the polar of P.

More generally, additional polarities can be defined by extending the definition of inversion to arbitrary conics.

Additional Euclidean Properties

While projective geometry is useful in its own right, it has significant application to Euclidean geometry as well. An earlier topic gave a taste of this, in which Ceva's theorem and Menelaus' theorem were shown to be equivalent. Additional results include the following:

Desargues' Theorem

Two triangles are in axial perspective if and only if they are in central perspective. Axial perspectivity means that, for triangles ABC and abc, $AB \cap ab, BC \cap bc$ and $CA \cap ca$ are collinear. Central perspectivity means that Aa, Bb, Cc all concur at a single point.

This theorem is intuitively understood as a theorem on art; more specifically, on perspective drawing:

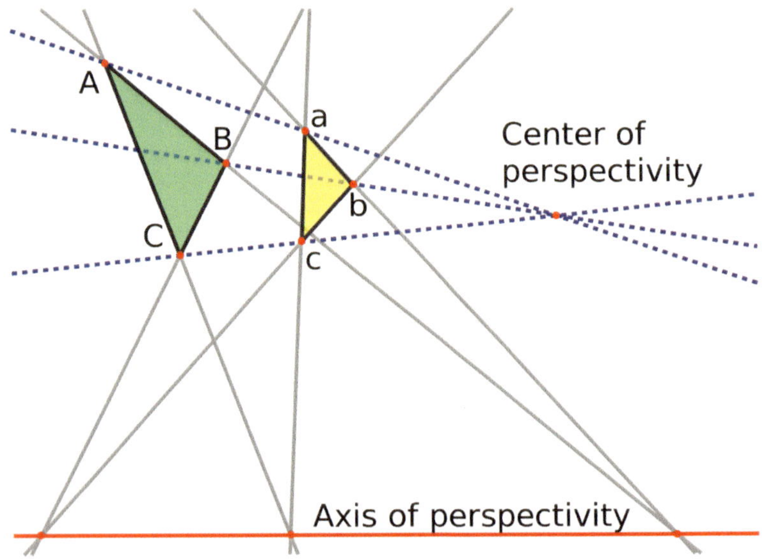

Interestingly, this result is not true in all projective planes, for instance in some finite projective planes of order. Nonetheless, it is true in most of them and certainly in the extended Euclidean plane. It is worth noting that the "iff" portion of this theorem is easily concluded from the principle of duality discussed above, so demonstrating either direction is sufficient to prove the theorem.

Pascal's Theorem

Let, *A, B, C, D, E, F* be six points on any conic (usually a circle). Then, $AB \cap DE, BC \cap EF, CD \cap FA$ are collinear.

Pascal's theorem is often useful even when there are fewer than 6 points on the conic, by considering a degenerate case where some of the points are equal. In this case, a line like *BB* is understood to be the tangent to the circle at B. It is also worth noting that Pascal's theorem has a converse: if $AB \cap DE, BC \cap EF, CD \cap FA$ are collinear, then A, B, C, D, E, F lie on some conic. This is not necessarily very useful information, but if five points are already known to lie on a conic, then the sixth must as well (as five points determine a conic). Thus the converse of Pascal's theorem can be useful in showing points lie on a certain circle, especially when combined with the degenerate strategy.

Example: Given a ΔABC triangle and a point *M*, a line passing through *M* intersects AB, BC and CA at $C_1, A_1, and\ B_1$ and B_1, respectively. The lines and $AM, BM\ and\ CM$ intersect the circumcircle ΔABC of repsectively at $A_2, B_2 and\ C_2$. Prove that the lines $A_1A_2, B_1B_2 and\ C_1C_2$ intersect in a point that belongs to the circumcircle of ΔABC.

Let A_1A_2 intersect the circumcircle of ABC for the second time at X, and let, $XB_2 \cap AC = B_1^{'}$. By Pascal's theorem on which is valid as all six of these points lie on a circle, the circumcircle of we get that $AC \cap B_2X = B_1^{'}CB \cap XA_2 = A_1$ and $BB_2 \cap A_2A = M$ are collinear. But then, since $B_1^{'}$ lies on A_1M and also AC, we must have $B_1^{'} = A_1M \cup AC = B_1$. So X, B_1, B_2 are collinear, and similarly X, C_1, C_2 are collinear, so X is the intersection of A_1A_2, B_1B_2, C_1C_2 which lies on the circumcircle as desired.

The projective dual of this theorem is important as well:

Example: Brianchon's Theorem

Let *ABCDEF* be a hexagon circumscribed about a conic section (again, usually a circle). Then concur at a point.

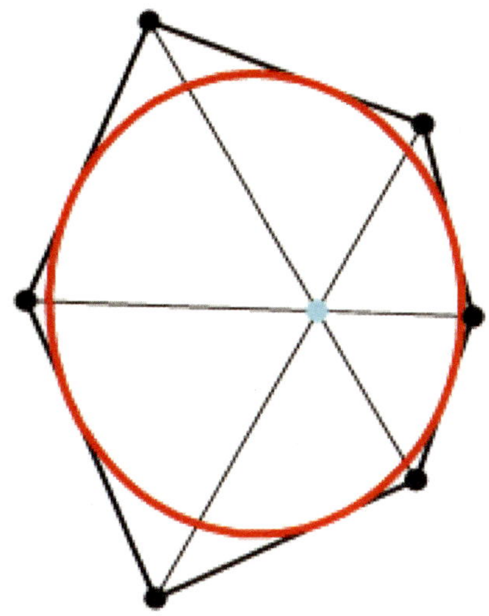

Illustration of Brianchon's theorem

References

- Ball, Simeon (2015), Finite Geometry and Combinatorial Applications, London Mathematical Society Student Texts, Cambridge University Press, ISBN 978-1107518438.
- Non-Euclidean-geometry, science: britannica.com, Retrieved 19 March 2018
- Riesz, Frigyes (1910), "Untersuchungen über Systeme integrierbarer Funktionen", Mathematische Annalen, 69 (4): 449–497, doi:10.1007/BF01457637
- Taxicabgeometry: mathematische-basteleien.de, Retrieved 31 March 2018
- Van Brummelen, Glen (2013). Heavenly Mathematics: The Forgotten Art of Spherical Trigonometry. Princeton University Press. ISBN 9780691148922. Retrieved 31 December2014
- Projective-geometry: brilliant.org, Retrieved 22 June 2018
- Ueberberg, Johannes (2011), Foundations of Incidence Geometry, Springer Monographs in Mathematics, Springer, doi:10.1007/978-3-642-20972-7, ISBN 978-3-642-26960-8

Permissions

All chapters in this book are published with permission under the Creative Commons Attribution Share Alike License or equivalent. Every chapter published in this book has been scrutinized by our experts. Their significance has been extensively debated. The topics covered herein carry significant information for a comprehensive understanding. They may even be implemented as practical applications or may be referred to as a beginning point for further studies.

We would like to thank the editorial team for lending their expertise to make the book truly unique. They have played a crucial role in the development of this book. Without their invaluable contributions this book wouldn't have been possible. They have made vital efforts to compile up to date information on the varied aspects of this subject to make this book a valuable addition to the collection of many professionals and students.

This book was conceptualized with the vision of imparting up-to-date and integrated information in this field. To ensure the same, a matchless editorial board was set up. Every individual on the board went through rigorous rounds of assessment to prove their worth. After which they invested a large part of their time researching and compiling the most relevant data for our readers.

The editorial board has been involved in producing this book since its inception. They have spent rigorous hours researching and exploring the diverse topics which have resulted in the successful publishing of this book. They have passed on their knowledge of decades through this book. To expedite this challenging task, the publisher supported the team at every step. A small team of assistant editors was also appointed to further simplify the editing procedure and attain best results for the readers.

Apart from the editorial board, the designing team has also invested a significant amount of their time in understanding the subject and creating the most relevant covers. They scrutinized every image to scout for the most suitable representation of the subject and create an appropriate cover for the book.

The publishing team has been an ardent support to the editorial, designing and production team. Their endless efforts to recruit the best for this project, has resulted in the accomplishment of this book. They are a veteran in the field of academics and their pool of knowledge is as vast as their experience in printing. Their expertise and guidance has proved useful at every step. Their uncompromising quality standards have made this book an exceptional effort. Their encouragement from time to time has been an inspiration for everyone.

The publisher and the editorial board hope that this book will prove to be a valuable piece of knowledge for students, practitioners and scholars across the globe.

Index

Q

R

S

T

U

V

Z